Fish Vaccines

This book is a timely reference text that highlights the role of vaccination in the fast-growing aquaculture industry. It discusses topics such as vaccine formulation, vaccine delivery, and enhancing the immune response of fish using nanoparticles. Information related to vaccine safety, ethical approval, and regulations is also discussed, together with dissemination of vaccines to fish farms across the globe. This cutting-edge book presents novel strategies to meet the growing demand for vaccines in finfish aquaculture. This book is useful to students, academics, clinicians, and professionals in the field of fisheries sciences, aquaculture, and veterinary sciences.

Fish Vaccines
Health Management for Sustainable Aquaculture

Edited by
Preetham Elumalai
Kim Thompson
Sreeja Lakshmi

CRC Press
Taylor & Francis Group
Boca Raton London New York

CRC Press is an imprint of the
Taylor & Francis Group, an **informa** business

First edition published 2023
by CRC Press
2385 Executive Center Drive, Suite 320, Boca Raton, Florida 33431-8530

and by CRC Press
4 Park Square, Milton Park, Abingdon, Oxon, OX14 4RN

CRC Press is an imprint of Taylor & Francis Group, LLC

ISBN: 978-1-032-48341-2 (hbk)
ISBN: 978-1-032-48344-3 (pbk)
ISBN: 978-1-003-38854-8 (ebk)

DOI: 10.1201/9781003388548

Typeset in Times
by codeMantra

Contents

SECTION III Adjuvants in Vaccination: An Underpinning

SECTION IV Technological Advancements and Efficacy of Fish Vaccination

SECTION V Vaccines in Aquaculture: Challenges and Perspectives

Acknowledgements

The editors cordially acknowledge 'International Veterinary Vaccinology Network' (IVVN) and 'BactiVac – The Bacterial Vaccines Network' for their excellent support for the successful production of the book. The authors are sincerely appreciated and acknowledged for their skillfull presentations of their respective chapters in articulating this book. I would like to thank Dr. Subramaniam Sivakumar, Associate Professor, Sri Sankara Arts and Science College, Enathur, Kancheepuram for offering his handful assistance and for technical support throughout the publishing process. The editors also greatfully acknowledge CRC Press/Taylor & Francis Group, especially Madhurima Kahali, Neha Bhatt and Vijay Shanker, for their patience and timely support towards the commendable task of organizing and integrating this compendium.

Foreword by Shri. Pinarayi Vijayan

As part of the Cochin University of Science and Technology's international project with the UK and Norway, on aquatic vaccination, this book has been jointly edited by Dr. Punnadath Preetham Elumalai, Dr. Kim Thomson, and Dr Sreeja Lakshmi. *Fish Vaccines: Health Management for Sustainable Aquaculture*, published by the CRC Press, USA, incorporates various chapters authored by experts from countries in which aquaculture is a prominent economic activity. Therefore, it will not merely be an academic document, but will also serve as a useful reference material for Kerala, especially as we are undertaking several measures to modernize and improve our fisheries sector.

Kerala's international collaborations in the fisheries sector date back to almost seven decades. Back in the early 1950s, we had collaborated with Norway to develop the capture fisheries sector in the State. Subsequently, the Neendakara Fishing Harbour was established under the Indo Norwegian Fisheries project, in 1953. Modern fishing techniques like trawling were learnt by our traditional fishing communities from the Norwegians. It is to further strengthen such collaboration, that an official delegation from Kerala recently visited Finland, Norway, and the UK.

Throughout the discussions and deliberations held by Kerala's delegation in these countries, the fisheries and aquaculture sector was featured prominently. Agreements were also made, with regard to co-operation in the sector. The Norwegian Institute of Marine Research has assured their support for research in fish nutrition, fish feed, and fish health management.

The NORD University that focusses on blue and green growth, and has a prominent Faculty of Biosciences and Aquaculture, has expressed its willingness to provide research orientation to students and teachers in Kerala. We have also decided to ensure more cooperation in student and faculty exchanges and research aimed at employment generation through offshore aquaculture, cage farming, and sustainable fish farming.

Fisheries and aquaculture play a significant role in the overall development of the world in terms of poverty alleviation, employment creation, and protein security. Unfortunately, owing to various reasons like climate change, over exploitation, marine pollution etc. our marine fish stocks are getting depleted and the production from the capture sector is showing dwindling trends for a few decades now. So, aquaculture has emerged as a sunrise sector to compensate this gap between production and demand.

The world has advanced significantly in aquaculture with varying production protocols and strategies, making this health food available in all the markets at a nominal price. Kerala's aquaculture production is low compared to countries like Norway and Vietnam and a few Indian States too. We aim to correct this situation through joint action and collaboration. In fact, both Finland and Norway have expressed interest in collaborating with the Maritime Cluster with a Sustainable Maritime Technology Hub, which is coming up in Kochi. It is in this larger backdrop of Kerala's focussed interventions toward emerging as a prominent maritime hub that this book assumes greater significance.

Apart from providing an introduction to vaccines in aquaculture, this book discusses various aspects of fish vaccines including their types and methods, technological advancements and challenges in the field, and so on. Certainly, it can serve as a much useful reference material for those who are engaged in improving the productivity of fisheries in general, and aquaculture in particular. Its contents would also go a long way in promoting cutting edge scientific and sustainable aquaculture. In that regard, this book benefits not only academia but also the general public who are keen to learn about and earn from aquaculture.

Kerala has the potential to emerge from a follower of aquaculture techniques developed around the world, to a creator of cutting edge technologies and techniques that can be followed around the world. In order to ensure that academic works such as this that balance research with experience are essential.

I congratulate the authors and wish the readers an enriching experience.

Shri. Pinarayi Vijayan
The Hon'ble Chief Minister of Kerala

Foreword by Prof Chris Secombes

Over the last couple of years, the importance of vaccination as a means of infectious disease control has been apparent to everyone. During the Covid pandemic, modern vaccination approaches were adopted and proven to be highly efficient in the fight against this disease. Such advances are likely to enable novel vaccine development in many areas of clinical and veterinary science, on a scale not previously seen. The need for vaccines to control diseases of fish in aquaculture is also readily apparent and crucial for the expansion of aquaculture in many developing countries, where fish are an important source of animal protein. Vaccination of fish is already highly successful in controlling many of the diseases of concern in the salmonid fish farming industry. Indeed, in Atlantic salmon culture, it is common for all fish to be vaccinated before transfer to seawater. This book, edited by Preetham Elumalai, Kim Thompson, and Sreeja Lakshmi, contains a number of review articles pertaining to the demand and ways in which vaccination can be used in aquaculture to improve fish health. There is an initial overview of the need for vaccination of fish, and the pathogens commonly encountered. It then describes the approaches taken to develop new and improved vaccines for aquaculture, using modern approaches and methodologies, including the requirement to ensure the safety as well as efficacy of those that are commercialized. The use of adjuvants to maximize (fish) vaccine efficacy is well known. For example, the benefit of incorporating adjuvants into salmonid vaccines against furunculosis (caused by the bacterium *Aeromonas salmonicida*) was realized at an early stage of their development. However, better adjuvants are needed, especially those that aid mucosal vaccination of fish, and their role and the advances on the horizon are covered in some of the later chapters. Such progress bodes well for the future. The challenges of vaccine commercialization, in terms of taking successful vaccines to market and increasing their use by fish farmers, is the final topic covered in this book. That there is an economic benefit as well as improved fish welfare is key to their uptake and should not be overlooked during the enthusiastic hunt for novel, highly effective vaccines to fish pathogens. Overall, this book highlights the vital role that vaccination can play in fish health management and the advances being made currently, which will help to ensure a robust and sustainable future for aquaculture.

Prof Chris Secombes
University of Aberdeen, Aberdeen, Scotland, UK

Foreword by Dr. Tim Connelley

The rapid growth of aquaculture as a primary source of providing the increasing needs for animal-derived protein in large parts of the world has been matched by a highly dynamic and progressive development in the science and application of vaccines. This has been achieved, in large part, by scientists from around the globe working collaboratively together. It has been a privilege to be able to support small-scale collaborations between some of these scientists as part of the portfolio of activities conducted by the International Veterinary Vaccinology Network (IVVN).

This book covers the broad spectrum of perspectives of vaccinology—from the foundational sciences such as immunology and microbiology, through the translational aspects of vaccine production and manufacture and onto the delivery of vaccines to farmers. As such it represents a useful text which considers and summarizes the current successes as well as the future challenges and opportunities in a comprehensive and holistic manner.

In India, aquaculture continues to increase as a source of both income and food security - growing at an annual rate of 8%, generating >USD6 billion in revenue and supporting the livelihoods of ~28 million fishermen and others involved in the aquaculture economy. With the expansion of the aquaculture sector comes increasing risks of disease outbreaks from a number of viral, bacterial, parasitic and fungal pathogens with the potential to lead to substantial morbidity and mortality for the farmed fish and economic loss to farmers. Vaccination is, and will continue to be, a principal method to mitigate and minimize the losses incurred from such diseases. As such it gives me great pleasure to provide a foreword to *Fish Vaccines: Health Management for Sustainable Aquaculture* edited by Dr. Preetham Elumalai, Dr. Kim D Thompson and Dr. Sreeja Lakshmi. I am certain the readers will find this book a useful reference by which a better understanding of the current status of fish vaccine development and deployment can be gained.

Dr. Timothy Connelley
Director of the International Veterinary Vaccinology Network,
The Roslin Institute, University of Edinburgh, United Kingdom

Editors

Preetham Elumalai, PhD, is an Associate Professor at the Department of Marine Biology, Cochin University of Science and Technology, Kochi, Kerala, India. He earned a Master's degree from the University of Madras and a PhD in biochemistry and molecular immunology from the Institute for Immunology, University of Regensburg, Germany. His research practice includes bioassay-guided identification of novel marine compounds, unveiling fish lectins in innate immune defense, aquatic vaccine development, evaluation of cost-effective feed additives and nutrigenomics, and effects of environmental pollutants on marine ecosystems. He has been a partner in numerous EU-, Indian-, and UK-funded projects (e.g., IVVN, BactiVac). He has written more than 70 peer-reviewed articles and has 2 patents in his name apart from editing 5 books (Springer, CRC Press) and has presented his work at more than 60 national and international conferences. He has been awarded the prestigious INSA fellowship (2018); MASTS, Fellowship (2019); IVVN award, UK (2020); FRSB, UK (2021); and BactiVac award, UK (2022).

Kim Thompson, PhD, is a Principal Investigator at Moredun Research Institute, where she heads the Aquaculture Research Group. She earned a BSc in biochemistry from the University of Stirling, an MSc in immunology from the University of Aberdeen, and a PhD in fish immunology from the University of Stirling. She has published over 160 articles in international journals relating to her work on fish health. Her research interests include vaccine development for fish pathogens (viral, bacterial, and parasitic), examining the immune response of fish to these pathogens and understanding the factors that influence this response. She is interested in developing alternative therapies for disease control in fish (e.g., functional feeds, immunostimulants, and probiotics).

Sreeja Lakshmi, PhD, is a Postdoctoral Research Scientist in collaboration with Moredun Research Institute (MRI), UK. She graduated from Calicut University and earned a PhD in biochemistry and functional genomics from the Institute for Molecular Biology, University of Regensburg, Germany. She has published research articles in peer-reviewed international journals and authored books and book chapters. She has been awarded prestigious research grants from the Bavarian Research Foundation (Bayerische Forschungsstiftung), Government of Bavaria, Germany; an HRD-Fellowship for Women Scientists from the Department of Health Research, Government of India; and a MASTS (Marine Alliance Science and Technology, Scotland) Award for Postdoctoral and Early Career Research Exchanges (PECRE), visited the University of Aberdeen, Scotland, and received IVVN Fellowship grant from the International Veterinary Vaccinology Network (IVVN), UK.

Contributors

Aneena Mariya Abraham
Department of Microbiology
Mar Athanasius College (Autonomous)
Kerala, India

Mohammad Noor Amal Azmai
Department of Biology
Faculty of Science, Universiti Putra Malaysia
Serdang, Selangor, Malaysia

Salleh Annas
Department of Veterinary Pathology and Microbiology
Faculty of Veterinary Medicine
Universiti Putra Malaysia
Serdang, Selangor, Malaysia

Aswathi Ashok
Department of Marine Biology, Microbiology and Biochemistry
School of Marine Sciences
Cochin University of Science and Technology (CUSAT)
Kochi, India

Adrian Astier
Aquaculture division
Virbac Asia Pacific Co. Ltd.
Bangkok, Thailand

Aruna Babu
Department of Microbiology
Mar Athanasius College (Autonomous)
Kerala, India

Joyal Elisabeth Baiju
Department of Microbiology
Mar Athanasius College (Autonomous)
Kerala, India

Alanta Maria Binu
Department of Microbiology
Mar Athanasius College (Autonomous)
Kerala, India

R. Cascione
Virbac Asia Pacific Co. Ltd.
Bangkok, Thailand

Hethesh Chellapandian
Department of Biotechnology and Microbiology
National College (Autonomous)
Tiruchirappalli, Tamil Nadu, India

Chou-Min Chong
Aquatic Animal Health and Therapeutics Laboratory (AquaHealth)
Institute of Bioscience
Universiti Putra Malaysia
Serdang, Selangor, Malaysia

Saurabh Dubey
Department of Production Medicine
Norwegian University of Life Sciences
Ås, Norway

A. P. Desbois
Institute of Aquaculture, University of Stirling
Stirling, Scotland, United Kingdom

Preetham Elumalai
Faculty of Fisheries
Kerala University of Fisheries and Ocean Studies
and
Department of Marine Biology, Microbiology and Biochemistry
School of Marine Sciences
Cochin University of Science and Technology
Kochi, India

Øystein Evensen
Norwegian University of Life Sciences
Ås, Norway

Uma Ganapathi
Aquatic Animal Health Laboratory, Centre for Marine Science and Technology (CMST)
Manonmaniam Sundaranar University
Tirunelveli, Tamil Nadu, India

Md Yasin Ina-Salwany
Aquatic Animal Health and Therapeutics Laboratory
Institute of Bioscience
Universiti Putra Malaysia
Serdang, Selangor, Malaysia

Sivakamavalli Jeyachandran
Department of Biotechnology and Microbiology
National College (Autonomous)
Tiruchirappalli, Tamil Nadu, India

Nayomi John
Department of Microbiology
Mar Athanasius College (Autonomous)
Kerala, India

Anu Joseph
Department of Microbiology
Mar Athanasius College (Autonomous)
Kerala, India

Jitendra Kumar
College of Fisheries
Acharya Narendra Deva University of Agriculture and Technology
Uttar, Pradesh, India

Kavitha Kuppuswamy
Department of Microbiology
Sri Sankara Arts and Science College
Kanchipuram, Tamil Nadu, India

Ihn-sil Kwak
Department of Fisheries
Chonnam National University
Yesou, South Korea

Sreeja Lakshmi
Aquaculture Research Group
Moredun Research Institute
Penicuik, United Kingdom

Po-Tsang Lee
Department of Aquaculture
National Taiwan Ocean University
Keelung, Taiwan

Philippe Mahl
Virbac Asia Pacific Co. Ltd.
Bangkok, Thailand

Tilusha Manchanayake
Department of Veterinary Pathology and Microbiology
Faculty of Veterinary Medicine
Universiti Putra Malaysia
Serdang, Selangor, Malaysia

Meril Mary Mathew
Faculty of Fisheries
Kerala University of Fisheries and Ocean Studies
Kochi, India

Mariavincent Michael Babu
Aquatic Animal Health Laboratory
Centre for Marine Science and Technology (CMST)
Manonmaniam Sundaranar University
Tirunelveli, Tamil Nadu, India

Aslah Mohamad
Aquatic Animal Health and Therapeutics Laboratory
Institute of Bioscience
Universiti Putra Malaysia
Serdang, Selangor, Malaysia

S. J. Monaghan
Institute of Aquaculture
University of Stirling
Stirling, Scotland, United Kingdom

Md. Shirajum Monir
Bangladesh Fisheries Research Institute Freshwater Station
Mymensingh, Bangladesh

Kizito K. Mugimba
Department of Biotechnical and Diagnostic Sciences
College of Veterinary Medicine
Animal Resources and Biosecurity
Makerere University
Kampala, Uganda

Bismi Mujeeb
Department of Microbiology
Mar Athanasius College (Autonomous)
Kerala, India

Hetron M. Munang'andu
Faculty of Biosciences and Aquaculture
Nord University
Bodø, Norway

Abdullah Bin Abdul Nazar
Faculty of Fisheries
Kerala University of Fisheries and Ocean Studies
Kochi, India

Neema
Department of Microbiology
Mar Athanasius College (Autonomous)
Kerala, India

Hoang Duc Nguyen
Center for Bioscience and Biotechnology
University of Science
and
Vietnam National University
Ho Chi Minh City, Vietnam

Kiyun Park
Department of Fisheries
Chonnam National University
Yeosu, South Korea

Nivya Mariam Paul
Department of Microbiology
Mar Athanasius College (Autonomous)
Kerala, India

Sanchu Prakash
Faculty of Fisheries
Kerala University of Fisheries and Ocean Studies
Kochi, India

Panneerselvam Punniyakotti
Department of Biochemistry
Kanchi Shri Krishna College of Arts and Science
Tamil Nadu, India

Amrutha Rajesh
Department of Microbiology
Mar Athanasius College (Autonomous)
Kerala, India

Sathish Kumar Ramamoorthy
Aquatic Animal Health Laboratory
Centre for Marine Science and Technology (CMST)
Manonmaniam Sundaranar University
Tirunelveli, Tamil Nadu, India

Malavika Biju Reni
Faculty of Fisheries
Kerala University of Fisheries and Ocean Studies
Kochi, India

Muhammed Salih Salim
Department of Biology
Norwegian University of Science and Technology (NTNU)
Trondheim, Norway

Panneerselvam Saravanan
Department of Biotechnology
Sri Sankara Arts and Science College
Kanchipuram, Tamil Nadu, India

Aparna Sasi
Department of Microbiology
Mar Athanasius College (Autonomous)
Kerala, India

Joseph Selvin
Department of Microbiology
Pondicherry University
Puducherry, India

Anuj Sharma
Faculty of Fisheries
Kerala University of Fisheries and Ocean Studies
Kochi, India

Muzammil Siddiqui
Faculty of Fisheries
Kerala University of Fisheries and Ocean Studies
Kochi, India

Wanna Sirimanapong
Faculty of Veterinary Science
Mahidol University
Nakhon Pathom, Thailand

Subramaniam Sivakumar
Department of Biochemistry
Sri Sankara Arts and Science College
Enathur, India

Chiranan Sudpraseart
Faculty of Veterinary Science
Mahidol University
Nakhon Pathom, Thailand

Vimal Sugumar
Aquatic Animal Health Laboratory
Centre for Marine Science and Technology (CMST)
Manonmaniam Sundaranar University
Tirunelveli, Tamil Nadu, India

Eleeswa Celin Syrus
Department of Microbiology
Mar Athanasius College (Autonomous)
Kerala, India

Prawporn Thaijongrak
Faculty of Veterinary Science
Mahidol University
Nakhon Pathom, Thailand

Citarasu Thavasimuthu
Aquatic Animal Health Laboratory
Centre for Marine Science and Technology (CMST)
Manonmaniam Sundaranar University
Tirunelveli, Tamil Nadu, India

Nam Tri
Center for Bioscience and Biotechnology and Laboratory of Molecular Biotechnology
University of Science
and
Vietnam National University
Ho Chi Minh City, Vietnam

Nhat HM Truong
Center for Bioscience and Biotechnology
University of Science
and
Vietnam National University
Ho Chi Minh City, Vietnam

Phuong Hong Vo
Center for Bioscience and Biotechnology
University of Science
Vietnam National University
and
Research Institute for Aquaculture
Ho Chi Minh City, Vietnam

Nam Tri Vo
Center for Bioscience and Biotechnology and Laboratory of Molecular Biotechnology
University of Science
and
Vietnam National University
Ho Chi Minh City, Vietnam

Eakapol Wangkahart
Laboratory of Fish Immunology and Nutrigenomics
Mahasarakham University
Kham Riang, Thailand

Boonsirm Withyachumnarnkul
Prince of Songkla University
Surat Thani Campus
Surat Thani, Thailand

Fernando Yamamoto
Mississippi Agriculture and Forestry Experiment Station
and
Department of Wildlife and Fisheries Sciences
Mississippi State University
Stoneville, Mississippi

Mohd Zamri-Saad
Department of Veterinary Pathology and Microbiology
Faculty of Veterinary Medicine
Universiti Putra Malaysia
Serdang, Selangor, Malaysia

Section I

Vaccines in Aquaculture

A Preface

1 Understanding Vaccine Development in Aquaculture

Nayomi John, Anu Joseph, Aparna Sasi, Bismi Mujeeb, Joyal Elisabeth Baiju, Eleeswa Celin Syrus, and Nivya Mariam Paul
Mar Athanasius College

CONTENTS

1.1 INTRODUCTION

Aquaculture is the breeding, rearing, and harvesting of fish, shellfish, algae, and different organisms in all forms of water environments. As the demand for seafood has multiplied, it has become feasible to develop foods in coastal marine waters and the open ocean. Aquaculture is a means to supply food and different industrial products, repair habitat and replenish wild shares, and rebuild populations of threatened and endangered species.

Fish farming has achieved an exponential growth over the last decades. The fast development of fish aquaculture has brought about a parallel growth in pathological situations affecting fish. Numerous pathogens, including microorganisms, viruses, parasites, and fungi, have had a negative effect on the popularity of farmed fish, inflicting critical financial losses. One of the most important challenges in sustainable aquaculture is the proper management of infectious diseases in fishes. Vaccination plays an important role in the community, environmental, and economic

DOI: 10.1201/9781003388548-2

sustainability in the global aquaculture fields. Although numerous commercial vaccines are presently available for extreme fish sicknesses, some of the fish are deprived of an efficient vaccination strategy. Despite various approaches to innovative therapy, fish diseases remain a major economic issue in commercial aquaculture worldwide. Due to the increasing demands in the discipline of fish vaccinology, this special review welcomes papers on latest vaccines and vaccination protocols for aquaculture.

Aquaculture has the ability to improve the environment on our planet and the fitness of our population, so long as it's carried out in a manner this is environmentally pleasant, socially responsible, and considers food protection and animal welfare. Not only is aquaculture necessary, but it is also a sustainable option for purchasers, especially in comparison to other farmed proteins. Seafood is particularly resource-efficient—it has the highest protein retention as compared to fowl, beef, and red meat. Aquaculture has lower greenhouse fuel emissions compared to other types of farming. Aquaculture is the major tool to fulfill the gap of seafood supply. Farming fish responsibly and sustainably has been the solution to provide healthy and environmentally pleasant protein alternatives to generations.

1.2 VACCINE

Vaccine is a biological technique that improves immunity against a particular ailment. Vaccines are used to reinforce the frame's immune system and prevent critical lifestyles-threatening disease. Vaccines were used almost solely to prevent or control infectious diseases with the intention of long-lasting immunity. The management of vaccines is referred to as vaccination. It is the direct method of preventing infectious diseases. The effect of vaccines is large and a long reaching, though not consistently quantifiable, analyzed, or communicated. Historically, the perceived benefits of vaccination had been to lessen morbidity and mortality from infections, and people were the drivers for the innovation of new vaccines, particularly during outbreaks of infections that afflict the most deprived in the society.

1.2.1 Ideal Vaccine

Certain features are to be looked for in a vaccine to qualify it as a "perfect" vaccine.

Foremost features include:

- **Safety:** The most critical characteristic that needs scrutiny in any vaccine is its safety.
- **Efficacy:** An excellent vaccine should be relatively efficacious in particular situations with 100% vaccine uptake.

Other features:

- Cost
- Ease of management
- Stability
- Multivalency
- Long-term immunity
- Economics of vaccination

1.2.2 History

Sickness prevention and management are crucial factors that allow for sustainable aquaculture, both economically and environmentally. Prophylactic measures based totally on stimulation of the immune system of the fish had been a powerful measure for attaining this aim. Immunoprophylaxis has become a vital part of the successful development of the fish-farming industry. When illnesses set in aquaculture operations involving fish, antibiotics or chemotherapeutics were used for sickness remedy, or even for disease prevention. The vaccine for aquaculture, a vaccine for the prevention of yersiniosis in salmonid fish, was licensed in the United States of America in 1976. Expanded use of vaccines in more locations and countries enhances the growth and development of aquaculture industries. At some stage in the first 20–30 years after World War II, there were a wide variety of reports about disease prevention in aquaculture by means of vaccination. The most vital reason for the disinterest in immunoprophylaxis in those years was possibly the provision of antimicrobial compounds. Snieszko and Friddle (1) concluded that chemotherapy with sulfamerazine turned into advanced oral administration of a vaccine to manipulate furunculosis. It was not till the seventies that vaccines were applied in business aquaculture.

Different fish species, which include carp and salmonid fish, had been included for immunological research, and elements of significance in vaccinology, like temperature and other environmental elements, had been additionally studied (2,3). The function of adjuvants in fish is important for the progress in immunology and vaccinology. Ambrosius and Lehman (4) discovered that adjuvants like aluminum hydroxide and Freund's adjuvant improved the amount of immunoglobulins—the previous slightly and the latter significantly.

The technique of management was additionally studied. Trout provided a terrific immune response while injected with antigens from *Aeromonas salmonicida* (5). The method of vaccine administration plays an important role in successful vaccination process. Oral immunization was considered as an important way for practical disease prevention (6). Based on immunization research, they concluded that injection of a concentrated vaccine might be a beneficial prophylactic way to manage vibriosis in rainbow trout. Three hundred and sixty-five days later, Ross and Klontz confirmed that enteric redmouth ailment (yersiniosis) can be prevented in fingerling rainbow trout fed pelleted food containing bacterial cells of the redmouth bacterium (*Yersinia ruckeri*) (7).

All through the 1970s, immunoprophylaxis was identified as a technique for the prevention of infection as a result of fish pathogenic species of *Vibrio* and *Yersinia* in aquaculture. In 1988, the maximum numbers of Atlantic salmon and rainbow trout in Norway were vaccinated, initially through immersion, but later on via injection, against this disorder, which was given the name cold water vibriosis. Antigen suspensions from *Vibrio anguillarum*, *V. salmonicida*, and *A. salmonicida* made in mineral oil adjuvants act as a powerful control over various fish diseases.

During the early years of aquaculture, major viral illnesses blanketed infectious pancreatic necrosis (IPN), viral hemorrhagic septicemia, and infectious hematopoietic necrosis (IHN). The first successful experiments on vaccination in opposition to those sicknesses were blanketed by vaccines, both avirulent and attenuated strains (8,9).

The history of fish vaccination is normally a tale of achievement. But, there have been additional barriers in the use of the fish immune system for ailment prevention. The history of fish vaccinology is a documentation of how the immune system of fish may be stimulated by way of vaccines to prevent unintended consequences of pathogenic microorganisms. Fish vaccinology has proven to be a brilliant improvement in recent years. Maximum vaccines are first-technology vaccines. Reliable scientific findings are very important to make environmental, social, and competitively priced sustainability in international aquaculture.

1.2.3 Requirement of Vaccine in Aquaculture

Aquaculture is currently the world's fastest growing agricultural sector producing food (10,11). It contributes to a great extent to food security and socioeconomic development in many countries. Global aquaculture practices have shifted from large-scale aquaculture systems to semi-intensive and intensive aquaculture systems, where high-value fish species are bred at higher breeding densities using commercial feed (12,13). But, there are many issues regarding the sustainable development of the aquaculture sector. Diseases are the most devastating threats to semi-intensive and intensive aquaculture systems. Bacterial infectious diseases are the most common illnesses or challenges in aquaculture while viral diseases are more difficult challenges due to lack of antiviral drugs and disease resistance mechanism in fish (14). The unavailability of controlled and efficient treatment modules to viral and bacterial diseases increased prerequisites for developing and implementing an effective prevention and management approach to these illnesses. In addition, the negative effects of infectious diseases have also prompted the strategic development of vaccines designed for the indiscriminate use of antibiotics in aquaculture, which may cause problems in developing bacterial resistance, food safety risks, and environmental issues (15). Treatment of many bacterial infections in fish with only antimicrobial agents is impossible.

Vaccination of fish has become very important in this situation—a simple and effective approach to prevent and fight against infections and illness of fish. Vaccination is a protective process by which the immune response is induced in the animal receiving the drug antigen derived from pathogens and made nonpathogenic by heat or other means. Vaccines stimulate the fish's immune system and strengthen protection against illness. Several important developments have been made in fish vaccines. However, so far, very few vaccines are on the market against infectious viral and bacterial diseases for aquaculture companies (10), Vaccines are currently available to economically important individual fish species used to treat bacterial and viral diseases.

1.2.4 Conventional and Modern Methods in Vaccine Development

Over the last 50 years, fish vaccination has been a crucial disease control tool that has contributed substantially to international aquaculture production. Nowadays, most licensed fish vaccines employed in aquaculture are developed with the use of traditional procedures, which primarily consist of inactivated whole microorganisms formulated with adjuvants and also live vaccines. Contemporary vaccine production involves targeting particular pathogen components and vaccines developed using such methods including subunits as well as nucleic acid vaccines.

1.2.4.1 Conventional Methods

1.2.4.1.1 Killed Vaccines/Inactivated

Presently, killed vaccines or inactivated vaccines are the principal vaccine types that are used commercially. The vaccines are prepared by killing the virulent agent and employing it as an antigen to stimulate the humoral antibody–related system of the immune response. In addition, the vaccine is found to be safe as it is incapable of replication when administered to the host (16). Killed vaccines usually generate short-range immunity because they do not survive in the vaccinated fish for a long time (17). As an alternative, the immunogenicity of these vaccines could be enhanced by the use of adjuvants or booster immunization doses (18,19). To cite an example, inactivated vaccines have been used for IPN in Atlantic salmon and grass carp hemorrhagic disease.

1.2.4.1.2 Live Vaccine

Live vaccines are developed by attenuating the pathogens using physical as well as chemical treatments and are prepared in the laboratory itself by the serial passages in cell cultures (20). These vaccines manipulate greater stimulation of both cellular and humoral immune responses due to their ability to proliferate inside the host (21). This will ensure a long-term immunity unlike killed vaccines. Since the vaccine itself is quite effective, the use of adjuvants is not typically required to boost the activity. A live vaccine containing *Arthrobacter* sp. has also been successfully demonstrated to cross protect against *Renibacterium salmoninarum*, a pathogen that causes bacterial kidney disease in salmonids (19,22).

1.2.4.2 Modern Methods

1.2.4.2.1 Subunit Vaccines

A subunit vaccine doesn't embody the whole pathogen but contains only the antigenic components that are needed to induce an immune response (23). Since they are found to have poor immunogenicity when compared to the killed or live vaccines, incorporation of adjuvants is necessary for a good response (10). Subunit vaccines can be prepared in several ways, which include direct extraction and purification of immunogenic proteins from the pathogens or by the use of recombinant vectors (24,25). Vaccine for IPN is an ideal successful subunit vaccine (26).

1.2.4.2.2 Nucleic Acid–Based Vaccines

Nucleic acid vaccines consist of DNA and RNA structures, which encode the antigen of interest. They are considered safe since reversion to their virulent state is prevented (27).

1.2.4.2.3 DNA-Based Vaccines

These vaccines are developed based on the advancement in molecular biology. It is a mode of vaccination that does not make use of antigens; instead, it consists of a plasmid that contains genes that code for the antigenic protein, which when expressed in the host produces a dominant immune response (28). DNA vaccines actively stimulate cellular and humoral immunity. Moreover, DNA vaccines do not comprise any impurities commonly found in other whole-cell vaccines (29). Although numerous DNA vaccines have been developed, only a few of these have been commercialized. An example is the salmonid alphavirus subtype 3 DNA vaccine against the pancreas disease virus in the trade name, Clynav (30).

1.2.4.2.4 Recombinant Vector Vaccine

The preparation of this vaccine involves the insertion and expression of the immunogenic part of the organism in the carriers. The resultant proteins can be purified and used as vaccines. Infectious salmon anemia and infective hematopoietic necrosis disease virus are expressed in carriers, and they serve as the best examples of this procedure (10).

1.2.5 Development of Vaccine for Specific Disease

Nowadays, many aquaculture systems, across the globe, face new challenges due to the emergence of new infectious diseases. Accordingly, the formulation of a promising vaccine that ensures long-term protection from disease outbreaks in the fisheries and aquaculture is inevitable. Several vaccines have been developed in aquaculture to fight against bacterial, viral and, parasitic infections.

1.2.5.1 Bacterial Fish Vaccines

The very first developed vaccines in aquaculture were bacterial whole-cell immersion vaccines that were introduced against red mouth disease and vibriosis (19). The advent of modern biological technologies enabled the discovery and commercialization of abundant vaccines like DNA vaccines, subunit vaccines, etc. (10). At present, some new approaches have been put forward for developing vaccines against significant bacterial infections that seriously affect the fish-farming industry (22). Mostly, these vaccines are based on inactivated whole-cell formulations that aid in preventing infections. For example, a simple inactivated bacterin vaccine has been found to be effective against vibriosis. The production is cost-effective making it an ideal method, but the existence of specifications such as the use of the right serotypes as well as starting vaccination in the larvae stage is very challenging (31). Another promising approach found to be more efficient is the use of live-attenuated isolates (32,33). In addition, further research in the development of recombinant protein-based vaccines is ongoing. For example, *Piscirickettsia salmonis* bacterins have to be replaced by recombinant vaccine as the former offer only a low efficacy (34).

1.2.5.2 Viral Fish Vaccines

Due to the presence of real challenges in regulating viral infections such as lack of information on the fish resistance mechanism, it is more difficult to control viral diseases in fish (10,35). Presently, most of the commercially available viral vaccines in aquaculture are inactivated pathogens or recombinant subunit proteins. In contrast, eco-safety aspects of live vaccines hinder their use as marketable vaccines. An example of an available vaccine is a koi herpes virus vaccine based on the attenuated strain of interstitial carp nephritis and gill necrosis virus (36). Disease outbreaks caused by IPN are a major threat to the salmonid industry partly due to the complex nature of this disease (26). Another challenge is the betanodavirus attack, which results in infectious outbreaks in marine fish farming (37). Although recombinant protein vaccines are available, administration of this vaccine in the early stage of life is required to prevent the illness. The inadequacy of viral vaccines is the prominent crisis dealing with fish vaccinology (19).

1.2.5.3 Fish Vaccines against Parasite

Fish farming is also affected by parasitic infections as well. However, no parasitic vaccines are made commercially available. Apart from that, various research and experiments found that live vaccines are more promising than dead vaccines (38). However, vaccine production is not feasible due to huge expenses and biosafety elements. Despite all of these problems, a successful vaccine against cattle tick (*Boophilus microplus*) has been developed after a lot of hardships (39) (Table 1.1).

TABLE 1.1
Overview of Fish Vaccines Used against Specific Diseases

Vaccine	Species of Administration	Illness Averted
1. Yersinia Ruckeri Bacterin	Salmonids	Yersiniosis
2. Vibrio Anguillarum-Ordalii	Salmonids	Vibriosis
3. Aeromonas Salmonicida Bacterin	Salmonids	Furunculosis
4. Vibrio Salmonicida Bacterin	Salmonids	Vibriosis
5. Edwardsiella Ictaluri Bacterin	Catfish	Enteric Septicemia
6. Flavobacterium Columnar Vaccine	Catfish	Columnaris Disease
7. Pasteurella Vaccine	Salmonids	Pasteurellosis
8. Streptococcus agalactiae Vaccine	Tilapia	Streptococcosis
9. Infectious Salmon Anemia Vaccine	Salmonids	Infectious Salmon Anemia
10. Arthrobacter Vaccine	Salmonids	Columnaris Disease
11. Nodavirus Vaccine	Seabass	Viral Nervous Necrosis
12. Pancreas Disease Virus Vaccine	Salmonids	Pancreas Disease
13. Betanodavirus	Grouper	Betanoda Virus Disease
14. Koi Herpes Virus	Koi carp	Koi Herpes Virus Disease
15. Carp Erythrodermatitis	Carp	Erythrodermatitis
16. Gaffkemia Vaccine	Lobsters	Gaffkemia
17. Piscirickettsia Salmonis Vaccine	Salmonids	Piscirickettsiosis
18. Inactivated Tenacibaculum	Turbot	Tenacibaculosis
19. Inactivated Yersinia Ruckeri	Salmonids	Enteric Red Mouth Disease
20. Inactivated L. graviae	Rainbow trout	Lactococciosis

1.2.6 Challenges in the Prevention and Control of Fish Diseases

In relative phrases, it is truthful to say that infectious sickness management in aquaculture is more complicated than terrestrial animal disorder control, because of the surroundings wherein the fish lives, and the character of the fish. Fish cannot be located near enough like we do in the case of terrestrial animals. Control of infectious diseases in aquaculture is much more complicated than terrestrial forms due to the environment where fish lives, the features of fish variety, etc., thus aquaculture diseases are often difficult to detect and characterize (40). The other important challenge is in diagnosing the disease of fish; in terrestrial animal disease diagnosis, the individual animal is the unit of interest. The scenario is not the same in aquaculture disease investigation because of the nature of the water where fish lives. A disease can transmit quickly and the whole tank may be the source of disease to healthy stock. In this case, the unit of interest is not a single fish rather the whole tank needs to be investigated and diagnosed. Samples should be collected not only of fish but also of water to measure important characteristics like pH, soil bottom conditions, and turbidity, which make aquatic diagnosis more complicated and challenging (41).

Advancing vaccination is the most crucial strategy to prevent and manage infectious ailments in fish. It is impossible to treat the various bacterial infections in aquatic animals with only antimicrobials (10). There are upgrades in fish vaccination lately. A number of the upgrades include immunization of massive stock at a time and the improvement of multivalent vaccines (11). Vaccination is broadly in use in almost all meals-producing animals. In aquaculture, it reduces the use of antibiotics and protects fish from infectious sicknesses. It also avoids the threat of drug resistance. Protection at stock degree due to herd immunity may be achieved and the need for licensing and registration of recent vaccine is an awful lot simpler than antibiotics (11).

The following are the few important issues or concerns that should be considered earlier than software of vaccination in fish: fish species to be vaccinated, immune status of fish, production cycle, existence history of the aquaculture machine, the illnesses to be screened in aquaculture, when do those illnesses occur (seasonal distribution of diseases in the aquarium), farming technology (coping with and mechanization), surroundings (temperature and salinity), strain factors, nutrients, and cost benefit (16).

1.2.7 Progress in Fish Vaccine Development

There has, however, been remarkable development in fish vaccine during the last four decades with 24 licensed fish vaccines now commercially available for use in a variety of fish species. These comprise killed, live-attenuated, peptide subunit, recombinant, DNA/RNA particle vaccines. Vaccination has emerged as a critical method for the prevention of infectious disorders in farmed fish, specifically salmonid species. Bacterial infections as a result of gram-poor microorganisms, which encompass Vibrio sp., *Aeromonas* sp., and *Yersinia* sp., have been correctly controlled for the usage of vaccination. With furunculosis, the success is attributed to the usage of injectable vaccines containing adjuvants. Vaccines against virus infections, together with IPN, have also been utilized in commercial fish farming. Vaccines against several bacterial and viral infections have been studied and determined to be technically possible. The overall positive effect of vaccination in farmed fish reduced mortality, but the future of the fish-farming industry is also important that vaccination contributes to sustainable biological production with the negligible intake of antibiotics.

Atlantic salmon and rainbow trout aquaculture in the United Kingdom, Norway, and the United States increased within the 1980s, accompanied by speedy growth in disease, especially with bacterial pathogens such are *Vibrio* sp., *Y. ruckeri*, and *A. salmonicida.* As a result, big quantities of antibiotics have been used and concerns grew about antibiotic resistance. This inspired the improvement of fish vaccines and led to the primary commercially available fish vaccines in opposition to Vibriosis, Enteric Red Mouth (ERM), and Furunculosis. Presently, 19 important corporations market fish vaccines globally and lots of small companies also exist (42).

The number of commercial vaccines for fish has expanded, from 2 in 1980 to 24 currently (43) with 1 vaccine additionally for lobsters (44). Fish vaccines are available for a wide variety of species (13,44,45) consisting of Atlantic salmon, Rainbow trout, sea bass (*Dicentrarchus labrax*), sea bream (*Sparus aurata*), tilapia (*Oreochromis niloticus/mossambicus*), amberjack (*Seriola dumerili*), and yellowtail (*Seriola quinqueradiata*) in Japan, and catfish (*Ictalurus punctatus*) and Vietnamese catfish (*Pangasianodon hypophthalmus*). Most are formalin killed whole-cell vaccines even though live-attenuated vaccines are certified in the United States of America to be used in catfish (46). A DNA vaccine against infectious IHN is licensed in Canada to be used in the Atlantic salmon (47), a subunit vaccine (peptide; VP2) is used in Norway (against infectious pancreatic necrosis virus, IPNV), and a recombinant vaccine toward infectious salmon anemia virus, ISAV, is used in Chile.

1.2.8 Challenges in Fish Vaccine Development

Even though there has been great development in the field of fish vaccinology, diverse demanding situations still are an obstacle for the development of fish vaccine. The most important issue that restricted the vaccine development system is the identification of protective antigens. Identifying protective antigens isn't easy and requires a variety of tactics, viz., pathogen type, fish species, administration method, antigen manufacturing method, and availability of challenge model for checking out the efficacy of the vaccine candidates (48).

1. **Pathogen type:** Piscine pathogens are heterogeneous in nature and have a variety of antigenic epitopes. Even though the majority of successful commercial fish vaccines are killed whole-cell bacteria preparations with intraperitoneal injection modalities, the applicability of this approach seems to be limited to other fish pathogens, mainly for intracellular or complicated pathogens (e.g., viruses and parasites); those are difficult or high priced to culture.
2. **Fish species:** The variety of fish species itself poses a challenge in vaccine development. As most fish pathogens have a wide variety of susceptible hosts and every fish species behaves differently to elucidate host–pathogen interactions, there aren't any regularly occurring formulae of developing a vaccine against a pathogen to be equally effective in each of its susceptible hosts.
3. **Administrative methods:** The technique of vaccine administration is the most challenging and needs to be determined. It's so far observed that a few novel vaccines being advanced are protective. However, recent administration methods and vaccination strategies are not suitable for the most fulfilling efficacy (e.g., may need high/booster vaccination). Even though the injection method is generally used for Atlantic salmon, administration of vaccines through the mucosal path is also practically possible for lower-valued fish species, for example, tilapia and Pangasius. However, various demanding situations have hampered the growth of mucosal vaccines, along with a lack of correlates of protection, lack of optimization of protective doses required, the possibility of oral tolerance, potential denaturation of oral vaccines in the stomach, and the capability of antigens to cross mucosal barriers to gain access to antigen-presenting cells (APCs) (49).
4. **Antigen production technique:** For developing an effective vaccine, the choice of protective antigen is of utmost importance. In addition to selection, the cost of manufacturing the antigen is also equally important as developed vaccines can be less affordable to low- to middle-income fish farmers.
5. **Challenge models:** Evaluation of vaccines' efficacy needs to be standardized in vivo disease challenge models that closely simulate the natural infection route. The bath and cohabitation challenge best fulfill the requirement of natural exposure, but they may be

more difficult to govern and standardize than the injection challenge methods. In addition, injection challenge isn't always an appropriate approach to test a mucosal vaccine (e.g., dip immersion) administered. Furthermore, a few pathogens are incapable of producing disease in experimental challenge models until some scarification or pressure is used (50). Within the absence of an experimental disease challenge model, the determination of vaccine efficacy taken into consideration is a problematic area, which needs to be sorted out in future.

1.2.9 Opportunities in Fish Vaccine Development

Regardless of the difficulties that hindered the improvement of the fish vaccine, opportunities also present for fish vaccinologists to apply novel technologies and vaccination techniques that may help in overcoming the demanding situations. With the advancement in the subject of bioinformatics, the cost of entire-genome sequencing of pathogens has decreased significantly, enabling targeted vaccine design for heterogeneous sp. The United Kingdom lately characterized more than 300 *Flavobacterium psychrophilum* sp. and developed an effective trivalent whole-cell vaccine, wherein undesirable immunosuppressive epitopes are removed, and only specific protective antigens are included thereby enhancing the efficacy of the vaccine. Likewise, in reverse vaccinology, the latest software programs are getting used nowadays to pick out highly immunogenic vaccine candidates for the development of protein subunit vaccines or DNA vaccines. After long pending legalization technique and substantial safety research, DNA vaccines are accredited to be used in Europe, consequently developing a huge opportunity for their growth in the future. Other alternatives to DNA vaccines, consisting of mRNA vaccination or safe to eat vaccine technologies, which have been confirmed to be enormously efficient and secure for different animals, can be used for prophylactic and therapeutic applications in fish also. In the case of live-attenuated vaccines, conventional attenuation techniques are being replaced by molecular methods, wherein genetically modified pathogens may be produced which permit higher control and protection than random mutations in live vaccines, but their classification as GMOs has restricted their use in aquaculture. However, with further safety measures, their use may be permitted in the future.

Vaccine administration strategies also provide huge scope for future research. Improved nanomaterials (<1000 nm), including virus-like particles (VLPs), liposomes, ISCOMs, polymeric, and nondegradable nanospheres, showed potential as antigen delivery vehicles, allowing the sustained release of antigens and additionally act as immune enhancer adjuvants (51). These delivery systems are being experimentally used for mucosal delivery of fish vaccines and are considered to be the immediate focus area for development in fish vaccinology. Furthermore, regarding vaccination strategies, prime booster vaccination with early immersion vaccination then IP booster vaccination or IP booster vaccination followed with the aid of oral booster vaccination has reported the stimulation of both mucosal and systemic immunity.

1.2.10 Effects and Benefits of Vaccine in Aquaculture

Over the last 2 decades, vaccination has become an important method for preventing infections in farmed fish, primarily salmon. To date, most commercial vaccines have been inactivated vaccines given by injection or immersion. Bacterial infections caused by *Vibrio sp.*, *Aeromonas Sp.*, and *Yersinia* sp. were triggered and effectively controlled by vaccination. In the case of boil, success is due to the use of an adjuvant and an injectable vaccine. Vaccines against viral infections, including IPN, have also been used in commercial aquaculture. Vaccines against several other bacterial and viral infections have been studied and have proven to be technically feasible. The overall positive effect of vaccination of farmed fish is a reduction in mortality. However, it is also important for the future of aquaculture that vaccination ignores antibiotic consumption and contributes to sustainable organic production (52).

Since the 1990s, vaccines have contributed significantly to improving fish health, well-being, and productivity. They are important factors in the prevention of various diseases. Fish develop immunity when exposed to infection and survive. So, they are usually fully or partially immune or resistant to other attacks by the same organism. Fish can get sick with the first infection and need to be treated. Vaccination mimics infection and provides immunity to the fish without giving in to the disease. Therefore, fish are resistant to disease before infection, and when fish are later exposed to infection, there is usually little or no evidence of disease. This makes the fish healthier and requires less processing. Fish, breeders, and consumers will now benefit. All fish are naturally immune to some diseases, and there is no risk of consuming food from previously vaccinated healthy fish to produce similar immunity.

1.3 LIMITATIONS

The predominant intention of vaccination is to set off a specific long-time period protection toward a certain sickness. It's been debated whether or not the effective long-term protection of oil-adjuvant injection vaccines is because of immunologic memory in the fish or steady stimulation from the antigen depot. Existing immersion vaccines are safely for a variety of aqua-cultured fish to protect against infectious diseases caused by bacteria and viruses. In evolutionary terms, fish are the first group of animals with simple elements of the immune system of higher vertebrates. Despite the fact that fish immune systems are primitive compared with mammals, there seem to be more similarities than differences (53,54).

In the restrained quantity of fish species studied, the primary type of antibody is an immunoglobulin Ig(M)-like isotope that usually exists as a tetramer in its secreted shape (55). Isotypes corresponding to mammalian IgG, IgA, or IgE have been diagnosed in fish and the secondary humoral immune reaction in fish is, if at present all, less outstanding than in mammals. Because of the ease of measurement, the presence and titer of specific antibodies is an often used immunologic assay. Assays to detect particular T-cell responses are still in their infancy; however, cellular responses from cloned fish showed prominent effects (56). For the commercial farming process of fish, an understanding of the immune response is important as it is for improved husbandry and health management of the species. The range of fish species is much more than the wide variety of mammalian species and the evolutionary distance between specific teleost (bony fish) households are significantly larger than between mammals.

Lack of precise expertise of the immune systems in distinctive fish species limits the opportunities to study each pathogen and vaccine-precipitated immunity. As with every veterinary vaccine, value effectiveness within the field is a critical hindrance to commercial vaccine development. Fish generally need a large antigen dose compared with terrestrial animals and cost-effective inactivated viral vaccines have proven difficult to develop. In a few species, all sorts of injection vaccines (or even immersion vaccines) are too steeply priced. Other species are too vulnerable to take care of the pressure prompted during the vaccination or may develop severe side effects after vaccination. Massive powerful vaccination strategies are essential for the aquaculture fields from the larval stages that have made an enormous contribution for the economic cultivation. The plain loss of maternal immunity in fish additionally limits the possibilities of shielding offspring through parental vaccination (57).

1.4 FUTURE PROSPECTS

There may be a tremendous ability for brand new achievements in fish vaccinology. Initially, there are several illnesses that are probably managed by means of vaccination. Bacterial kidney ailment infections as a result of motile aeromonads and piscirickettsiosis are examples of bacterial sickness. There also are several viral and parasitic diseases. So, to develop fish vaccinology, cooperation between the immunologist and the vaccinologist has to be developed. Better knowledge on the immune status of fishes greatly promotes the development of the aquaculture industry and many aquatic vaccine developments are in the clinical trial stage, and it is believed that there will be many commercial vaccines available in the coming years. The promising results of DNA vaccines imply that better fish vaccines can be produced in the future. Subsequently, immunopharmacology, a discipline covering adjuvants, immunostimulants, and antigen carriers, is fundamental for the fulfillment of fish vaccinology.

1.5 CONCLUSION

A perfect fish vaccine is one that is safe for the animal and environment, incurs a low cost for large-scale production, clean to manage, capable of inducing robust immunity all through durations of finest susceptibility, and demonstrates minimum facet effects. New and alternative fish vaccines are adapting advanced technologies often developed based on needs in animal or human medicines. Those who meet the standards for a powerful aquaculture vaccine will provide the maximum advantage and have the finest capacity for commercialization. New fish vaccines using opportunity technologies (beyond just cellular preparations) can be expensive to expand. However, given the limited success of conventional methods for brand new illnesses issues, it is crucial to discover such approaches. As aquaculture keeps to grow globally, there will be a need for new vaccines long into the future, and the application of all available biotechnology toward solving emerging disease issues will be critical.

REFERENCES

1. Snieszko, S. F., & Friddle, S. B. (1949). Prophylaxis of furunculosis in brook trout (Salvelinus fontinalis) by oral immunization and sulfamerazine. *The Progressive Fish-Culturist*, *11*(3), 161–168. https://doi.org/10.1577/1548-8640(1949)11[161:POFIBT]2.0.CO;2
2. Snieszko, S. F. (1974). The effects of environmental stress on outbreaks of infectious diseases of fishes. *Journal of Fish Biology*, *6*(2), 197–208. https://doi.org/10.1111/j.1095-8649.1974.tb04537.x
3. Avtalion, R. R., & Clem, L. W. (1981). Environmental control of the immune response in fish. *Critical Reviews in Environmental Science and Technology*, *11*(2), 163–188. https://doi.org/10.1080/10643388109381687
4. Ambrosius, H., & Lehmann, R. (1965). Beitrage Zur Immunbiologie Poikilothermer Wirbeltiere. 3. Der Einfluss Von Adjuvantien Auf Die Antikorperproduktion Von Knochenfischen (Teleostei). *Acta Biologica et Medica Germanica*, *14*(6), 830–944.
5. Krantz, G. E., Reddecliff, J. M., & Heist, C. E. (1963). Development of antibodies against Aeromonas salmonicida in trout. *The Journal of Immunology*, *91*(6), 757–760.
6. Snieszko, S. F. (1970). Immunization of fishes: A review. *Journal of Wildlife Diseases*, *6*(1), 24–30. https://doi.org/10.7589/0090-3558-6.1.24
7. Ross, A. J., & Klontz, G. W. (1965). Oral immunization of rainbow trout (*Salmo gairdneri*) against an etiologic agent of "Redmouth disease". *Journal of the Fisheries Board of Canada*, *22*(3), 713–719. https://doi.org/10.1139/f65-063
8. Fryer, J. L., Rohovec, J. S., Tebbit, G. L., McMichael, J. S., & Pilcher, K. S. (1976). Vaccination for control of infectious diseases in Pacific salmon. *Fish Pathology*, *10*(2), 155–164. https://doi.org/10.3147/jsfp.10.155
9. Hill, B. J., Dorson, M., & Dixon, P. F. (1980). Studies on immunization of trout against IPN. In *Fish diseases* (pp. 29–36). Springer, Berlin, Heidelberg. https://doi.org/10.1007/978-3-642-67854-7_6
10. Dadar, M., Dhama, K., Vakharia, V. N., Hoseinifar, S. H., Karthik, K., Tiwari, R., Khandia, R., Munjal, A., Salgado-Miranda, C., & Joshi, S. K. (2017). Advances in aquaculture vaccines against fish pathogens: Global status and current trends. *Reviews in Fisheries Science & Aquaculture*, *25*(3), 184–217. https://doi.org/10.1080/23308249.2016.1261277
11. Plant, K. P., & LaPatra, S. E. (2011). Advances in fish vaccine delivery. *Developmental & Comparative Immunology*, *35*(12), 1256–1262. https://doi.org/10.1016/j.dci.2011.03.007
12. Gudding, R., & Van Muiswinkel, W. B. (2013). A history of fish vaccination: Science-based disease prevention in aquaculture. *Fish & Shellfish Immunology*, *35*(6), 1683–1688. https://doi.org/10.1016/j.fsi.2013.09.031

13. Assefa, A., & Abunna, F. (2018). Maintenance of fish health in aquaculture: Review of epidemiological approaches for prevention and control of infectious disease of fish. *Veterinary Medicine International, 2018*, 5432497. https://doi.org/10.1155/2018/5432497
14. Hatha, M., Vivekanandhan, A. A., & Joice, G. J. (2005). Antibiotic resistance pattern of motile aeromonads from farm raised fresh water fish. *International Journal of Food Microbiology, 98*(2), 131–134. https://doi.org/10.1016/j.ijfoodmicro.2004.05.017
15. Cabello, F. C., Godfrey, H. P., Buschmann, A. H., & Dölz, H. J. (2016). Aquaculture as yet another environmental gateway to the development and globalisation of antimicrobial resistance. *The Lancet Infectious Diseases, 16*(7), e127–e133. https://doi.org/10.1016/S1473-3099(16)00100-6
16. Monaghan, S. J., Thompson, K. D., Smith, P. D., & Adams, A. (2016). Potential of DIVA vaccines for fish. In *Fish vaccines* (pp. 143–173). Springer, Basel, Switzerland. https://doi.org/10.1007/978-3-0348-0980-1_7
17. Baxter, D. (2007). Active and passive immunity, vaccine types, excipients and licensing. *Occupational Medicine, 57*(8), 552–556. https://doi.org/10.1093/occmed/kqm110
18. Dhar, A. K., & Allnutt, F. (2011). Challenges and opportunities in developing oral vaccines against viral diseases of fish. *Journal of Marine Science: Research and Development, 1*. http://dx.doi.org/10.4172/2155-9910.S7-001
19. Sommerset, I., Krossøy, B., Biering, E., & Frost, P. (2005). Vaccines for fish in aquaculture. *Expert Review of Vaccines, 4*(1), 89–101. https://doi.org/10.1586/14760584.4.1.89
20. Ma, J., Bruce, T. J., Jones, E. M., & Cain, K. D. (2019). A review of fish vaccine development strategies: Conventional methods and modern biotechnological approaches. *Microorganisms, 7*(11), 569. https://doi.org/10.3390/microorganisms7110569
21. Levine, M. M., & Sztein, M. B. (2004). Vaccine development strategies for improving immunization: The role of modern immunology. *Nature Immunology, 5*(5), 460–464. https://doi.org/10.1038/ni0504-460
22. Håstein, T., Gudding, R., & Evensen, O. (2005). Bacterial vaccines for fish–An update of the current situation worldwide. *Developments in Biologicals, 121*, 55–74. https://doi.org/10.1038/ni0504-460
23. Hansson, M., Nygren, P. A. K., & Ståhl, S. (2000). Design and production of recombinant subunit vaccines. *Biotechnology and Applied Biochemistry, 32*(2), 95–107. https://doi.org/10.1042/BA20000034
24. Kelly, D. F., & Rappuoli, R. (2005). Reverse vaccinology and vaccines for serogroup B Neisseria meningitidis. In *Hot topics in infection and immunity in children II* (pp. 217–223). Springer, Boston, MA. https://doi.org/10.1007/0-387-25342-4_15
25. Rappuoli, R., Pizza, M., Masignani, V., & Vadivelu, K. (2018). Meningococcal B vaccine (4CMenB): The journey from research to real world experience. *Expert Review of Vaccines, 17*(12), 1111–1121. https://doi.org/10.1080/14760584.2018.1547637
26. Biering, E., Villoing, S., Sommerset, I., & Christie, K. E. (2005). Update on viral vaccines for fish. *Developments in Biologicals, 121*, 97–113.
27. Ulmer, J. B., Mason, P. W., Geall, A., & Mandl, C. W. (2012). RNA-based vaccines. *Vaccine, 30*(30), 4414–4418. https://doi.org/10.1016/j.vaccine.2012.04.060
28. Lorenzen, N., & LaPatra, S. E. (2005). DNA vaccines for aquacultured fish. *Revue scientifique et technique (International Office of Epizootics), 24*(1), 201–213.
29. Dhar, A. K., Manna, S. K., & Thomas Allnutt, F. C. (2014). Viral vaccines for farmed finfish. *Virusdisease, 25*(1), 1–17. https://doi.org/10.1007/s13337-013-0186-4
30. Anderson, E. D., Mourich, D. V., Fahrenkrug, S. C., LaPatra, S., Shepherd, J., & Leong, J. A. (1996). Genetic immunization of rainbow trout (Oncorhynchus mykiss) against infectious hematopoietic necrosis virus. Molecular Marine Biology and Biotechnology, *5*, 114–122.
31 Shoemaker, C. A., Klesius, P. H., Evans, J. J., & Arias, C. R. (2003, April). Modified live vaccines against *Edwadsiella ictaluri* and *Flavobacterium culumnare*. In *Proceedings of the 28th Annual Eastern Fish Health Workshop*.
32. Klesius, P. H., & Shoemaker, C. A. (1999). Development and use of modified live *Edwardsiella ictaluri* vaccine against enteric septicemia of catfish. *Advances in Veterinary Medicine, 41*, 523–537.
33. Arias, C. R., Shoemaker, C. A., Evans, J. J., & Klesius, P. H. (2003). A comparative study of Edwardsiella ictaluri parent (EILO) and E. ictaluri rifampicin-mutant (RE-33) isolates using lipopolysaccharides, outer membrane proteins, fatty acids, Biolog, API 20E and genomic analyses. *Journal of Fish Diseases, 26*(7), 415–421. https://doi.org/10.1046/j.1365-2761.2003.00475.x
34. Kuzyk, M. A., Burian, J., Machander, D., Dolhaine, D., Cameron, S., Thornton, J. C., & Kay, W. W. (2001). An efficacious recombinant subunit vaccine against the salmonid rickettsial pathogen Piscirickettsia salmonis. *Vaccine, 19*(17–19), 2337–2344. https://doi.org/10.1016/S0264-410X(00)00524-7

35. Muktar, Y., Tesfaye, S., & Tesfaye, B. (2016). Present status and future prospects of fish vaccination: A review. *Journal of Veterinary Science and Technology*, *7*(2), 299. https://doi.org/10.4172/2157-7579.1000299
36. Ronen, A., Perelberg, A., Abramowitz, J., Hutoran, M., Tinman, S., Bejerano, I., & Kotler, M. (2003). Efficient vaccine against the virus causing a lethal disease in cultured Cyprinus carpio. *Vaccine*, *21*(32), 4677–4684. https://doi.org/10.1016/S0264-410X(03)00523-1
37. Munday, B. L., Kwang, J., & Moody, N. (2002). Betanodavirus infections of teleost fish: A review. *Journal of Fish Diseases*, *25*(3), 127–142. https://doi.org/10.1046/j.1365-2761.2002.00350.x
38. ickerson, H. W., Brown, J., Dawe, D. L., & Gratzek, J. B. (1984). *Tetrahymena pyriformis* as a protective antigen against *Ichthyophthirius multifiliis* infection: Comparisons between isolates and ciliary preparations. *Journal of Fish Biology*, *24*(5), 523–528. https://doi.org/10.1111/j.1095-8649.1984.tb04822.x
39. Willadsen, P., McKenna, R. V., & Riding, G. A. (1988). Isolation from the cattle tick, Boophilus microplus, of antigenic material capable of eliciting a protective immunological response in the bovine host. *International Journal for Parasitology*, *18*(2), 183–189. https://doi.org/10.1016/0020-7519(88)90059-8
40. Peeler, E. J., & Taylor, N. G. (2011). The application of epidemiology in aquatic animal health-opportunities and challenges. *Veterinary Research*, *42*(1), 1–15. https://doi.org/10.1186/1297-9716-42-94
41. Ninawe, A. S., Hameed, A. S., & Selvin, J. (2017). Advancements in diagnosis and control measures of viral pathogens in aquaculture: An Indian perspective. *Aquaculture International*, *25*(1), 251–264. https://doi.org/10.1007/s10499-016-0026-9
42. Brudeseth, B. E., Wiulsrød, R., Fredriksen, B. N., Lindmo, K., Løkling, K. E., Bordevik, M., ... & Gravningen, K. (2013). Status and future perspectives of vaccines for industrialised fin-fish farming. *Fish & Shellfish Immunology*, *35*(6), 1759–1768. https://doi.org/10.1016/j.fsi.2013.05.029
43. Shefat, S. H. T. (2018). Vaccines for infectious bacterial and viral diseases of fish. *Journal of Bacteriology and Infectious Diseases*. 2018; *2*(2): 1, 5.
44. Evensen, Ø. (2016). Development of fish vaccines: Focusing on methods. In *Fish vaccines* (pp. 53–74). Springer, Basel, Switzerland. https://doi.org/10.1007/978-3-0348-0980-1_3
45. Adams, A., & Subasinghe, R. (2019). Use of fish vaccines in aquaculture (including methods of administration). Veterinary vaccines for livestock (1st Edition), published by The Food and Agriculture Organization of the United Nations.
46. Adams, A. (2019). Progress, challenges and opportunities in fish vaccine development. *Fish & Shellfish Immunology*, *90*, 210–214. https://doi.org/10.1016/j.fsi.2019.04.066
47. Alonso, M., & Leong, J. A. (2013). Licensed DNA vaccines against infectious hematopoietic necrosis virus (IHNV). *Recent Patents on DNA and Gene Sequences*, *7*(1), 62–65.
48. Lillehaug, A., Lunestad, B. T., & Grave, K. (2003). Epidemiology of bacterial diseases in Norwegian aquaculture a description based on antibiotic prescription data for the ten-year period 1991 to 2000. *Diseases of Aquatic Organisms*, *53*(2), 115–125. https://doi.org/10.3354/dao053115
49. Kim, C. H., Johnson, M. C., Drennan, J. D., Simon, B. E., Thomann, E., & Leong, J. A. C. (2000). DNA vaccines encoding viral glycoproteins induce nonspecific immunity and Mx protein synthesis in fish. *Journal of Virology*, *74*(15), 7048–7054. https://doi.org/10.1128/JVI.74.15.7048-7054.2000
50. Madetoja, J., Nyman, P., & Wiklund, T. (2000). Flavobacterium psychrophilum, invasion into and shedding by rainbow trout Oncorhynchus mykiss. *Diseases of Aquatic Organisms*, *43*(1), 27–38. https://doi.org/10.3354/dao043027
51. Becker, R. S. (2014). Vaccine technologies & the rationale for new nanoparticle formulations. Drug Development & Delivery, 14(7), 36–40.
52. Arun Sudhagar, S., Khan, F., & Linga Prabu, D. Fish vaccination: A health management tool for aquaculture. *Fish & Shellfish Immunology*, *111*, 145–151. https://doi.org/10.1016/j.fsi.2020.10.011
53. Manning, M. J. (1996). The specific immune system: Cellular defences. *The Fish Immune System*. 1570009750647018880
54. Van Muiswinkel, W. B., & Nakao, M. (2014). A short history of research on immunity to infectious diseases in fish. *Developmental & Comparative Immunology*, *43*(2), 130–150. https://doi.org/10.1016/j.dci.2013.08.016
55. Pilström, L., & Bengtén, E. V. A. (1996). Immunoglobulin in fish—Genes, expression and structure. *Fish & Shellfish Immunology*, *6*(4), 243–262. https://doi.org/10.1006/fsim.1996.0026
56. Nakanishi, T., Fischer, U., Dijkstra, J. M., Hasegawa, S., Somamoto, T., Okamoto, N., & Ototake, M. (2002). Cytotoxic T cell function in fish. *Developmental & Comparative Immunology*, *26*(2), 131–139. https://doi.org/10.1016/S0145-305X(01)00055-6
57. Lillehaug, A., Sevatdal, S., & Endal, T. (1996). Passive transfer of specific maternal immunity does not protect Atlantic salmon (*Salmo salar*L.) fry against yersiniosis. *Fish & Shellfish Immunology*, *6*(7), 521–535. https://doi.org/10.1006/fsim.1996.0049

2 Antigens and Immune Responses in Fishes

An Overview

Nivya Mariam Paul, Aruna Babu, Amrutha Rajesh, Neema, Alanta Maria Binu, Anu Jose, Aneena Mariya Abraham, and Nayomi John
Mar Athanasius College (Autonomous)

CONTENTS

DOI: 10.1201/9781003388548-3

2.1 INTRODUCTION

Fishes are a divergent group of vertebrate animals, originating during the Devonian period. Despite certain differences, the immune system of fish physiologically resembles that of higher vertebrates. The heterogeneous group of fishes is the evident link between innate immunity and the first outcome of the adaptive immune response. Importantly, fishes have immune organs comparable to that of mammalian immune systems. In contrast to higher vertebrates, fishes lived freely in their environment from the early embryonic stage. During that time, they are frequently dependent on nonspecific immune systems for their hold up. In the fishes, nonspecific immunity is the basic defense mechanism, therewith acquired immunity also plays a key role in preserving homeostasis by activation through a system of receptor proteins, which identify pathogen-related molecular patterns typical of pathogenic microorganisms containing lipopolysaccharides, peptidoglycans, DNA, RNA, and other molecules that are typically not found on the surface of a multicellular organism. There are many external factors like environmental factors, biological factors, stress, and internal factors like genetic makeup, sex, age, maternal effect, etc. that can change the immunological defense capabilities of the fishes. Evolution has created many genetic and physiological innovations in animal phyla containing alterations in the immune mechanism. The immune system of fish is a subject that gives a unique insight into the evolution of the shielding system in the vertebrate lineage. As the earliest vertebrate in evolutionary history, Fish has an apparent pattern of immune morphogenesis compared to other higher vertebrates. They are a heterogeneous group of poikilothermic animals that contain jawless fish (e.g., lamprey) and jawed fish of class Chondrichthyes and Osteichthyes. Their physiology and immune system formation vary among them and it is greatly influenced by environmental parameters, dissimilar to warm-blooded vertebrates. External parameters like photoperiodism, the oxygen concentration of water, and temperature influence the growth and functioning of both innate (e.g., complement, lysozyme activity) and adaptive immunity. Aside from environmental influence, some deviations are inherited through genetic alterations. In adaptive immune mechanism, the genetic recombination is the key to the variation of receptors of lymphocyte-based antigen recognition sites. The role of various genes and organs involved in the defense mechanism of jawed and jawless fishes are commented upon here to provide complete information on the progress or innovation in the fish immune system (1).

The functions of the fish immune system are: defense against disease and protection against neoplastic cells. Fishes, the largest vertebrate class, are divided into jawless fish and jawed fish. It can be further subdivided into cartilaginous fish and bony fish (2).

The basal position of fish in vertebrate phylogeny makes them very enchanting for genomic and functional comparative immunity studies. The immune molecules, like Ag-recognizing lymphocytes, immunoglobulins, major histocompatibility complex (MHC), recombination activation genes (RAG) 1 & 2, and the recombination mechanisms are identical in fishes and mammals. These molecules and their immune reaction mechanisms piece together the primordial vertebrate immune system repertoire and adaptive radiations (3). In this article, we are discussing the immune system, antigens, immune modulation, and regulations.

2.2 IMMUNE SYSTEM IN FISHES

Evolution has introduced many genetic and physiological improvements in animal phyla including alterations in immune mechanisms. Fish as the earliest vertebrate in evolutionary records has a different sample of immune morphogenesis in comparison to other higher vertebrates. They are the heterogeneous organization of poikilothermic animals, which include jawless fish and jawed fish. Their physiology and immune system development vary among them, and it is particularly influenced via environmental parameters, not like warm-blooded vertebrates. Outside parameters like photoperiodism, temperature, and concentration of oxygen in water influences the maturation and functioning of each innate (e.g., supplement, lysozyme pastime) and adaptive immunity in fish (4). Apart from antimicrobial immune reactions, environmental factors have an impact on some of the changes that are inherited and evolved via genetic changes. The primary mechanism through which the adaptive immunity diversifies its repertoires of lymphocyte-based antigen receptor is genetic recombination (5). The roles of various genes and organs involved in the defense mechanism of jawed and jawless fishes are discussed right here so that they will provide complete information on the development of the fish immune system.

2.2.1 Immunity of Agnathans

No matter the diversification, many features of fish immunity, i.e., immune gene expression, inflammation, wound recovery, antigen sample receptors, signaling, and trafficking of lymphocytes, remain conserved across the vertebrate lineage. These features are by and large performed through the cell and humoral factors of immunity. The Agnathans lack hematopoietic organs, i.e., spleen, thymus, or kidney. However, they have a unique strip of medullary tissue present in the trunk referred to as the immune body (6). The dedicated organs for immunity have not been detected so far. However, a number of areas of lamprey typhlosole and renal folds carry hematopoietic stem cells (HSC) and lymphoid-like cells and differentiated cells including thrombocyte-, granulocyte-, monocyte-, and lymphocyte-like cells have additionally been detected (7). The humeral elements like antimicrobial peptide coding genes, i.e., cathelicidin genes have been detected in Atlantic hagfish (Myxine glutinosa) (8). Other innate immunity-associated genes, such as reactive oxygen species modulator I, peroxiredoxin coding gene, and NFκB inhibitor gene, are being detected in the immune body and other tissues, which imply the presence of a well-advanced innate protection mechanism (9). The lamprey oral gland is additionally determined to secrete many defenses related to functional proteins, e.g., interferon-prompted lethality protein-19 and disintegrins. The components present in the supplement activation pathway have been detected in lamprey (10). The homologous components like C3, MBL, and MBL associated serum protease (MAPS) of the lectin pathway and the alternative pathway had been identified from lamprey and/or hagfish, but the cytolysis in terms of serum protein is named "lamprey pore-forming protein." The biomarkers of adaptive immunity (11), i.e., MHC, T cell receptors (TCRs), and B cell receptors (BCRs), are absent in primitive forms, but in spaces, there are a lot of leucine-rich repeats indicating an alternative pathway. Some of the research has located precise agglutinin-based reminiscence for antigen reputation in Atlantic lamprey and agglutinin-secreting cells within the gut. The lamprey has particular lymphocytes-expressing genes encoding B-cell signaling additives. The VDJ recombination system, which is required for developing a diversified repertoire of Ig, is absent in Agnathans. The lymphoid cells have determined a specific complex leucine rich repeats (LRR) molecule known as variable lymphocyte receptors (VLR), which undergoes subsequent assembly through an entirely novel genomic mechanism in which massive banks of LRR cassettes are used to build the "diversity" area of the receptor molecules (12). The fundamental composition of these VLR consists of a conserved sign peptide, an N-terminal LRR (LRRNT), followed by nine variable and tremendously numerous LRRs, a connecting peptide, a C-terminal LRR (LRRCT), a conserved C terminus (GPI)-anchor site, and a hydrophobic tail. Upon antigen induction, there is the marked proliferation of hematopoietic lymphoid cells and increased VLR protein receptors for variable antigen detection.

2.2.2 Immunity of Osteichthyes

Fishes have a resilient innate immunity which helps them to continue to exist and adapt to the unfavorable conditions. The kidney, thymus, and spleen are important lymphoid organs in fish (13). Head kidney plays a chief role in hematopoiesis and antimicrobial activity. The pattern of fish lymphoid organs is unique with the kind of fish and its development also varies among marine and freshwater fishes (14). For example, the kidney, thymus, and spleen are the largest lymphoid organs in teleost fishes. In the case of freshwater teleost, the kidney is the primary lymphoid organ and the spleen is the secondary organ. In marine fishes, it develops in the order of kidney, spleen, and thymus. In marine water teleost fishes, the anterior kidney is the first organ to appear followed by spleen and thymus. But, in each case, the thymus is the primary organ to have lymphoid cells accompanied via the kidney and spleen.

2.2.3 Fish Innate Immunity

Nonspecific immunity is observed in all residing organisms and is the primary line of defense against all pathogens. It also plays a crucial function in the activation of the adaptive immune response. The cell of the innate body recognizes and generically responds to pathogens. It also possesses memory as the host evolves its innate immune components based on the evolutionary experience of its ancestors encountering comparable pathogens. Innate immunity is commonly divided into three compartments: surface barrier, humoral factors, and cellular elements. As in the first line of protection, it is not astonishing that most people of the wide-spectrum parameters of innate immunity are conserved across species and taxa. In all jawed vertebrates, the innate immune system features a rapid reaction toward invading pathogens and tissue damage. But, it cannot provide perfectly directed, specific protection from individual pathogens or long-term period immunological reminiscence.

2.2.3.1 Surface Barrier

Mucus, skin, gills, and gastrointestinal (GI) tract act as the first line of barrier to any diseases. The layer of mucus found in pores and skin, gills, and GI tract entrap microorganisms with the aid of continuously sloughing and inhibiting colonization. The mucus of fish is toxic to certain microorganisms because of the presence of a few humoral factors. The rate of mucus production will increase in reaction to infection or physical or chemical irritants (15). The dermis of fish skin consists of nonkeratinized living cells and the integrity of those cells plays an important function in maintaining osmotic stability and excluding microorganisms. Speedy healing is likewise observed in the epidermis of fishes (16). The large surface area of delicate gill epithelium is taken into consideration as a crucial course of pathogen entry. The gills are covered via mucus production and highly responsive epithelium, resulting in hyperplasia, often seen in numerous gill infections. Phagocytic cells line the branchial capillaries and lymphoid cells on the caudal edge of the intrabronchial septum. GI tract is lined by the mucus membrane and additionally the digestive enzymes, bile, and low pH of the stomach give an extremely adverse environment for pathogens, typically based on their pattern recognition specificities or effector functions.

2.2.3.2 Humoral Factors

They have a protective function that inhibits microorganisms and neutralizes the host enzymes on which pathogens depend. The classification of humoral parameters is based on their pattern recognition factors or effector functions.

2.2.3.2.1 Growth Inhibitors

Growth inhibitors act either on crucial vitamins or interfere with microbial metabolism. Transferrin (TF) in serum exerts bacteriostatic and fungistatic effects. TF, protein with high iron-binding capability, is a crucial element for the growth of microorganisms and preventing the utilization of iron by microorganisms (17). Pathogenic bacteria can produce siderophores to overcome this defense mechanism and the hyperferremic activity acts as a counter-reaction established in fish species. TF is also an acute-phase protein cited in inflammatory reaction to get rid of iron from damaged tissues (18) and an activator of fish macrophages (19). Interferons are virus-inducible cytokines that induce the expression of different antiviral proteins (20). The lysozyme variants (types I and II), purified from the head kidney of rainbow trout, has antibacterial impact gram –ve bacterial fish pathogens (1). INF α and β have nonspecific antiviral characteristics that cause the inhibition of nucleic acid replication. They induce the expression of interferon-stimulated genes inducing Mx, Viperin, ISG 15, and PKR. IFN-γ promotes Th 1 responses via CD4+Th 1 and natural killer (NK) cells. It provides defense against intracellular pathogens. Fish IFN modulates cytokine and chemokine expression, and induces proinflammatory cytokines (IL-1, IL-6, IL-12, and TNF).

2.2.3.3 Cellular Factors

Cellular factors include monocytes, macrophages, granulocytes, neutrophils, dendritic cells, and NK cells. When an innate immune system recognizes a pathogen through its pathogen-associated molecular pattern, these immune cells get activated and cause phagocytosis and destruction of pathogens (21).

2.2.3.3.1 Macrophages

Macrophages are differentiated via circulating monocytes or through tissue-resident macrophages like kupffer cells in the liver, glial cells in the brain, and so on. Its differentiation is controlled by the CSF receptor 1 (22). It was first recognized inside the elephant shark (*Callorhinchus milii*) genome. Macrophages in teleost play a role in both innate and adaptive immunity. In the innate immune reactions, they destroy pathogens via phagocytosis, reactive oxygen species, nitric oxide production, and the release of anti-inflammatory cytokines and chemokines. In teleost fish species, classically activated macrophages (M1) produce TNF, IL-1b, ROS, and NO (23), and kill pathogens through engulfment and production of toxic reactive intermediates, phagolysosomal acidification, and limiting nutrient availability. M2 are alternatively activated macrophages and are immunosuppressant and anti-inflammatory.

2.2.3.3.2 Fish Gill

Diseases associated with gills harm and cause enormous losses in the aquaculture industry not only via an extended mortality rate among fish but additionally through impaired growth and also with the aid of increased remedial and sanitation costs. Damage to gill tissues is especially characterized by infection and increased epithelial cells hyperplasia or hypertrophy. A gill epithelium of salmonids has a better variety of MHC class II positive cells (24), whereas a low wide variety of macrophages-like cells has been detected in the gill epithelium of presumably healthy salmonid fish (25) cytokines including interleukin (IL)-10.

2.3 IMMUNE ORGANS OF FISH

Fishes can be classified under the phylum chordate and subphylum vertebrate. Fishes are a very fascinating and diverse group. Fishes originated during the Devonian period. Fishes have their own clear immunological mechanisms when they are affected by pathogens. In spite of certain dissimilarities, fishes have an immune system similar to that of higher vertebrates (1). The immune response in fishes of two types is as follows:

1. An innate (inborn) defense system
2. An acquired (adaptive) defense system

Fishes have immune organs that are similar to that of the mammalian immune system. In the early embryonic stage, fishes have innate immunity. Different organs in fish act as immune barriers. Increased awareness of fish immune system will ease the buildout of novel vaccination strategies in fish (26).

2.3.1 Lymphoid Organs of Fish

Lymphoid organs are organs of the lymphatic system involved in the immune mechanism of fish. Lymphoid organs of fish are of two main types – primary lymphoid and secondary lymphoid. Primary lymphoids are the thymus and anterior kidney that produce and mature stem cells. The secondary lymphoid organs are head kidney, spleen, and mucosa. These are the main organs that take part in the immune mechanism, whereas liver, skin, intestine, and heart are important organs that take part in the defense. In the case of teleost fishes, kidney, spleen, and thymus are the largest lymphoid organ. Features are different between marine water and freshwater species (27).

2.3.1.1 Primary Lymphoid Organs

The thymus and anterior kidney are the primary lymphoid organs.

2.3.1.1.1 Thymus

The thymus is the most crucial lymphoid organ and is found in all vertebrates. In fish, thymus is the first organ to be lymphoid. Thymus is the paired organ situated in the dorsal lateral region of the gill chamber close to the opercular cavity. To protect from harsh environment, it is covered by the pharyngeal epithelium (27). T-cells are involved in the stimulation of phagocytosis and B cells produce antibodies (28). The thymus starts developing 24 hours after fertilization, and it is the lymphoid tissue that contains lymphocytes during development. Once the thymocytes are developed, they migrate from thymus to the spleen and kidney. The development of thymus in fish is more dependent on the hormonal cycles. The morphology of thymus changes with age (2).

2.3.1.1.2 Anterior Kidney

The fish kidney works similar to that of bone marrow in the vertebrates and anatomy of the kidney varies from species to species. In the case of teleost fish, kidney not only functions as the excretory organ but also is the largest site of hematopoiesis (29). Immune cells are present over the entire kidney. In fish, kidney is the main organ for phagocytosis, antigen processing, and formation of IgM. Antibody production mainly takes place in the anterior kidney (23). It has a role in the development and differentiation of HSC. Fish immune system is similar to higher vertebrates because after migration of precursor cells from fetal liver, pro-myeloid cell formation occurs in bone marrow for lifetime. It also contains the cortical adrenal homologs, and medullary and hematopoietic tissues. Excretory function of anterior kidney stops when it becomes mature (30).

2.3.1.2 Secondary Lymphoid Organs

Secondary lymphoid organs are the head kidney, spleen, and mucus.

2.3.1.2.1 Head Kidney

The head kidney functions as the secondary lymphoid organ. Head kidney is the site where antibody-producing cells are formed. In studies, it has been shown that the head kidney has antigen-sampling abilities and antigen retention (30). The proliferating B-cell precursors and plasma cells are mostly present in the head kidney. It is the site for HSC development and differentiation, and it has the highest concentration of developing lymphoid cells (27).

2.3.1.2.2 Spleen

The spleen functions as a secondary immune organ. It is formed in the form of mesenchymal cell aggregates surrounded by blood capillaries. The spleen is the third important hematopoietic organ to be lymphoid. However, for a short period of time, it carries erythroid cells only. The zebra fish contains ellipsoids. They appear at 3 months after hatching and are involved in plasma filtration and blood-borne antigens (31). The size of the spleen is used as a simple measurable immune parameter with potential role in immune response against parasite infections (32).

2.3.1.2.3 Mucus

The mucus is present in the skin, gills, and GI tract, and it acts as the first line of defense against the invading pathogen. When the fish body comes in contact with pathogen skin, gills and GI tract have the ability to fight against it. In some cases, the mucus is toxic for some microorganism due to some humoral factors. For example, if the infection rate is high in a fish, the mucus production also increases. Gills are the major route for the entry of pathogens, so the gills are protected by the mucus and epithelium. Likewise, skin also contains keratinized living cells, which help in mucus production, preventing the entry of pathogen; gastrointestine containing mucus membrane; and digestive enzymes (15).

2.4 ANTIGENS IN FISH

External invaders, such as pathogens, chemicals, pollutants, and pollens, can be antigens. Under pathological conditions, normal mobile proteins can turn out to be self-antigens. Fish can reject allografts (33) and synthesize particular antibodies after antigenic undertaking. Infectious and non-infectious diseases can affect fishes. Infectious diseases are caused by pathogenic organisms present in the environment or are spread by other fishes. They are contagious and treatment is necessary to control them. Noninfectious diseases are caused by environmental factors, nutritional deficiencies, or genetic problems.

They are not contagious and cannot be cured by medicines. Infectious diseases are again classified into parasitic, bacterial, viral, or fungal diseases. Parasitic diseases infect the gills and skin of fishes causing irritation, loss in weight, and eventually leading to death. They can be easily controlled using chemicals like copper sulfate, formalin, etc. Viral diseases are difficult to cure and no specific medications are available. Infectious pancreatic necrosis (IPN), infectious hematopoietic necrosis (IHN), and viral hemorrhagic septicemia (VHS) are predominant viral sicknesses affecting fishes (34). Fungal spores are common among the aquatic habitat. Fungi colonize the damaged exterior tissues and these areas appear to be cottony or brown matted. Formalin or potassium permanganate is effective against most fungal infections.

Bacterial diseases are internal infections and require medications containing antibodies. Fishes infected with such infections have hemorrhagic spots or ulcers along the body wall, and around the eyes and mouth.

Examples for some bacterial infections are given below:

1. *Aeromonas salmonicida* is the infectious agent of furunculosis, a common and critical sickness in the fisheries sector globally. The disease is essentially managed by using commercial oil adjuvanted vaccines containing bacterins. However, the mechanisms leading to immune response remain poorly understood.
2. *Vibrio anguillarum* is the infectious agent of vibriosis. For many years, basic bacterin vaccines had been used to defend cultured fish from vibriosis. The vaccines incorporate two essential serotypes, namely, O1 and O2. This provides protection against *V. anguillarum* (35) and *V. anguillarum* (36).
3. *Yersinia ruckeri* is the infectious agent of enteric red mouth (ERM). As a crude bacterin, the vaccine has been administered through all routes of management, namely, injection, bathtub, spray, and orally.
4. *Edwardsiella ictaluri* is the reason for large losses in channel catfish, and vaccination using a bacterin has been found to be effective—best results were obtained while injected with freund's whole adjuvant (37). Immunity is obtained by stimulation of CMI, as this pathogen can multiply in macrophages of infected fish. The bacterium facilitates immune serum killing (38) and resists phagocytosis through catfish neutrophils (39).

5. *Photobacterium damselae* is the causative agent of pseudotuberculosis in yellow tail fish in Japan and Mediterranean countries. Virulent strains are resistant to killing via fish serum and this is due to the presence of a polysaccharide pill (40).

Noninfectious environmental diseases are common in commercial aquaculture. They include low dissolved oxygen, high ammonia and nitrate, or natural and manmade causes. Water-quality management will prevent such diseases. Noninfectious nutritional diseases are very difficult to diagnose. One example is broken back disease in catfish (due to vitamin C deficiency). It leads to improper bone development and deformation of the spinal column. Another example is "no blood disease" in catfish (folic acid deficiency). Affected fishes become anemic and die. Genetic abnormalities include conformational oddities like lack of a tail or presence of an extra tail. Fish disease outbreaks are often complex and includes both infectious and noninfectious processes. Proper therapy includes medication and changes in husbandry practices. Assistance from UF/IFAS aquaculture extension specialists is available to manage disease outbreaks and develop management programs to prevent them.

2.5 ELEMENTS OF IMMUNE RESPONSES

The fish immune system is of two types:

- Innate Immunity
- Adaptive Immunity

Innate immunity will invade pathogens by recognizing the pathogens. Toll-like receptors (TLRs) and phagocytosis are the key elements of the innate immunity that protect the organism against invaders by recognizing and destroying the pathogens. Adaptive immunity recognizes pathogens through molecules that are developed by somatic mechanism by humoral and cellular responses through B- and T-lymphocytes.

2.5.1 Innate Immunity

Innate immunity is the first layer of defensive mechanism against pathogens. It is divided into CMIR and humoral immune response (HIR). In cellular immunity, epithelial cells and mucus act as a barrier that is lining the stomach, skin, and gills. If the pathogens cross these barriers, macrophages, granulocytes, monocytes, and nonspecific cytotoxic cells (NCCs) destroy and digest these pathogens. They are retained at the site of infection by inflammatory cytokines. Humoral immune reactions have a variety of proteins and glycoproteins that are capable of destroying the antigens.

2.5.2 Nonspecific Cellular Immunity

Leukocytes and white blood cells (WBCs) are involved in the innate nonspecific cellular immunity of fish. Toll-like receptors, NCCs, macrophages, and granulocytes are part of it. TLRs have the capability to recognize the epitopes in pathogens. Granulocytes and macrophages are the phagocytic cells in the circulating blood and the 2° lymphoid organs. Host cells infected by pathogens are the targets for NCCs making them appear in the mucosal area, blood circulation, lymphoid tissues, and killing them through apoptosis and necrosis (41).

2.5.2.1 Toll-Like Receptors

TLRs are vital components of innate immunity reactions and are type I transmembrane protein with an extracellular domain (leucine-rich repeats) and a cytoplasmic tail region. In 1985, Christiane

Nusslein Volhard identified toll gene in Drosophila. Pathogen associated molecular pattern molecules (PAMP) has a role in the initiation of the innate immune receptors and have a role in the revealing of microbial infection (42).

2.5.2.2 Macrophages

Macrophages are mononuclear. They are peroxidase negative leukocytes. They are also nonspecific esterase positive. They are phagocytic and emit nitrogen free radicals and O_2 that kill microbes. Macrophages have both complementary and antibody receptors and are expressed in class II MHC molecules.

Macrophages are produced in fish during primary hematopoiesis. Monocytes circulate in the blood to mature as tissue macrophages. They are found in both layers of fish thymus—the cortex and medulla. Macrophages are also found between the inner and outer zones of the pharyngeal epithelium, lymphoid organ, blood, and in the perineal cavity (43).

2.5.2.3 Granulocytes

Granulocytes are polymorphonuclear leukocytes in fish species. Three types of granulocytes are found in fishes such as neutrophil, eosinophil, and basophil. They are differentially distributed and can be isolated from blood, peritoneal cavity, and lymphoid tissues of fishes. The distribution of granulocytes in fish depends on season of the year, diseases, environmental pollutants, and various stressors that greatly enhance the number of granulocytes in blood within 24 hours of stressing the fish (44).

2.5.3 Nonspecific Cytotoxic Cells

NCCs are like NK cells and show several similarities such as competent lytic cycle, target cells for lysis, effectors to lyse the infectious microbes, and also have differences such as kinetics of killing, morphology, and specificity of the target cell. They are active in the kidney of fish but in spleen and peripheral blood lymphocytes (PBL) have cytotoxic abilities (44).

2.5.4 Lysozymes

Lysozymes play an important role in the defensive mechanism against invading pathogens or microorganisms. Neutrophils and monocytes can produce lysozymes in fish. They are mainly distributed at the sites of microbial invasions such as GI tract, gills, skin, eggs, and in the leukocyte-rich organs. They can be identified from the body mucus, blood, and various tissues of the fishes, and they play an important role in preventing vertical transmission of bacterial pathogens in fishes. They have antibacterial activity against gram +ve and gram –ve bacteria due to the disruption of lipid polysaccharide layer, which leads to damage of the outer cell membrane allowing additional lysozymes to reach and injure deep structures by increasing permeability, which will lead to the loss of viability of cells without lysis (45).

2.5.5 Alkaline Phosphatase

Alkaline phosphatase is found in various body secretions such as mucus, blood, and its concentration will increase during stress. In fish, high activity of alkaline phosphatase occurs during skin regeneration in wound healing and has a role in protection.

2.5.6 Complements

Complements are involved in both initiation of the innate immune reactions and mounting of an adaptive immune reaction using more than 35 soluble proteins and a combination of alternative, lectin, and classical pathways (41).

2.5.7 Interferons

Interferons have antiviral activity by inhibiting viral replication. In 2003, the first interferon gene of fish was cloned. Fish cells produce IFN-α and β in response to virus attack. Type I IFN has a defensive role against virus attack and has five axons and four intron genes (46).

2.5.8 C-Reactive Protein

C-reactive protein is the first protein that exists in the blood plasma of humans and most species in response to tissue damage, infection, and inflammation. It was discovered that C polysaccharide of Pneumococcus in the serum results in acute swelling and so it is named as C-reactive protein. The fat cells synthesize CRP and are members of the pentraxin family of proteins and were the first PRR to be identified, and they were isolated from many fish species (47).

2.5.9 Transferrin

TF is a bi-lobed monomeric iron-binding glycoprotein actively involved in iron metabolism closely associated with an immune response, which helps in transporting iron around the body. Iron acts as a constituent for growth and survival, and surplus free iron radicals show a virulent effect on the cells. Regulation of iron metabolism maintains the balance between beneficial and toxic effects by iron transporter transferrin, which also responds to infection. TF can be ascertained from nearly all fish species such as lamprey and catfish (48).

2.5.10 Lectins

Lectins are primitive molecules that have multiple functions and were initially identified as hemagglutinins as they bind carbohydrates and agglutinate cells. Lectins can be divided into C- and S-type lectins; C-type lectin is calcium dependent. Lectins have been isolated from many species of fish eggs, which provide protection to the developing egg and prevent the transmission of pathogenic organisms (49).

2.5.11 Adaptive Immunity

Adaptive immunity is also known as specific immune response that comprises specialized cells that eliminate pathogens or prevent their growth. It consists of antibodies and lymphocytes, often called as humoral, and cell-mediated immune response. Cells of adaptive immunity include B lymphocyte, which is derived from the bone marrow. They produce antibodies and T-lymphocytes mature in thymus and participate in lymphocyte maturation or kill virus infected cells. Humoral immunity varies in fish and mainly depends on external conditions and species of fish. Several characteristics of fish humoral immunity include Ig structure, cellular requirement for stimulation of Ab and function of Ab in neutralization, opsonization, and complement fixation. Cellular immunity or cell-mediated immunity involves macrophages, NK cells, mast cells, neutrophils, basophils, and eosinophils (50).

2.6 WHAT IS IMMUNITY MODULATION?

The substance stimulates or suppresses the immune device and surely assists the body in combatting most cancers, contamination, or different sicknesses. Specific immune machine modulators, compendious of monoclonal antibodies, cytokines, and vaccines, affect precise sections of the immune device (1).

2.6.1 How Does the Fish Immune System Work?

Fishes possess a properly evolved nonspecific immune system. The skin, lateral line, and gills are the primary line of defense against pathogens in fishes. The skin of fish is lined through mucus, which is continuously secreted through goblet cells and incorporates antibodies besides the lysozyme (51).

2.6.2 Why Is Immune Regulation Important?

The immune machine is built of a complex community of cells that protect us against contamination from other life-threatening pathogens. It's so far an effective force, which requires to be carefully regulated to prevent harm to the host. Stress and immune modulation in fish is something that most animals experience, and it generates several responses associated with all three regulatory structures, namely, neural, endocrine, and immune. While the stressor is acute and short term, the reaction structure is stimulatory and the fish immune reaction displays an activating section that particularly improves innate responses. If the stressor is persistent, the immune reaction indicates suppressive effects and consequently, the possibilities of contamination can be high. Furthermore, coping with the stressor puts an allostatic value, which can intrude on the needs of the immune reaction (52).

In the following years, it has become more obvious that the mammalian immune system and the neuroendocrine system are linked together which is essential for maintaining the homeostatic feature. General studies have focused on the interaction between the two structures through cytokines and neuropeptides and as a result, it has become difficult to define many of those molecules exactly in terms of the function with which they were related at the beginning. Much later, immune±endocrine interactions in nonmammalian vertebrates, specifically fish, have obtained multiplied interest.

2.6.3 Immunomodulatory Effects of Hormones

2.6.3.1 Cortisol

The primary corticosteroid formed in the teleost interregnal gland is cortisol (53). The immunomodulatory effects of stress and corticosteroids in mammals (54) are similar to that in fish. Stress or cortisol administration reduces the resistance of fish to microbial infections (55). Cortisol administration and catecholamines release have an effect on the fish immune system (56). Cortisol administration decreases the circulating T-cells and B-like circulating phagocytes (neutrophils and macrophages) may increase. Thus, acute stress or cortisol treatment in juvenile coho salmon reduces the number of circulating leukocytes but raises the number of these cells in the thymus and head kidney. Cortisol, however, has also been shown to induce apoptosis in B cells and this could be responsible for their removal from the blood following stress (57). Cortisol has depressive effects on phagocytosis, lymphocyte mitogenesis, and reduces the antibody production and circulating IgM titers (58). The effects of cortisol are mediated through the glucocorticoid receptor (GR), which has been recognized in leukocytes from coho salmon and carp (59).

2.6.3.2 Growth Hormone and Prolactin

The polypeptides growth hormone and prolactin are synthesized in the adenohypophysis of the pituitary gland and share numerous structural and functional characteristics. They are involved in the regulation of postnatal somatic growth in all vertebrates and also influence osmoregulation and increase gonadal steroidogenesis in fish. Prolactin is a versatile peptide with a wide range of functions that can be classified into three broad categories: growth and development, osmoregulation, and replication (54). In animals, GH and PRL promote thymocyte maturation, differentiation, and phagocytes formation (54). They inhibit glucocorticoid-induced T cell progenitor death in gilthead sea bream and silver sea bream. GH has been shown to increase lymphopoiesis and phagocytosis (60). In salmon, it increases leukocyte mitogenesis, phagocytosis, NK cells, and antibody production. The peptide enhances the resistance of rainbow trout to the *Vibrio anguillarum* (61). Prolactins have been shown to stimulate leukocyte mitogenesis (62), breathing burst interest, phagocytosis, and increasing IgM titers (63).

2.7 CONCLUSION

Fishes have properly advanced immune structures, with complete representation of all acknowledged fundamental components of innate and adaptive immunity, although with a few specializations and specific capabilities. The knowledge of the structure and function of fish immune system is important. However, on understanding about the immune system, there is a possibility of promoting appropriate stimuli, such as immune stimulants and prebiotics administration in order to prevent losses and to avoid the unwanted use of antibiotics. The study on the modification and manipulation of the innate immune system is a successful development for the expansion of sustainable aquaculture. Antigenic stimulation in fish evokes responses that are similar to those in warm-blooded vertebrates. The presence of an effective innate immune system and acquired immune responses show the expected characteristics of specificity and memory. There is a clear influence of environmental factors, like temperature and stress conditions. Our knowledge of the immune system of fish is useful to evaluate the health status of fish under different conditions, and can also be used for vaccination and breeding for disease resistance in aquaculture. While commercial vaccines, which can induce protection against bacterial diseases, are available, only little is understood about the nature of the protective mechanisms involved. There are several other economically important fish diseases for which no effective vaccines exist, for example, rainbow trout fry syndrome (RTFS), pasteurellosis, edwardsiellosis, etc. In the case of vibriosis and ERM, simple bacterins have proved to be powerful vaccines. However in the case of other diseases, antibody response is directed in opposition to antigens, which are most effective under particular conditions. Much more information is required to know the host–pathogen relationships in these diseases to understand potential protective mechanisms. With such records, more directed processes for the production and presentation of pertinent antigens can allow similar improvement of vaccination of fish in opposition to many economically vital bacterial diseases of fish.

REFERENCES

1. Sahoo, S., Banu, H., Prakash, A., & Tripathi, G. (2021). Immune system of fish: An evolutionary perspective. *Antimicrobial Immune Response*, 1.
2. Hitzfeld, B. (2005). Fish immune system. In *Encyclopedic reference of immunotoxicology*. Springer, Berlin, Heidelberg.
3. Rauta, P. R., Nayak, B., & Das, S. (2012). Immune system and immune responses in fish and their role in comparative immunity study: A model for higher organisms. *Immunology Letters*, *148*(1), 23–33.
4. Bowden, T. J. (2008). Modulation of the immune system of fish by their environment. *Fish & Shellfish Immunology*, *25*(4), 373–383.
5. Cooper, M. D., & Alder, M. N. (2006). The evolution of adaptive immune systems. *Cell*, *124*(4), 815–822.
6. Yang, X. U., Si-Wei, Z. H. U., & Qing-Wei, L. I. (2016). Lamprey: A model for vertebrate evolutionary research. *Zoological Research*, *37*(5), 263.
7. Bajoghli, B., Guo, P., Aghaallaei, N., Hirano, M., Strohmeier, C., McCurley, N.,... & Boehm, T. (2011). A thymus candidate in lampreys. *Nature*, *470*(7332), 90–94.
8. Uzzell, T., Stolzenberg, E. D., Shinnar, A. E., & Zasloff, M. (2003). Hagfish intestinal antimicrobial peptides are ancient cathelicidins. *Peptides*, *24*(11), 1655–1667.
9. Sun, J., Liu, X., & Li, Q. (2010). Molecular cloning, expression and antioxidant activity of a peroxiredoxin 2 homologue from Lampetra japonica. *Fish & Shellfish Immunology*, 28(5–6), 795–801.
10. Matsushita, M. (2018). The complement system of Agnathans. *Frontiers in Immunology*, *9*, 1405.
11. Pancer, Z., Amemiya, C. T., Ehrhardt, G. R., Ceitlin, J., Larry Gartland, G., & Cooper, M. D. (2004). Somatic diversification of variable lymphocyte receptors in the Agnathan sea lamprey. *Nature*, *430*(6996), 174–180.
12. Velikovsky, C. A., Deng, L., Tasumi, S., Iyer, L. M., Kerzic, M. C., Aravind, L., ... & Mariuzza, R. A. (2009). Structure of a lamprey variable lymphocyte receptor in complex with a protein antigen. *Nature Structural & Molecular Biology*, *16*(7), 725–730.
13. Fijan, N. (2002). Composition of main haematopoietic compartments in normal and bled channel catfish. *Journal of Fish Biology*, *60*(5), 1142–1154.

14. Zapata, A., Diez, B., Cejalvo, T., Gutierrez-de Frias, C., & Cortes, A. (2006). Ontogeny of the immune system of fish. *Fish & Shellfish Immunology*, *20*(2), 126–136.
15. Ángeles Esteban, M. (2012). An overview of the immunological defenses in fish skin. *International Scholarly Research Notices*, *2012*, 853470. https://doi.org/10.5402/2012/853470
16. Sveen, L., Karlsen, C., & Ytteborg, E. (2020). Mechanical induced wounds in fish–A review on models and healing mechanisms. *Reviews in Aquaculture*, *12*(4), 2446–2465.
17. Bayne, C. J., & Gerwick, L. (2001). The acute phase response and innate immunity of fish. *Developmental & Comparative Immunology*, *25*(8–9), 725–743.
18. Stafford, J. L., & Belosevic, M. (2003). Transferrin and the innate immune response of fish: Identification of a novel mechanism of macrophage activation. *Developmental & Comparative Immunology*, *27*(6–7), 539–554.
19. Ellis, A. E. (2001). Innate host defense mechanisms of fish against viruses and bacteria. *Developmental & Comparative Immunology*, *25*(8–9), 827–839.
20. Beck, G., & Habicht, G. S. (1996). Immunity and the invertebrates. *Scientific American*, *275*(5), 60–66.
21. Bayne, C. J., & Gerwick, L. (2001). The acute phase response and innate immunity of fish. *Developmental & Comparative Immunology*, *25*(8–9), 725–743.
22. Smith, N. C., Rise, M. L., & Christian, S. L. (2019). A comparison of the innate and adaptive immune systems in cartilaginous fish, ray-finned fish, and lobe-finned fish. *Frontiers in Immunology*, *10*, 2292.
23. Hodgkinson, J. W., Grayfer, L., & Belosevic, M. (2015). Biology of bony fish macrophages. *Biology*, *4*(4), 881–906.
24. Olsen, M. M., Kania, P. W., Heinecke, R. D., Skjoedt, K., Rasmussen, K. J., & Buchmann, K. (2011). Cellular and humoral factors involved in the response of rainbow trout gills to Ichthyophthirius multifiliis infections: Molecular and immunohistochemical studies. *Fish & Shellfish Immunology*, *30*(3), 859–869.
25. Goldes, S. A., Ferguson, H. W., Daoust, P. Y., & Moccia, R. D. (1986). Phagocytosis of the inert suspended clay kaolin by the gills of rainbow trout, *Salmo gairdneri* Richardson. *Journal of Fish Diseases*, *9*(2), 147–151.
26. Bjørgen, H., & Koppang, E. O. (2022). Anatomy of teleost fish immune structures and organs. *Principles of Fish Immunology*, *73*(1), 53–63.
27. Rauta, P. R., Nayak, B., & Das, S. (2012). Immune system and immune responses in fish and their role in comparative immunity study: A model for higher organisms. *Immunology Letters*, *148*(1), 23–33.
28. Zapata, A., Diez, B., Cejalvo, T., Gutierrez-de Frias, C., & Cortes, A. (2006). Ontogeny of the immune system of fish. *Fish & Shellfish Immunology*, *20*(2), 126–136.
29. Whyte, S. K. (2007). The innate immune response of finfish–A review of current knowledge. *Fish & Shellfish Immunology*, *23*(6), 1127–1151.
30. Bowden, T. J., Cook, P., & Rombout, J. H. W. M. (2005). Development and function of the thymus in teleosts. *Fish & Shellfish Immunology*, *19*(5), 413–427.
31. Brattgjerd, S., & Evensen, Ø. (1996). A sequential light microscopic and ultrastructural study on the uptake and handling of Vibrio salmonicida in phagocytes of the head kidney in experimentally infected Atlantic salmon (Salmosalar L.). *Veterinary Pathology*, *33*(1), 55–65.
32. Lefebvre, F., Mounaix, B., Poizat, G., & Crivelli, A. J. (2004). Impacts of the swimbladder nematode *Anguillicola crassus* on *Anguilla anguilla*: Variations in liver and spleen masses. *Journal of Fish Biology*, *64*(2), 435–447.
33. Botham, J. W., Grace, M. F., & Manning, M. J. (1980). Ontogeny of first set and second set alloimmune reactivity in fishes. In *Phylogeny of immunological memory* (pp. 83–92). Elsevier, North-Holland, Amsterdam.
34. McAllister, P. E. (1979). Fish viruses and viral infections. In *Comprehensive virology* (pp. 401–470). Springer, Boston, MA.
35. Viele, D., Kerstetter, T. H., & Sullivan, J. (1980). Adoptive transfer of immunity against *Vibrio anguillarum* in rainbow trout, *Salmo gairdneri* Richardson, vaccinated by the immersion method. *Journal of Fish Biology*, *17*(4), 379–386.
36. Harrell, L. W., Etlinger, H. M., & Hodgins, H. O. (1976). Humoral factors important in resistance of salmonid fish to bacterial disease. II. Anti-*Vibrio anguillarum* activity in mucus and observations on complement. *Aquaculture*, *7*(4), 363–370.
37. Thune, R. L., Hawke, J. P., Fernandez, D. H., Lawrence, M. L., & Moore, M. M. (1997). Immunization with bacterial antigens: Edwardsiellosis. *Developments in Biological Standardization*, *90*, 125–134.

38. Waterstrat, P. R., Ainsworth, A. J., & Capley, G. (1991). In vitro responses of channel catfish, Ictalurus punctatus, neutrophils to Edwardsiella ictaluri. *Developmental & Comparative Immunology, 15*(1–2), 53–63.
39. Ainsworth, A. J., & Dexiang, C. (1990). Differences in the phagocytosis of four bacteria by channel catfish neutrophils. *Developmental & Comparative Immunology, 14*(2), 201–209.
40. Romalde, J. L., & Magariños, B. (1997). Immunization with bacterial antigens: Pasteurellosis. *Developments in Biological Standardization, 90*, 167–177.
41. Firdaus-Nawi, M., & Zamri-Saad, M. (2016). Major components of fish immunity: Review. *Pertanika Journal of Tropical Agricultural Science, 39*(4), 393–420.
42. Roach, J. C., Glusman, G., Rowen, L., Kaur, A., Purcell, M. K., Smith, K. D., …, Aderem, A. (2005). The evolution of vertebrate toll-like receptors. *Proceedings of the National Academy of Sciences, 102*(27), 9577–9582.
43. Firdaus-Nawi, M., & Zamri-Saad, M. (2016). Major components of fish immunity: a review. *Pertanika Journal of Tropical Agricultural Science, 39*(4), 393–420.
44. Evans, D. L., & Jaso-Friedmann, L. (1992). Nonspecific cytotoxic cells as effectors of immunity in fish. *Annual Review of Fish Diseases, 2*, 109–121.
45. Demers, N. E., & Bayne, C. J. (1997). The immediate effects of stress on hormones and plasma lysozyme in rainbow trout. *Developmental & Comparative Immunology, 21*(4), 363–373.
46. Castro, R., Martin, S. A., Bird, S., Lamas, J., & Secombes, C. J. (2008). Characterisation of γ-interferon responsive promoters in fish. *Molecular Immunology, 45*(12), 3454–3462.
47. Mitra, S., & Bhattacharya, S. (1992). Purification of C-reactive protein from Channa punctatus (Bloch). *Indian Journal of Biochemistry & Biophysics, 29*(6), 508–511.
48. Neves, J. V., Wilson, J. M., & Rodrigues, P. N. (2009). Transferrin and ferritin response to bacterial infection: The role of the liver and brain in fish. *Developmental & Comparative Immunology, 33*(7), 848–857.
49. Russell, S., & Lumsden, J. S. (2005). Function and heterogeneity of fish lectins. *Veterinary Immunology and Immunopathology, 108*(1–2), 111–120.
50. Uribe, C., Folch, H., Enríquez, R., & Moran, G. J. V. M. (2011). Innate and adaptive immunity in teleost fish: A review. *Veterinarni Medicina, 56*(10), 486–503.
51. Uzzell, T., Stolzenberg, E. D., Shinnar, A. E., & Zasloff, M. (2003). Hagfish intestinal antimicrobial peptides are ancient cathelicidins. *Peptides, 24*(11), 1655–1667.
52. Whyte, S. K. (2007). The innate immune response of finfish–A review of current knowledge. *Fish & Shellfish Immunology, 23*(6), 1127–1151.
53. Kime, Y. J., Driscoll, D. C., & Dowben, P. A. (1987). The stability of the carbon tetrahalide ions. *Journal of the Chemical Society, Faraday Transactions 2: Molecular and Chemical Physics, 83*(2), 403–410.
54. Harris, J., & Bird, D. J. (2000). Modulation of the fish immune system by hormones. *Veterinary Immunology and Immunopathology, 77*(3–4), 163–176.
55. Wiik, R., Andersen, K., Uglenes, I., & Egidius, E. (1989). Cortisol-induced increase in susceptibility of Atlantic salmon, *Salmo salar*, to *Vibrio salmonicida*, together with effects on the blood cell pattern. *Aquaculture, 83*(3–4), 201–215.
56. Narnaware, Y. K., & Baker, B. I. (1996). Evidence that cortisol may protect against the immediate effects of stress on circulating leukocytes in the trout. *General and Comparative Endocrinology, 103*(3), 359–366.
57. Weyts, F. A. A., Cohen, N., Flik, G., & Verburg-van Kemenade, B. M. L. (1999). Interactions between the immune system and the hypothalamo-pituitary-interrenal axis in fish. *Fish & Shellfish Immunology, 9*(1), 1–20.
58. Schreck, C. B. (1996). Immunomodulation: Endogenous factors. *The Fish Immune System: Organism, Pathogen, and Environment, 15*, 311–338.
59. Weyts, F. A. A., Verburg-van Kemenade, B. M. L., & Flik, G. (1998). Characterisation of glucocorticoid receptors in peripheral blood leukocytes of Carp, Cyprinus carpioL. *General and Comparative Endocrinology, 111*(1), 1–8.
60. Yada, T., Azuma, T., & Takagi, Y. (2001). Stimulation of non-specific immune functions in seawater-acclimated rainbow trout, Oncorhynchus mykiss, with reference to the role of growth hormone. *Comparative Biochemistry and Physiology Part B: Biochemistry and Molecular Biology, 129*(2–3), 695–701.

61. Rawson, R. B., Zelenski, N. G., Nijhawan, D., Ye, J., Sakai, J., Hasan, M. T., … & Goldstein, J. L. (1997). Complementation cloning of S2P, a gene encoding a putative metalloprotease required for intramembrane cleavage of SREBPs. *Molecular Cell*, *1*(1), 47–57.
62. Sakai, T. (1996). *Riemannian geometry* (Vol. 149). American Mathematical Society. Charles street Providence, Rhode Island, USA.
63. Dake, M. D., Kato, N., Mitchell, R. S., Semba, C. P., Razavi, M. K., Shimono, T., … & Miller, D. C. (1999). Endovascular stent–graft placement for the treatment of acute aortic dissection. *New England Journal of Medicine*, *340*(20), 1546–1515.

3 Aquaculture
Facing an Array of Pathogens

Wanna Sirimanapong, Prawporn Thaijongrak, and Chiranan Sudpraseart
Mahidol University

Boonsirm Withyachumnarnkul
AquaAcademy Farm and Prince of Songkla University

CONTENTS

3.1 INTRODUCTION

The growth in the human population requires increasing sources of protein. Sources of protein are plants and animals, that is, both terrestrial and aquatic animals. Since the production of aquatic animals is more economical than the production of terrestrial animals (Engelmann 1966; Golley 1968; McNeill and Lawton 1970; Humphreys 1979), the world is moving toward aquaculture as a way of producing food for humans. As predicted by FAO (2020), global aquaculture production is expected to grow at 5% per year. Thus far, world aquaculture is mainly focused on high-density monoculture, which employs high-level technology and aims at maximum profit for the producers. In the past decades, intensive aquaculture activities aimed at the production of a single species have resulted in increasing aquatic products for human consumption. However, unless the activities are carried out under strict regulations, intensive aquaculture can lead to environmental deterioration.

Strict regulations to avoid environmental deterioration caused by intensive aquaculture practices cannot be strictly applied by all farmers. Not only does intensive aquaculture have a negative impact on the environment, but it is also unsustainable. History has shown aquaculture to have a "boom and bust" phenomenon; farmers usually face alternative success and failure crops. The shrimp farming industry in Thailand is a good example of this phenomenon. In the past 40 years of intensive land-based marine shrimp culture, the production of farmed shrimp in Thailand has been up and down (Phornprapha 2020). The country's farmed shrimp production has increased from when it first started in the 1980s to becoming the world's number one shrimp exporter in the 1990s. Behind this

DOI: 10.1201/9781003388548-4

successful story, there is a saga of several crop failures from disease outbreaks (Lightner 2011). As a result of severe financial problems, several farmers stopped farming, while others started new farms in different locations, changed husbandry techniques, and gained profit for a few years before facing crop failures again. These cycles of events have continued up to the present time (2022), resulting in the country having many deserted shrimp farms. It is likely that Thailand is not the only country that has experienced this problem, with many other shrimp-production countries probably having experienced the same phenomenon.

Many different pathogens affect the fish. The main reason for mortality and economic loss in commercial aquaculture is diseases caused by bacterial, viral, or parasitic infection. Disease control is usually by using antibiotics and chemicals. However, the use of antibiotics can lead to the development of antimicrobial resistance in aquatic animals and human pathogens (Lulijwa, Rupia, and Alfaro 2020). Disease prevention strategies used during fish production include biosecurity to minimize disease risks. This includes quarantining of fish; screening for the disease before fish are introduced into their new aquaculture facilities; reducing the risk of wild animals acting as a pathogen reservoir that can infect the farm population (Oidtmann et al. 2011); water management, that is, pre- and post-treatment by filter, ultraviolet, ozone sterilization, and magnetic field reorganizing the water molecules (Sharrer and Summerfelt 2007; Irhayyim et al. 2020); and immunostimulants, that is, prebiotics, symbiotic, and probiotics, which stimulate the innate immune system to help protect against disease and act as a growth promoter (Amenyogbe et al. 2020). Moreover, vaccines can also help protect the host against disease and are one of the most effective approaches for preventing disease in fish.

In this chapter, we would like to challenge the traditional shrimp farming method in Thailand with a relatively new farming methodology, land-based integrated multi-trophic aquaculture (IMTA), accompanied by re-circulatory aquaculture (RAS) systems (Chopin 2006; Knowler et al. 2020; de Oliveira Costa et al. 2021). This method has been practiced for a few years in a private farm in Thailand, with promising results. In this system, marine shrimp is the main product, with other aquatic species as the secondary product. It is a polyculture system that is ecologically balanced and environmentally friendly.

3.2 TURNING INTENSIVE SHRIMP FARMING INTO LAND-BASED IMTA

As transmission of pathogens from one host to another is facilitated in intensive monoculture (Pulkkinen et al. 2010), epidemics and pandemics of various shrimp diseases causing frequent mass morbidity and mortality have occurred along the history of shrimp farming. Despite a large volume of research and accumulated scientific data on shrimp diseases and recommendations on using new technologies to mitigate disease problems, in reality, this knowledge has not really helped much. As a result, the shrimp culture industry in Thailand, for example, has faced decline in production since 2009 (Phornprapha 2020; Rubel et al. 2019), with more than 50% of the shrimp-rearing area in the country deserted and awaiting some future remedy to disease problems.

To be fair, it should be mentioned that "proper" intensive shrimp culture does exist and has been successful, but it must be equipped with suitable biosecurity measures and modern science-based technologies (Ulhaq et al. 2022). These activities require high investment and scientific knowledge not affordable for the majority of shrimp farming operators. However, the land-based IMTA with RAS is likely to be affordable for most farmers. In IMTA, the main products (finfish or shrimp) are provided with commercial pellets; they are the "fed species." Water discharged from the fed species is distributed to the area stocked with other species that feed on nutrients present in the water. These can be divided into suspended (e.g., phytoplankton, bioflocs) and soluble (e.g., ammonia, nitrite, nitrate, phosphate) nutrients, with the species that consume these nutrients termed "extractive species." The extractive species feeding on the suspended part, the suspension feeders, are bivalves and fish. The extractive species that consume inorganic material, inorganic feeders, is mainly seaweed (Buck et al. 2017). IMTA can be set up either in the sea, near-shore or offshore (marine IMTA), or inland (land-based IMTA). An example of marine IMTA is the one with salmon (fed species)

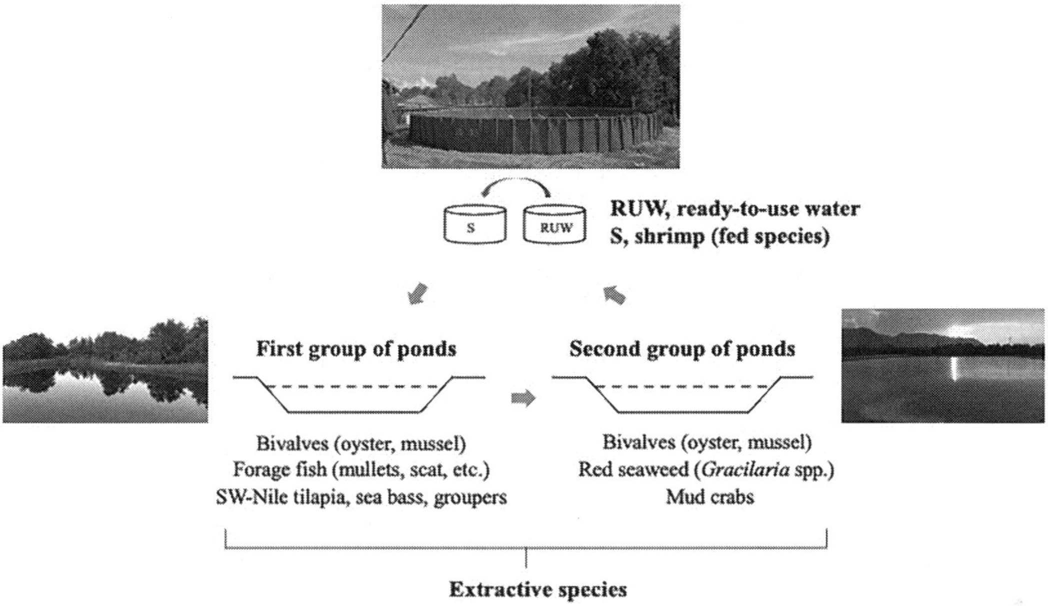

FIGURE 3.1 Land-based IMTA uses marine shrimp as the fed species and bivalves, fish, and seaweed as extractive species.

cage culture, surrounded by facilities for oyster culture (suspension feeder) and seaweed (inorganic feeder) situated in the outer rows (Chopin 2006). An example of the land-based IMTA is described herein (Figure 3.1); marine shrimp is the fed species, and oysters and seaweed being suspension feeder and inorganic feeder, respectively.

3.3 BIOECONOMY OF LAND-BASED IMTA WITH SHRIMP AND FISH AS THE FED SPECIES

At present, examples of land-based IMTA are scarce, although farmers might have already practiced parts of the process without knowing it. In an attempt to find a new way to improve shrimp production, a group of Thai farmers have grown marine shrimp and cockles in separate ponds with water flowing between the cockle- and shrimp-stocking ponds (Withyachumnarnkul, personal observation). Apparently, farmers have been successful in this farming method, at least within the past 2 years (2021–2022), in the farms that previously failed in shrimp rearing. The farmers also have additional income from cockle production, which is considerably a high-priced commodity in Asian markets. This practice, though being one of a few examples, holds promise for the future practice of land-based IMTA with marine shrimp being the fed species.

The operation cost for the production of extractive species in the land-based IMTA is usually less than the cost for the production of these species under the traditional monoculture since the feed cost is minimized. The extractive species feed on a natural diet. However, the volume of production for the primary product, marine shrimp, may not fit with the current commercial set-up, which requires a big volume of shrimp for export. Under current practice, at harvesting time, a processing plant company is responsible for the harvesting process. The volume of the shrimp harvested should be of cost-benefit for the company, which in most cases would be 4 tons or more per a single harvest. Therefore, shrimp production per crop should be high enough to make the business feasible. Besides, income from the sale of the extractive species alone is unlikely to cover the farm expenses. This bioeconomic consideration is probably the key factor to make land-based IMTA a reality.

To solve this problem, two options are available: one is to produce enough volume of marine shrimp for export and the other is to produce a low quantity and sell the shrimp to the domestic market. Sale price of the marine shrimp at farm gate in the domestic market is usually higher than the sale price for export, due to the low number of middle men involved.

3.4 CANVAS-POND CULTURE OF MARINE SHRIMP

In shrimp culture, the main problem is diseases caused by viruses, bacteria, fungi, and parasites (Thitamadee et al. 2016). Since shrimp have only innate defense (Wang et al. 2014), vaccination against pathogens to enhance their acquired defense is not an option. Preventive measures are the only option that farmers have to avoid shrimp disease outbreaks. Therefore, under the current situation where shrimp pathogens exist in the surrounding environment, one way to avoid shrimp diseases is to rear the shrimp in confined, smooth-surfaced areas, such as concrete or canvas tanks (Figure 3.1). In that kind of rearing tanks, pathogens coming in with incoming water can be completely eliminated by chemical treatment. In addition, it is best to minimize water exchange during rearing to avoid incoming pathogens from intake water; and the rearing water must be exchanged with pathogen-free water. An available method of shrimp culture that employs minimum- or zero-water exchange is through employing biofloc technology (Khanjani and Sharifinia 2020), by which soluble nitrogen waste (especially ammonia and nitrites) are kept at safe, low levels by applying nitrifying bacteria.

In addition, the shrimp is provided with additives that stimulate their defense mechanisms, such as feed supplemented with extracts from seaweed (Boonsri et al. 2017; Rudtanatip et al. 2017), peptidoglycans (Song et al. 2013), chelated minerals (Katya et al. 2016), and organic acids (He et al. 2017). All chemicals applied must have been regarded as safe for human consumption. Water discharge from the shrimp tanks contain both suspended particles (friendly aerobic bacteria, phytoplankton, flagellates, zooplankton, shrimp exuviates, etc.) and soluble inorganic substances (ammonia, nitrite, phosphate, etc.). This discharge becomes nutrients for oysters and seaweed.

3.5 FEEDING ON DIFFERENT TROPHIC LEVELS

In this example, the marine shrimp (the Pacific white shrimp, *Litopenaeus vannamei* or the black tiger shrimp, *Penaeus monodon*) are reared in canvas ponds (Figure 3.2). The ponds are divided into two groups: one for stocking treated, stored seawater ready for use; and another for stocking shrimp (Figure 3.1). The shrimp are fed with commercial pellets. Discharged water from shrimp-rearing tanks is then drained into two groups of earthen ponds for rearing extractive species.

The first group of ponds are stocked with bivalves, for instance, oysters (*Crassostrea belcheri* and *C. lugubris*), mussels (*Mytilus edulis*), foraging fish, and commercial fish (seawater-acclimated Nile tilapia, *Oreochromis niloticus*; Asian sea bass, *Lates calcarifer*; and groupers, *Epinephelus* spp.). The second group of ponds are stocked with red seaweed (*Gracilaria fisheri*) and mud crab (*Scylla serrata*).

The bivalves feed on phytoplankton and flagellates, and produce biodeposits as feces and pseudofeces (Haven, Morales-Alamo, and Oceanography 1966). The biodeposits on the external surface of the bivalves set up conditions and nutrients for amphipods, isopods, polychaete, and other small aquatic epibionts (Smith et al. 2018). Several species of forage fish (scat, mullets, crescent grunter, rabbitfish, etc.) prey on biodeposits and epibionts (Figure 3.2). While feeding, they also shake the baskets containing the bivalves, making the bivalves healthier (Solomon and Ahmed 2016).

The oysters and three groups of commercial fish stocked in the first group of ponds become additional income for the operator. The seawater-acclimated *O. niloticus* have been specially bred for several generations and continuously cultured in the facility since 2017; they survived and grow well under 30-ppt seawater (Withyachumnarnkul et al. 2017, 2022). The three commercial groups of fish (seawater-acclimated Nile tilapia, Asian sea bass, and groupers) prey on forage fish and grow

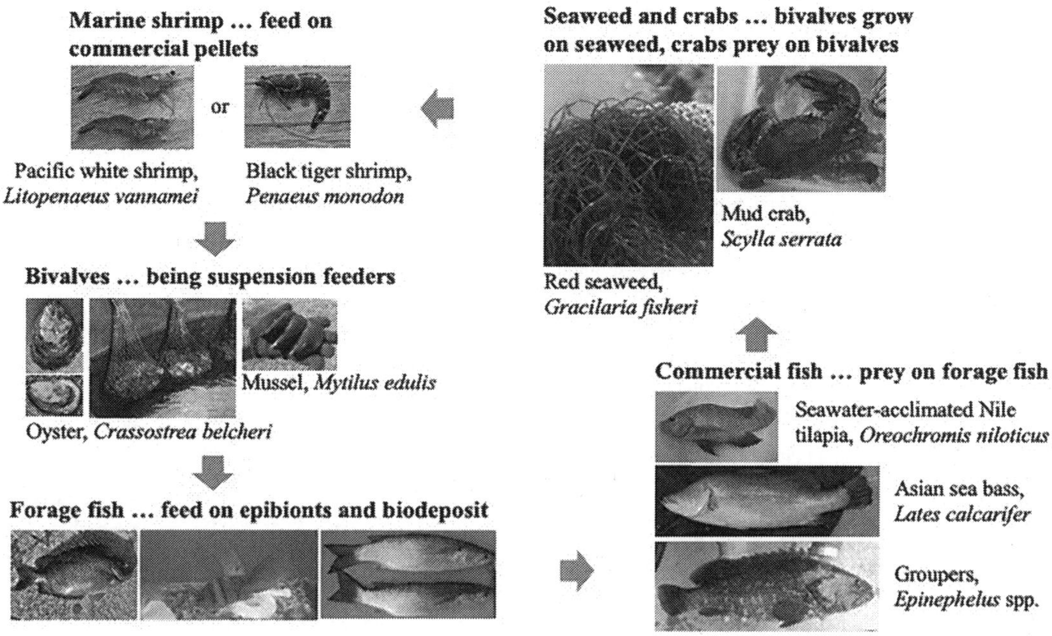

FIGURE 3.2 Aquatic species in one example of land-based IMTA, with marine shrimp as the fed species. The extractive species are reared under balanced ecology, feeding on different trophic levels.

to marketable size without receiving commercial feed. Water discharged from this group of ponds is pumped out to the second group of ponds stocking red seaweed, swimming crabs, and mud crabs.

The red seaweed, *G. fisheri*, consumes soluble inorganic waste in the water, that is, nitrogen and phosphate waste (Azis, Tresnati, and Tuwo 2020). The seaweed is also a substrate for small bivalves, such as false mussel, *Mytella strigata*, and horse mussel, *Musculus senhousia*, on which the swimming and mud crabs prey. The reason for growing the seaweed *G. fisheri*, besides for bioremediation, is that a special ingredient in the seaweed extracts helps protect the shrimp from infections by certain pathogens (Boonsri et al. 2017; Rudtanatip et al. 2017).

Seawater from the second group of ponds is pumped into the ready-to-use water canvas ponds for shrimp culture. The water is treated and pumped into the shrimp-rearing canvas ponds. The water is then re-circulated through the ponds. No water from the system is drained out into the environment. However, seawater from public waterways is pumped in at a minimum volume occasionally to replace evaporated water from the system.

The stocking density of the fed species is at 80–100 individuals/m^3 of water for *L. vannamei*, or at 30–40 individuals/m^3 of water for *P. monodon*. These stocking densities are within the density range of marine shrimp culture in Thailand. For the extractive species, the most effective and profitable stocking density is still being determined, but the levels shown in Table 3.1 have been proven to make the system run without problems.

3.6 THE HEALTH OF EXTRACTIVE FISH SPECIES

Since the extractive fish species, namely, seawater-acclimated Nile tilapia, Asian sea bass, and groupers, are stocked at low density (Table 3.1), they are expected to be much healthier than those reared under intensive culture. Since the manifestation of disease results from a complex interaction between fish health, the environment, and pathogens, the chance of having disease outbreaks in the land-based IMTA is likely to be much lower than intensive culture systems. It has been shown that pathogens existing in the water or even in the fish's body do not necessarily cause disease, but when

TABLE 3.1
Stocking Density of the Fed and Extractive Species in the Land-Based IMTA, with Marine Shrimp as the Fed Species

Fed Species	Stocking Density
Pacific white shrimp, *L. vannamei*	80–100 ind/m^3
Black tiger shrimp, *P. monodon*	30–40 ind/m^3
Extractive Species, Group 1	
Oyster, *Crassostrea belcheri* and *C. lugubris*	5–10 kg/m^3
Mussel, *Mytilus edulis*	5–10 kg/m^3
Forage fish	20–30 ind/m^3
Seawater-acclimated Nile tilapia, *O. niloticus*	15–20 ind/m^3
Asian sea bass, *Lates calcarifer*	1–2 ind/m^3
Groupers, *Epinephelus* spp.	1–2 ind/m^3
Extractive Species, Group 2	
Red seaweed, *Gracilaria fisheri*	500–700 kg/m^3
Mud crabs, *Scylla serrata*	0.5-1.0 ind/m^3

Ind, individuals.

the fish are under stress, the pathogens may increase in number, turn virulent, and cause diseases (Bricknell 2017; Withyachumnarnkul et al. 2017). However, certain types of pathogens are naturally virulent and may cause disease outbreaks in healthy fish.

Under low stocking densities and when circulating with other aquatic species, pathogens are likely to have a slower capacity to spread. Theoretically, biodiversity of aquatic species within the aquatic environment balances the food chain and causes ecological stability. In nature, a positive spatiotemporal interaction among species exists, which helps sustain food webs and control certain epidemic diseases. Intensive monoculture in aquaculture leads to high productivity per unit area but also eliminates the interactions among species that occur in ecosystems. Once individuals in the pond carry pathogens, severe disease outbreaks might quickly occur due to the high density of hosts and lack of species interactions. Although the theory behind this is reasonably sound, it still requires scientific support. In marine shrimp culture, white-spot disease was controlled by polyculture of shrimp with carps and other fish species that preyed on dead and weakened shrimp infected by white-spot syndrome virus, the causative agent of the disease (Wang et al. 2021). Land-based IMTA would provide more opportunities for researchers to confirm these points conclusively.

In the following example of a land-based IMTA, three fish groups are raised in the same pond: seawater Nile tilapia, Asian sea bass, and several species of groupers. There is a possibility that cross infections may occur among the fish of different species. An example is the disease streptococcosis, which is one of the most common infectious bacterial diseases of tilapia and sea bass. The disease is caused by *Streptococcus spp.* Infection, resulting in severe mortality. Streptococcosis outbreaks may occur in the ponds that stock both fish species.

Certain pathogens are considered highly virulent and possibly cause diseases in the fish despite the fish being healthy and living in a suitable environment. A good example of that kind of pathogens is Tilapia Lake Virus (TiLV), a globally emerging virus associated with high mortality in tilapia farms and in natural environments (Aich et al. 2021). In this example, the outbreak of TiLV infection may occur due to the seawater-acclimated *O. niloticus* in the IMTA ponds, despite the fish being stocked under low density and apparently being healthy. This, however, has not happened, at least within the past 2 years of the land-based IMTA practice. Besides, it was shown that TiLV has long been discovered in tissues of a population of *O. niloticus* several years before the first report of its outbreaks (Senapin et al. 2018). This finding agrees well with a highly acceptable concept that fish diseases are manifested though the combination of pathogens, environment, and fish health, not simply the existence of pathogens (Bowden 2008; Withyachumnarnkul et al. 2017).

Despite these arguments, for safety precaution, the fish in the land-based IMTA should be vaccinated against known and common pathogens present in the local area.

The diseases commonly reported in *L. calcarifer* are related to infections caused by iridovirus, nervous necrosis virus (NNV), *Vibrio tenacibaculum*, *Streptococcus agalactiae*, *V. harveyi*, and *Aeromonas hydrophila* (for review, see Gibson-Kueh, 2012). For groupers, vibriosis caused by *V. harveyi* and *V. alginolyticus* is the most common, followed by NNV (Hoihuan et al. 2021).

3.7 A POSSIBLE VACCINATION PROGRAM FOR FISH IN LAND-BASED IMTA SYSTEMS

Diseases in Nile tilapia, Asian sea bass, and groupers have been reviewed elsewhere (Shoemaker, Klesius, and Evans 2000; Kasornchandra 2002). Several types of vaccines for diseases of known pathogens are listed in Table 3.2. For effective protections, the vaccination programs should be for protection against diseases frequently observed in the local areas.

TABLE 3.2
Vaccines for Protection against Common Pathogens in Land-Based IMTA Commercial Fish

Pathogen	Type	Administration Route	References
	For Seawater Nile Tilapia (*Oreochromis niloticus*)		
S. agalactiae	Attenuated vaccine	Intraperitoneal	Liu et al. (2019)
	Recombinant vaccine	Oral	Nur-Nazifah, Sabri, and Siti-Zahrah (2014)
	DNA vaccine	Oral	Zhu et al. (2017)
	Inactivated vaccine (whole-cell, formalin-killed)	Intraperitoneal	Evans, Klesius, and Shoemaker (2004)
S. iniae	Attenuated vaccine	Intraperitoneal	Miccoli et al. (2021)
	DNA vaccine	Intramuscular	Kayansamruaj et al. (2017)
	Attenuated vaccine	Intracoelomic	Heckman et al. (2022)
A. hydrophila	Inactivated whole bacteria	Intraperitoneal	Ruangpan, Kitao, and Yoshida (1986)
	Attenuated vaccine	Intraperitoneal	Pridgeon, Klesius, and Yildirim-Aksoy (2013)
S. iniae and V. vulnificus	Inactivated whole bacteria	Intraperitoneal	Shoemaker, LaFrentz, and Klesius (2012)
V. vulnificus	Inactivated whole bacteria	Intraperitoneal	Shoemaker, LaFrentz, and Klesius (2011)
V. harveyi	Inactivated whole bacteria	Intraperitoneal	Abu Nor et al. (2020)
E. tarda	Recombinant vaccine	Intraperitoneal	Pridgeon, Klesius, and Yildirim-Aksoy (2013)
F. Noatunensis subsp. Orientalis (Fno)	Inactivated vaccine (formalin-killed *Fno* vaccine)	Intraperitoneal	Pulpipat et al. (2020)
	Inactivated vaccine (whole-cell)	Intraperitoneal	Shahin et al. (2019)
TiLV	Inactivated vaccine (formaldehyde- and β-propiolactone-inactivated vaccines)	Intramuscular	Zeng et al. (2021)
	DNA vaccines	Intramuscular	Yu et al. (2022)

(Continued)

TABLE 3.2 (*Continued*)
Vaccines for Protection against Common Pathogens in Land-Based IMTA Commercial Fish

Pathogen	Type	Administration Route	References
	For Asian Sea Bass (*Lates calcarifer*)		
S. iniae	Inactivated vaccine (formalin-killed bacterins)	Intraperitoneal	Aviles et al. (2013)
V. anguillarum	DNA vaccine	Intramuscular	Kumar et al. (2007)
V. harveyi, S. agalactiae and A. hydrophila	Polyvalent vaccine (formalin-killed vaccine)	Oral	Mohamad et al. (2021)
S. agalactiae and S. iniae	Inactivated vaccine (oil-based formalin-killed bivalent)	Intraperitoneal	Lan et al. (2021)
NNV	DNA vaccine	Intramuscular	Vimal et al. (2016)
	For Grouper (*Epinephelus* spp.)		
V. vulnificus	Inactivated whole bacteria	Intraperitoneal	Hoihuan et al. (2021)
V. harveyi	Glutathione peroxidase (GPx) DNA vaccine	Intramuscular	Wang et al. (2017)
	Recombinant vaccine (outer-membrane protein)	Intramuscular	Zhu et al. (2019)
	Attenuated vaccine	Intraperitoneal	Bai et al. (2020)
V. alginolyticus	Attenuated vaccine	Intraperitoneal	Pang et al. (2018, 2022)
NNV	DNA vaccine	Intramuscular	Chen, Peng, and Chiou (2015)
	β-Propiolactone (BPL) inactivated virus and formalin-inactivated virus	Intraperitoneal	Ou-yang et al. (2012)

3.7.1 Vaccines for Seawater Nile Tilapia

Streptococcosis, one of the most common infectious bacterial diseases of farmed Nile tilapia with high mass mortality, is caused by *Streptococcus spp.* infections. Vaccines protecting tilapia (and other kinds of fish) against a number of bacterial diseases have been developed and commercialized, for example, *S. iniae* (AQUAVAC® Strep *Si*, MSD; Vaxxinova®, International BV; Himmvac Agilban-S Plus® and Himmvac Agilban S-3 Plus®, KBNP Inc.) and *S. agalactiae* serotypes Ib and III (AQUAVAC® Strep *Sa*1, MSD; Vaxxinova®, International BV). In one report, tilapia receiving the vaccine against *S. iniae* had a significant rise in specific antibody levels and relative percent survival (RPS) of 100% by intraperitoneal injection, and 88% by immersion vaccination (Miccoli et al. 2021). Likewise, an attenuated erythromycin-resistant *S. agalactiae* vaccine produced a significant rise in antibody titers, which protected the fish for at least 16 weeks after vaccination (Liu et al. 2019). This period of protection is long enough for the fish to grow to marketable size without the need for a booster vaccination. DNA vaccine against the disease was tested but not yet commercialized and was found to stimulate both innate and adaptive immunity (Evensen and Leong 2013; Kayansamruaj et al. 2017; Mondal and Thomas 2022). The feed-based recombinant vaccine against *S. agalactiae* in red tilapia displayed a high IgM antibody titer in serum, mucus, and gut of the vaccinated fish, with RPS at 70% following the challenge (Nur-Nazifah, Sabri, and Siti-Zahrah 2014). Using recombinant proteins to produce the vaccine would probably lower the cost of vaccines; however, its effectiveness also needs to be considered.

Vaccines against two other types of bacteria, *Aeromonas hydrophila* and *Edwardsiella tarda*, have been developed with high protective efficiency (Pridgeon, Klesius, and Yildirim-Aksoy 2013). Both vaccines were designed for intraperitoneal administration and 100% protection was reported (Pridgeon and Klesius 2011).

Vibriosis has been rarely reported in farmed tilapia in Thailand, and when it does occur, it is mostly red tilapia that is affected. A formalin-killed *V. harveyi* vaccine was found to promote strong IgM antibody titers and lysozyme activities in red tilapia, which showed higher rate of survival upon challenging the fish with the pathogen, when compared to unvaccinated control fish (Abu Nor et al. 2020). Red tilapia vaccinated with a whole-cell vaccine against *V. vulnificus* infection had RPS values of 88% following bacterial challenge (Shoemaker, LaFrentz, and Klesius 2011).

Another type of bacteria that infects tilapia, especially red tilapia, is the intracellular bacteria, *Francisella* spp. This group of bacteria has attracted attention, since it was found in a large population of tilapia without causing any obvious disease manifestations. Red tilapia infected with *F. noatunensis* were found to be more susceptible to other diseases than the noninfected fish (Sirimanapong et al. 2018). At present, there is no commercial vaccine against these bacteria.

The main viral disease in tilapia is TiLV, but effective vaccine for TiLV is still in the process of being developed (Lertwanakarn et al. 2021). A DNA vaccine under development against TiLV had produced some promising results (Criollo Joaquin et al. 2019).

3.7.2 Vaccines for the Asian Sea Bass

Several bacterial diseases affecting tilapia also affect Asian sea bass: *Streptococcus* spp., *Vibrio* spp., and *A. hydrophila*. Therefore, vaccines used against pathogens in tilapia are also applicable to Asian sea bass. The formalin-killed *S. iniae* bacterin vaccines provided 100% protection against the vaccine strain in vaccinated *L. calcarifer*, whereas significantly reduced protection was observed when the fish were challenged with a heterologous strain of the bacterium (Aviles et al. 2013; Caipang et al. 2014). An oil-based formalin-killed bivalent vaccine containing *S. agalactiae* and *S. iniae* promoted strong systemic and mucosal antibody responses in vaccinated Asian sea bass, with RPS of 85% obtained following challenge with both bacteria (Lan et al. 2021). Live attenuated *V. harveyi* provided RPS of 68% to *L. calcarifer* fingerlings (Chin et al. 2020). The feed-based polyvalent vaccine against *V. harveyi*, *A. hydrophila*, and *S. agalactiae* showed a significant rise in IgM antibody levels as well as the RPS of 75%, 80%, and 80%, respectively, after the challenge with the three pathogens (Mohamad et al. 2021). Likewise, DNA vaccines encoding an outer-membrane protein of *V. anguillarum* provided RPS of 56% after the challenge with the bacteria (Kumar et al. 2007).

A commercial vaccine for *L. calcarifer* is available for *V. anguillarum* (serotype 01) and *Photobacterium damsela* subsp. *piscicida* (causing pasteurellosis) (for immersion, ALPHA DIP® 2000, AQUAVET S.A; for intraperitoneal injection, ALPHA JECT® 2000, AQUAVET S.A.). In addition, an autogenous vaccine against *Aeromonas veronii* has also been developed by the same company.

For viral infections in *L. calcarifer*, a DNA vaccine against NNV in *L. calcarifer* was tried with pFNCPE42-DNA, resulting in a RPS value of 77%, and a significant increase in the antibody titer in vaccinated fish (Vimal et al. 2016). An injectable vaccine against NNV in red grouper is also effective against NNV in *L. calcarifer* (ALPHA JECT micro® 1 noda, AQUAVET S.A.).

3.7.3 Vaccines for Groupers (*Epinephelus* spp.)

The most common disease in groupers is vibriosis, which is usually caused by *V. harveyi* and *V. alginolyticus* infections (Mohamad et al. 2019). Viral infections have also been reported, due to iridovirus and NNV. A mixed formalin-killed cell vaccine of three subgroups of *V. vulnificus* biotype 1 was effective in protecting the brown-marbled grouper, *E. fuscoguttatus*, in Thailand (Hoihuan et al. 2021). In China, one study produced a RPS value of 78% with a live attenuated vaccine following being challenged with a wild type *V. alginolyticus* strain HY9901ΔvscB in pearl-gentian or hybrid grouper (♀*E. fuscoguttatus* × ♂*E. lanceolatus*) (Pang et al. 2022; Pang et al. 2018). A glutathione peroxidase DNA vaccine was tested in China in 2017 for intramuscular administration, and it gave 78% RPS after being *V. harveyi* challenged (Wang et al. 2017), while a recombinant vaccine in the hybrid grouper, constructed from the outer-membrane protein of *V. harveyi*, showed the protein to be a potential vaccine candidate (Zhu et al. 2019).

For viral infections, β-propiolactone- and formalin-inactivated virus vaccines were produced against Singapore grouper iridovirus (SGIV) that showed promising results for protecting orange-spotted grouper against the virus when challenged at 30 days post-vaccination with RPSs values of 92% and 100% obtained, respectively (Ou-yang et al. 2012). The oil-adjuvanted vaccine is commercially available (AQUAVAC® IridoV, MSD). In addition, mass mortality in grouper larvae caused by NNV infection could be prevented by DNA vaccine with modulated CpG oligodeoxynucleotide (Viral Nervous Necrosis Vaccine, Nisseiken Co., Ltd.) (Chen, Peng, and Chiou 2015). However, NNV inactivated with 0.4 mM binary ethylenimine (BEI) or 0.1%–0.2% formalin in orange-spotted grouper larvae resulted in 95% RPS and 43% RPS, respectively, in vaccinated fish (Kai and Chi 2008). When the vaccine from the viral supernatant was injected intramuscularly into the potato grouper, *E. tukula*, broodstock, NNV-specific antibodies were found in eggs from vaccinated broodstock within 5-month post-vaccination. Moreover, NNV was detected in the eggs of the nonvaccinated fish, but not in the vaccinated fish. Therefore, the vaccination looked like it protected against the vertical transmission of the pathogen from the grouper broodstock (Kai et al. 2010).

It is interesting to find out the levels of innate and acquire defenses for the fish reared under the land-based IMTA, compared to those of the same species reared under intensive monoculture. It is possible to set up the two groups of animals for this study in a scientific way. A net cage stocked with the fish at a commercial density, for example, at 30 individuals/m^3, is placed in the IMTA ponds. The same species of fish outside the cage, at a density of 2 individuals/m^3, are those under IMTA conditions. The fish in the cage are fed with commercial pellets but the fish outside the cage swim freely, interact with other aquatic species, and prey on their natural diet. But yet, the water conditions of the two groups are mostly identical.

To compare the efficacy of vaccines on the fish under intensive monoculture and those under the land-based IMTA, the same set-up as described, that is, a net cage in the IMTA pond, can be employed as well. At this point, it may be helpful to explain how the fish are prepared for stocking in the IMTA ponds. They are hatchery-produced and nursed in canvas tanks before being released into the pond. Seawater-acclimated *O. niloticus* are released at the size of 50 g; *L. calcarifer*, at the length of 6 in; and *Epinephelus* spp., at the size of 200 g. Therefore, vaccinations can be carried out at the time before releasing the fish into the pond. In these experiments, one group of the vaccinated fish should be released into the net cages and the other group outside the cages. The choice of vaccine administration is preferably by intramuscular or intraperitoneal injections as these two methods of vaccine administration are most effective (Ismail et al. 2016; Silva et al. 2009). In this scenario, booster doses, if required, can be accomplished only through oral administrations since it would not be feasible to catch all the experimental fish in the pond for the booster-dose injection.

3.8 CONCLUSIONS

Intensive aquaculture faces problems worldwide, especially with respect to environmental issues. In order to achieve both high levels of production and reduced environmental impact from aquaculture, high investment and scientific knowledge are required. But most farmers cannot reach those requirements. Besides, it is still under debate as to whether high production of industrial aquaculture is sustainable; for instance, high stocking density is prone to disease outbreaks and any small error in the operation could result in a catastrophic loss of the crop. Land-based IMTA with RAS is a good alternative for the average farmer, as it generates sufficient profit for farmers, is environmentally friendly, and is likely to be more sustainable than industrial types of aquaculture production. In addition, the practice could be a platform for future research and development; for instance, research on the interactions among different aquatic species, ecological balance, feeding behavior, and vaccination programs.

REFERENCES

Abu Nor, N., Zamri-Saad, M., Md Yasin, I.-S., Salleh, A., Mustaffa-Kamal, F., Matori, M.F., and Amal Azmai, M.N. 2020. "Efficacy of whole cell inactivated *Vibrio harveyi* vaccine against vibriosis in a marine red hybrid Tilapia (*Oreochromis niloticus*× *O. mossambicus*) model." *Vaccines*. *8*(4):734.

Aich, N., Paul, A., Choudhury, T.G., and Saha, H. 2021. "Tilapia lake virus (TiLV) disease: Current status of understanding." *Aquaculture and Fisheries*. *7*(1):7–17.

Amenyogbe, E., Chen, G., Wang, Z., Huang, J., Huang, B., and Li, H. 2020. "The exploitation of probiotics, prebiotics and synbiotics in aquaculture: present study, limitations and future directions: A review." *Aquaculture International*, *28*(3):1017–1041.

Aviles, F., Zhang, M.M., Chan, J., Delamare-Deboutteville, J., Green, T.J., Dang, C., and Barnes, A.C. 2013. "The conserved surface M-protein SiMA of *Streptococcus iniae* is not effective as a cross-protective vaccine against differing capsular serotypes in farmed fish." *Veterinary Microbiology*, *162*(1):151–159.

Azis, H.Y., Tresnati, J., and Tuwo, A. 2020. "Seaweed *Gracilaria changii* as a bioremediator agent for ammonia, nitrite and nitrate in controlled tanks of Whiteleg Shrimp *Litopenaeus vannamei*." *IOP Conference Series: Earth and Environmental Science*, *564*:1–9. doi:10.1088/1755-1315/564/1/012059

Bai, J.-Y., Long, H., Cui, J., Zhang, X., Cai, X.-N., and Xie, Z.-Y. 2020. "Characterization of a pathogenic *Vibrio harveyi* strain from diseased *Epinephelus coioides* and evaluation of different methods to control its infection." *Aquaculture*, *526*:735371. doi:10.1016/j.aquaculture.2020.735371

Boonsri, N., Rudtanatip, T., Withyachumnarnkul, B., and Wongprasert, K. 2017. "Protein extract from red seaweed *Gracilaria fisheri* prevents acute hepatopancreatic necrosis disease (AHPND) infection in shrimp." *Journal of Applied Phycology*, *29*(3):1597–1608.

Bowden, T.J. 2008. "Modulation of the immune system of fish by their environment." *Fish & Shellfish Immunology*, *25*(4):373–383.

Bricknell, I. 2017. "Types of pathogens in fish, waterborne diseases." Fish Diseases Prevention and Control Strategies, 53–80. doi:10.1016/B978-0-12-804564-0.00003-X

Buck, B.H, Nevejanm N., Wille, M., Chambers, M.D., and Chopin, T. 2017. "Offshore and multi-use aquaculture with extractive species: Seaweeds and bivalves." Aquaculture Perspective of Multi-Use Sites in the Open Ocean, pp. 23–69. Cham, Switzerland: Springer.

Caipang, C.M.A., Lucanas, J.B., and Lay-yag, C.M. 2014. "Updates on the vaccination against bacterial diseases in tilapia, *Oreochromis* spp. and Asian seabass, *Lates calcarifer*." *Aquaculture, Aquarium, Conservation & Legislation*, *7*(3):184–193.

Chen, S.-P., Peng, R.-H., and Chiou, P.P. 2015. "Modulatory effect of CpG oligodeoxynucleotide on a DNA vaccine against nervous necrosis virus in orange-spotted grouper (*Epinephelus coioides*)." *Fish & Shellfish Immunology*, *45*(2):919–926. doi: 10.1016/j.fsi.2015.06.013

Chin, Y.K., Al-saari, N., Zulperi, Z., Mohd-Aris, A., Salleh, A., Silvaraj, S., Mohamad, A., Lee, J.Y., Zamri-Saad, M., and Ina-Salwany, M.Y. 2020. "Efficacy of bath vaccination with a live attenuated *Vibrio harveyi* against vibriosis in Asian seabass fingerling, *Lates calcarifer*." *Aquaculture Research*, *51*(1):389–399.

Chopin, T. 2006. "Integrated multi-trophic aquaculture. What it is and why you should care and don't confuse it with polyculture." *North Aquaculture*, *12*(4):4.

de Oliveira Costa, L.C., da Silva Poersch, L.H., and Abreu, P.C. 2021. "Biofloc removal by the oyster *Crassostrea gasar* as a candidate species to an integrated multi-trophic aquaculture (IMTA) system with the marine shrimp *Litopenaeus vannamei*." *Aquaculture*, *540*:736731.

Engelmann, M.D. 1966. "Energetics, terrestrial field studies, and animal productivity." *Advances in Ecological Research*, *3*:73–115. doi: 10.1016/S0065–2504(08)60310-4

Evans, J.J, Klesius, P.H., and Shoemaker, C.A. 2004. "Efficacy of *Streptococcus agalactiae* (group B) vaccine in tilapia (*Oreochromis niloticus*) by intraperitoneal and bath immersion administration." *Vaccine*, *22*(27–28):3769–3773.

Evensen, Ø., and Leong, J.-A.C. 2013. "DNA vaccines against viral diseases of farmed fish." *Fish & Shellfish Immunology*, *35*(6):1751–1758.

Gibson-Kueh, S. 2012. "Diseases of Asian seabass (or barramundi), *Lates calcarifer* Bloch." Murdoch University. Last modified June 14, 2022. http://researchrepository.murdoch.edu.au/id/ eprint/14817

Golley, F.B. 1968. "Secondary productivity in terrestrial communities." *American Zoologist*, *8*(1):53–59.

Haven, D.S., and Morales-Alamo, R. 1966. "Aspects of biodeposition by oysters and other invertebrate filter feeders 1." *Limnology and Oceanography*, *11*(4):487–498.

He, W., Rahimnejad, S., Wang, L., Song, K., Lu, K., and Zhang C. 2017. "Effects of organic acids and essential oils blend on growth, gut microbiota, immune response and disease resistance of Pacific white shrimp (*Litopenaeus vannamei*) against *Vibrio parahaemolyticus*." *Fish & Shellfish Immunology*, *70*:164–173.

Heckman, T.I., Shahin, K., Henderson, E.E., Griffin, M.J., and Soto, E. 2022. "Development and efficacy of *Streptococcus iniae* live-attenuated vaccines in Nile tilapia, *Oreochromis niloticus*." *Fish & Shellfish Immunology*, *121*:152–162.

Hoihuan, A., Soonson, P., Bunlipatanon, P., Thawonsuwan, J., Tanasomwang, V., Areechon, N., and Unajak, S. 2021. "Molecular genotyping and phenotyping of *Vibrio vulnificus* isolated from diseased, brown-marbled grouper (*Epinephelus fuscoguttatus*) in Thailand with preliminary vaccine efficacy analysis." *Aquaculture*, 545:737188. doi:10.1016/j.aquaculture.2021.737188

Humphreys, Ws. F. 1979. "Production and respiration in animal populations." *The Journal of Animal Ecology*, *48*:427–453.

Irhayyim, T., Beliczky, G., Havasi, M., and Bercsényi, M. 2020. "Impacts of magnetic water treatment on water quality, feeding efficiency and growth performance of common carp in integrated recirculating aquaculture systems." *Journal of Central European Agriculture*, *21*(2):246–255.

Ismail, M.S., Siti-Zahrah, A., Syafiq, M.R.M., Amal, M.N.A., Firdaus-Nawi, M., and Zamri-Saad, M. 2016. "Feed-based vaccination regime against streptococcosis in red tilapia, *Oreochromis niloticus* × *Oreochromis mossambicus*." *BMC Veterinary Research*, *12*(1):1–6.

Joaquin, C., Paola, M., Motte, E., Salvatierra, M., Medina, J., Diringer, B., Sandoval, G., and Mialhe, E. 2019. "Design and evaluation of the expression of a potential DNA vaccine against Tilapia lake virus (TiLV)." *Revista peruana de biología*, *26*(3): 301–310. doi: 10.15381/rpb.v26i3.15516

Kai, Y.-H., and S.-C., Chi. 2008. "Efficacies of inactivated vaccines against betanodavirus in grouper larvae (*Epinephelus coioides*) by bath immunization." *Vaccine*, *26*(11):1450–1457. doi:10.1016/j.vaccine.2007.12.043

Kai, Y.-H., H.-M., Su, Kun-Tsai, T., and Shau-Chi, C. 2010. "Vaccination of grouper broodfish (*Epinephelus tukula*) reduces the risk of vertical transmission by nervous necrosis virus." *Vaccine*, *28*(4):996–1001. doi:10.1016/j.vaccine.2009.10.132

Kasornchandra, J. 2002. "Major viral and bacterial diseases of cultured seabass and groupers in Southeast Asia." Diseases in Asian aquaculture IV: Proceedings of the Fourth Symposium on Diseases in Asian Aquaculture, November 22–26, 1999, Cebu City, Philippines, pp. 205–212.

Katya, K., Lee, S., Yun, H., Dagoberto, S., Browdy, C.L., Vazquez-Anon, M., and Bai, S.C. 2016. "Efficacy of inorganic and chelated trace minerals (Cu, Zn and Mn) premix sources in Pacific white shrimp, *Litopenaeus vannamei* (Boone) fed plant protein based diets." *Aquaculture*, *459*:117–123.

Kayansamruaj, P., Dong, H.T., Pirarat, N., Nilubol, D., and Rodkhum, C. 2017. "Efficacy of α-enolase-based DNA vaccine against pathogenic *Streptococcus iniae* in Nile tilapia (*Oreochromis niloticus*)." *Aquaculture*, *468*:102–106.

Khanjani, M.H., and Sharifinia, M. 2020. "Biofloc technology as a promising tool to improve aquaculture production." *Reviews in Aquaculture*, *12*(3):1836–1850.

Knowler, D., Chopin, T., Martinez-Esplineira, R., Neori, A., Nobre, A., Noce, A., and Reid, G. 2020. The economics of integrated multi-trophic aquaculture: Where are we now and where do we need to go? *Reviews in Aquaculture*, *12*:1579–1594.

Kumar, S.R., Parameswaran, V., Ishaq Ahmed, V.P., Musthaq, S.S., and Sahul Hameed, A.S. 2007. "Protective efficiency of DNA vaccination in Asian seabass (*Lates calcarifer*) against *Vibrio anguillarum*." *Fish & Shellfish Immunology*, *23*(2):316–326.

Lan, N.G.T., Salin, K.R., Longyant, S., Senapin, S., and Dong, H.T. 2021. "Systemic and mucosal antibody response of freshwater cultured Asian seabass (*Lates calcarifer*) to monovalent and bivalent vaccines against *Streptococcus agalactiae* and *Streptococcus iniae*." *Fish & Shellfish Immunology*, *108*:7–13.

Lertwanakarn, T., Trongwongsa, P., Yingsakmongkol, S., Khemthong, M., Tattiyapong, P., and Surachetpong, W. 2021. "Antiviral activity of ribavirin against tilapia tilapinevirus in fish cells." *Pathogens*, *10*(12):1616.

Lightner, D.V. 2011. "Status of shrimp diseases and advances in shrimp health management." Diseases in Asian Aquaculture VII: Proceedings of the Seventh Symposium on Diseases in Asian Aquaculture, June 20–26, 2008, Taipei, Taiwan. pp. 211–134.

Liu, L., Lu, D.-Q., Xu, J., Luo, H.-L., and Li, A.-X. 2019. "Development of attenuated erythromycin-resistant *Streptococcus agalactiae* vaccine for tilapia (*Oreochromis niloticus*) culture." *Journal of Fish Diseases*, *42*(5):693–701.

Lulijwa, R., Rupia, E.J., and Alfaro, A.C. 2020. "Antibiotic use in aquaculture, policies and regulation, health and environmental risks: A review of the top 15 major producers." *Reviews in Aquaculture*, *12*(2):640–663.

McNeill, S, and J.H., Lawton. 1970. "Annual production and respiration in animal populations." *Nature*, 225(5231):472–474.

Miccoli, A., Manni, M., Picchietti, S., and Scapigliati, G. 2021. "State-of-the-art vaccine research for aquaculture use: The case of three economically relevant fish species." *Vaccines*, *9*(2):140.

Mohamad, A., Zamri-Saad, M., Azmai Amal, M.N., Al-Saari, N., Monir, M.S., Chin, Y.K., and Md Yasin, I.-S. 2021. "Vaccine efficacy of a newly developed feed-based whole-cell polyvalent vaccine against vibriosis, streptococcosis and motile aeromonad septicemia in Asian Seabass, *Lates calcarifer*." *Vaccines*, *9*(4):368.

Mohamad, N., Mohd Roseli, F.A., Amal Azmai, M.N., Zamri Saad, M., Md Yasin, I-S., Zulkiply, N.A., and Nasruddin, N.S. 2019. "Natural concurrent infection of *Vibrio harveyi* and *V. alginolyticus* in cultured hybrid groupers in Malaysia." *Journal of Aquatic Animal Health*, *31*(1):88–96.

Mondal, H., and Thomas, J. 2022. "A review on the recent advances and application of vaccines against fish pathogens in aquaculture." *Aquaculture International*, *30*(4):1971–2000. doi: 10.1007/s10499-022-00884-w

Nur-Nazifah, M., Sabri, M.Y., and Siti-Zahrah, A. 2014. "Development and efficacy of feed-based recombinant vaccine encoding the cell wall surface anchor family protein of *Streptococcus agalactiae* against streptococcosis in *Oreochromis* sp." *Fish & Shellfish Immunology*, *37*(1):193–200.

Oidtmann, B.C., Thrush, M.A., Denham, K.L., and Peeler, E.J. 2011. "International and national biosecurity strategies in aquatic animal health." *Aquaculture*, *320*(1–2):22–33.

Ou-yang, Z., Wang, P., Huang, X., Cai, J., Huang, Y., Wei, S., Ji, H., Wei, J., Zhou, Y., and Qin, Q. 2012. "Immunogenicity and protective effects of inactivated Singapore grouper iridovirus (SGIV) vaccines in orange-spotted grouper, *Epinephelus coioides*." *Developmental & Comparative Immunology*, *38*(2):254–261.

Pang, H., Chang, Y., Zheng, H., Tan, H., Zhou, S., Zeng, F., Hoare, R., Monaghan, S.J., Wang, N., and Ding, Y. 2022. "A live attenuated strain of HY9901ΔvscB provides protection against *Vibrio alginolyticus* in pearl gentian grouper (♀*Epinephelus fuscoguttatus* × ♂*Epinephelus lanceolatu*)." *Aquaculture*, *546*:737353. doi: 10.1016/j.aquaculture.2021.737353

Pang, H., Qiu, M., Zhao, J., Hoare, R., Monaghan, S.J., Song, D., Chang, Y., and Jian, J. 2018. "Construction of a Vibrio alginolyticus hopPmaJ (hop) mutant and evaluation of its potential as a live attenuated vaccine in orange-spotted grouper (*Epinephelus coioides*)." *Fish & Shellfish Immunology*, *76*:93–100. doi: 10.1016/j.fsi.2018.02.012

Phornprapha, W. 2020. "Shrimp Farming in Thailand: A pathway to Sustainability." Pomona Senior Theses, Pomona College, Claremont, CA, USA, pp. 208.

Pridgeon, J.W., and Klesius, P.H. 2011. "Virulence of *Aeromonas hydrophila* in the presence or absence of extracellular products to channel catfish fingerlings." *Disease of Aquatic Organism*, *95*:209–215.

Pridgeon, J.W., Klesius, P.H., and Yildirim-Aksoy, M. 2013. "Attempt to develop live attenuated bacterial vaccines by selecting resistance to gossypol, proflavine hemisulfate, novobiocin, or ciprofloxacin." *Vaccine*, *31*(18):2222–2230.

Pulkkinen, K., Suomalainen, L.-R., Read, A.F., Ebert, D., Rintamäki, P., and Valtonen, E.T. 2010. "Intensive fish farming and the evolution of pathogen virulence: The case of columnaris disease in Finland." *Proceedings of the Royal Society B: Biological Sciences*, *277*(1681):593–600.

Pulpipat, T., Maekawa, S., Wang, P.-C., and Chen, S.-C. 2020. "Immune responses and protective efficacy of a formalin-killed *Francisella noatunensis* subsp. *orientalis* vaccine evaluated through intraperitoneal and immersion challenge methods in *Oreochromis niloticus*." *Vaccines*, *8*(2):163.

Ruangpan, L., Kitao, T., and Yoshida, T. 1986. "Protective efficacy of *Aeromonas hydrophila* vaccines in Nile tilapia." *Veterinary Immunology and Immunopathology*, *12*(1–4):345–350.

Rubel, H., Woods, W., Pérez, D., Unnikrishnan, S., Zum Felde, A.M., Zielcke, S., Lidy, C., and Lanfer, C. 2019. "A strategic approach to sustainable shrimp production in Thailand: The case for improved economics and sustainability." Boston, MA: Boston Consulting Group.

Rudtanatip, T., Boonsri, N., Asuvapongpatana, S., Withyachumnarnkul, B., and Wongprasert, K. 2017. "A sulfated galactans supplemented diet enhances the expression of immune genes and protects against *Vibrio parahaemolyticus* infection in shrimp." *Fish & Shellfish Immunology*, *65*:186–197.

Senapin, S., Shyam, K.U., Meemetta, W., Rattanarojpong, T., and Dong, H.T. 2018. "Inapparent infection cases of tilapia lake virus (TiLV) in farmed tilapia." *Aquaculture*, *487*:51–55.

Shahin, K., Shinn, A.P., Metselaar, M., Ramirez-Paredes, J.G., Monaghan, S.J., Thompson, K.D., Hoare, R., and Adams, A. 2019. "Efficacy of an inactivated whole-cell injection vaccine for Nile tilapia, *Oreochromis niloticus* (L), against multiple isolates of *Francisella noatunensis* subsp. *orientalis* from diverse geographical regions." *Fish & Shellfish Immunology*, *89*:217–227.

Sharrer, M.J., and Summerfelt, S.T. 2007. "Ozonation followed by ultraviolet irradiation provides effective bacteria inactivation in a freshwater recirculating system." *Aquacultural Engineering*, *37*(2):180–191.

Shoemaker, C.A., Klesius, P.H., and Evans, J.J. 2000. "Diseases of tilapia with emphasis on economically important pathogens." Proceedings of the 5th International Symposium on tilapia Aquaculture. Last modified June 14, 2022. https://www.ars.usda.gov/research/publications/publication/?seqNo115=163551

Shoemaker, C.A., LaFrentz, B.R., and Klesius, P.H. 2011. "Vaccination of sex reversed hybrid tilapia (*Oreochromis niloticus*× *O. aureus*) with an inactivated *Vibrio vulnificus* vaccine." *Biologicals*, *39*(6):424–429.

Shoemaker, C.A., LaFrentz, B.R., and Klesius, P.H. 2012. "Bivalent vaccination of sex reversed hybrid tilapia against *Streptococcus iniae* and *Vibrio vulnificus*." *Aquaculture*, 354:45–49.

Silva, B.C., Martins, M.L., Jatobá, A., Buglione Neto, C.C., Vieira, F.N., Pereira, G.V., Jerônimo, G.T., Seiffert, W.Q., and Mouriño, J.L.P. 2009. "Hematological and immunological responses of Nile tilapia after polyvalent vaccine administration by different routes." *Pesquisa Veterinária Brasileira*, *29*(11):874–880.

Sirimanapong, W., Thompson, K.D., Shinn, A.P., Adams, A., and Withyachumnarnkul, B. 2018. "*Streptococcus agalactiae* infection kills red tilapia with chronic *Francisella noatunensis* infection more rapidly than the fish without the infection." *Fish & Shellfish Immunology, 81*:221–232.

Smith, C.S., Ito, M., Namba, M., and Nakaoka, M. 2018. "Oyster aquaculture impacts Zostera marina epibiont community composition in Akkeshi-ko estuary, Japan." *PLoS One*, *13*(5):1–17. doi:10.1371/journal.pone.0197753

Solomon, O.O., and Ahmed, O.O. 2016. "Ecological consequences of oysters culture: A review." *International Journal of Fish Aquatic Studies*, *4*:1–6.

Song, X., Zhang, Y., Wei, S., and Huang, J. 2013. "Effects of different enzymatic hydrolysis methods on the bioactivity of peptidoglycan in *Litopenaeus vannamei*." *Chinese Journal of Oceanology and Limnology*, *31*(2):374–383.

Thitamadee, S., Prachumwat, A., Srisala, J., Jaroenlak, P., Salachan, P.V., Sritunyalucksana, K., Flegel, T.W., and Itsathitphaisarn, O. 2016. "Review of current disease threats for cultivated penaeid shrimp in Asia." *Aquaculture*, *452*:69–87.

Ulhaq, I., Pham, N.T.A., Le, V., Pham, H.-C., and Le, T.C. 2022. "Factors influencing intention to adopt ICT among intensive shrimp farmers." *Aquaculture*, *547*:737407.

Vimal, S., Farook, M.A., Madan, N., Abdul Majeed, S., Nambi, K.S.N., Taju, G., Sundar Raj, N., Venu, S., Subburaj, R., and Thirunavukkarasu, A.R. 2016. "Development, distribution and expression of a DNA vaccine against nodavirus in Asian Seabass, *Lates calcarifier* (Bloch, 1790)." *Aquaculture Research*, *47*(4):1209–1220.

Wang, H., Zhu, F., Huang, Y., Ding, Y., Jian, J., and Wu, Z. 2017. "Construction of glutathione peroxidase (GPx) DNA vaccine and its protective efficiency on the orange-spotted grouper (*Epinephelus coioides*) challenged with *Vibrio harveyi*." *Fish & Shellfish Immunology*, *60*:529–536. doi:10.1016/j.fsi. 2016.11.018

Wang, M., Chen, Y., Zhao, Z., Weng, S., Yang, J., Liu, S., Liu, C., Yuan, F., Ai, B., and Zhang, H. 2021. "A convenient polyculture system that controls a shrimp viral disease with a high transmission rate." *Communications Biology*, *4*(1):1–11.

Wang, P.-H., Huang, T., Zhang, X., and He, J.-G. 2014. "Antiviral defense in shrimp: From innate immunity to viral infection." *Antiviral Research*, *108*:129–141.

Withyachumnarnkul, B., Foroutan, B., Kerdmusik, C., Vanijviriyakij, R., and Withyachumnarnkul, B. 2022. "Oogenesis, spermatogenesis and hatching rate of seawater-acclimated Nile Tilapia, *Oreochromis niloticus*." *Songklanakarin Journal of Science & Technology*, *44*(1):21–25.

Withyachumnarnkul, B., Palang, I., Reungsri, J., Sirithammajak, S., Jitrakorn, S., Kiatpathomchai, W., Saksmerprome, V., Pongtippatee, P., and Withyachumnarnkul, B. 2017. "Nile tilapia reared under full-strength seawater: Hemato-immunological changes and susceptibility to pathogens." *Aquaculture*, *480*:42–50.

Yu, N.-t., Zeng, W.-W., Xiong, Z., and Liu, Z.x. 2022. "A high efficacy DNA vaccine against tilapia lake virus in Nile tilapia (*Oreochromis niloticus*)." *Aquaculture Reports*, *24*:101166.

Zeng, W., Wang, Y., Hu, H., Wang, Q., Bergmann, S.M., Wang, Y., Li, B., Lv, Y., Li, H., and Yin, J. 2021. "Cell culture-derived tilapia lake virus-inactivated vaccine containing Montanide adjuvant provides high protection against viral challenge for tilapia." *Vaccines*, *9*(2):86.

Zhu, L., Yang, Q., Huang, L., Wang, K., Wang, X., Chen, D., Geng, Y., Huang, X., Ouyang, P., and Lai, W. 2017. "Effectivity of oral recombinant DNA vaccine against *Streptococcus agalactiae* in Nile tilapia." *Developmental & Comparative Immunology*, *77*:77–87.

Zhu, Z., Dong, C., Weng, S., and He, J. 2019. "Identification of outer membrane protein TolC as the major adhesin and potential vaccine candidate for *Vibrio harveyi* in hybrid grouper, *Epinephelus fuscoguttatus* (♀) × *E. lanceolatus* (♂)." *Fish & Shellfish Immunology*, *86*:143–151. doi:10.1016/j.fsi.2018.11.037

Section II

Vaccination in Fishes

Types and Methods

4 Concepts and Types of Vaccines

Subramaniam Sivakumar
Sri Sankara Arts and Science College

Panneerselvam Punniyakotti
Kanchi Shri Krishna College of Arts and Science

Kuppuswamy Kavitha and Panneerselvam Saravanan
Sri Sankara Arts and Science College

CONTENTS

DOI: 10.1201/9781003388548-6

4.1 INTRODUCTION

One of the most rapidly developing food industries is aquaculture. Different fish varieties are formed throughout the world. The latest *State of World Fisheries and Aquaculture* edited by the Food and Agriculture Organization reported that the global fish production accounted for 171 million tons in 2016 (1). The economic losses due to infectious diseases in aquaculture exceeded 20 billion yen before the 21st century because of the limited strategies for disease prevention in the aquaculture industry (2).

The aquaculture industry has become an important producer of seafoods as the fishing capture industry has declined and wild marine stocks have diminished. It is one of the fastest growing food production sectors, with an average annual growth of 5.8% during the period 2001–2016 (3).

Worldwide, Atlantic salmon is one of the most intensively farmed fish and the major producers are from Norway, Chile, The United Kingdom, and Canada. Depending on the development of the aquaculture system, farming of high-value marine species such as European seabass and seabream, halibut, flounder, cod, tuna, eel and amberjack/yellowtail will increase. Limited stress and proper hygienic conditions with proper fish management are key factors in the prophylaxis of infectious diseases (4).

Chemotherapeutic agents and antibiotics are often used for the control and prevention of diseases. Inappropriate use and overdose of antibiotics may cause side effects in fish and ultimately in human consumers, as well as increase the risk of developing antibiotic-resistant microbes in the environment (5). Thus, the importance of vaccine development was realized by the fishing industry. It is also necessary for the optimal functioning of vaccines (4).

The development of fish vaccines is an ongoing interaction among the academia, regulatory authorities, and the pharmaceutical industry. Infectious diseases are a major issue for the commercial fishing industries. Reduced usage of antibiotics is essential for the food industry. In this context, development of vaccines for aquaculture is the need of the hour.

At present, vaccines are used to treat more than 6 viral diseases and 22 different bacterial diseases. About 30 different types of vaccines are available worldwide (6). In global aquaculture, vaccination efforts contribute to social, environmental, and economic sustainability (7). Therefore, the development of effective and versatile vaccines is required both for preventing infections and for avoiding the use of antibiotics in the aquaculture industry (8).

4.2 VACCINES

A vaccine is an agent that can be used to induce immune response against infection and/or disease that provides protection on a subsequent exposure to the respective pathogen (9).

The fish vaccine mainly serves as an antigen, which either contains or produces an antigenic substance (7). At the end of 2019, there are 29 vaccines approved against 10 fish pathogens including both single and combined vaccines in Korea (10). Vaccination is considered as the most effective strategy to control infectious diseases in fish with adaptive immunity (11).

4.2.1 Fish Immune System

To have a specific long-term protection against diseases is a major goal of vaccination. Fish are the initial primitive group of animals with the basic immune system of higher vertebrates with respect to the evolutionary system. There are more similarities than differences although fish immune systems are primitive compared to mammals. Furthermore, there are more than 25,000 species of fish living in habitats ranging from muddy fresh water to the ocean, from polar regions to the tropics, from dark high-pressure depths to the bright low-pressure surface areas. Therefore, their response to vaccines and combating mechanisms against pathogens differ for different fish species. The study of both pathogen and vaccine-induced immunity is limited due to the lack of detailed knowledge of the immune systems in different fish species (4).

4.2.1.1 Fish Immunity

Innate immunity is more vital in fish than mammals against invading pathogens because the adaptive immunity of fish is not as diversified; there are only three isotypes of immunoglobulins and some fish lack few major histocompatibility complexes (MHC) (12).

More than 550 million years ago, the adaptive immune system emerged along with the evolution of jawless vertebrates. Only variable lymphocyte receptors are used to recognize antigens in jawless vertebrates due to lack of T- and B-cells, while in jawed vertebrates, immunoglobulins (Ig) are involved in the recognition of pathogens in the process of immune exclusion (13). Immunoglobulin T is one of the key molecules of jawed vertebrate's adaptive immunity (14).

To date, five immunoglobulin isotypes have been characterized in mammals, namely, Ig M, IgD, IgG, IgE, and IgA, whereas in fishes only three isotypes were well characterized, namely, IgM, IgD, and IgT or IgZ (15). IgT plays important roles in mucosal immunity in the gut (16), skin, and gills (17,18) of fish (19). Nayak and Nakanishi (2013) demonstrated the direct antibacterial activity of $CD8^+$ and $CD4^+$ T-cells and surface IgM^+ cells in fish (20). The interbranchial lymphoid tissue (ILT) of Atlantic salmon originates from an embryological location that in higher vertebrates acted as origin for the development of primary and secondary lymphoid tissues (21). The skin-associated lymphoid tissue (SALT) is a mucosal lymphoid tissue in fish that is different from that of mammals that is the outermost cell layer alive and actively dividing. The SALT protection is efficient due to the high concentration of immunoglobulin M (22).

The mucosal surface of the fish digestive tract is covered with immunoglobulins, complement factors, antimicrobial peptides, and other surface defensins beneath which several types of immune cells are found in fish gut, which provides innate immunity to fish (23). The CD4 cell surface marker plays an important role in distinguishing between T-helper cell and cytotoxic T-cell (24).

Oral vaccines induce fish immune system selectively by activating CCL25/CCR9 ligand/receptor system in innate immunity (25,26). In fish, central and peripheral immune tolerance is important in the gills, where the intimate contact between gill tissue and the aqueous environment would otherwise lead to continual immune stimulation by innocuous antigens (27).

4.2.1.1.1 Fish Immune Organs

The organization of immune organs in fish is slightly different from higher vertebrates. The main difference is that fishes lack bone marrow and lymph nodes and the primary lymphoid organs are the thymus, the head kidney, or pronephros for teleosts, and the leydig and epigonal organs for chondrichthyes, while secondary lymphoid organs are the spleen, the kidney, and mucosa-associated lymphoid tissue (MALT) present in peripheral immune tissues (28).

Bony fish (teleosts) have a complete set of lymphoid organs except for bone marrow, and they can mount both innate and adaptive immune responses. Their lymphoid organs are very different from those in mammals. For example, their thymus may involute in response to hormones or season. Age involution is inconsistent, and a thymus may be found in old fish. Fish kidneys differentiate into two sections. The opisthonephros or posterior kidney is similar to the mammalian kidney.

In contrast, the pronephros or anterior kidney is a lymphoid organ containing antibody-forming cells and phagocytes. Its function is analogous to mammalian bone marrow and lymph nodes. Fish have a spleen with a structure and function similar to that in mammals. Aggregates of lymphocytes are prominent in the fish intestinal tract. Fish also have clusters of macrophages containing melanin and hemosiderin. These melanomacrophage centers are found in the spleen, liver, and kidney. Antigens may persist in these centers for long periods, and they appear to be precursors of the germinal centers found in more evolved vertebrates. Teleosts also have dendritic-like cells that can present antigens to T-cells.

Fish lymphocytes resemble those of mammals. B-cells can be found in the thymus, anterior kidney, spleen, leydig organ, and blood, and their surface immunoglobulins act as antigen receptors. These B-cells can mature into plasma cells. Unlike mammalian B-cells, however, teleost B-cells can phagocytose particles, generate phagolysosomes, and kill ingested microbes. Both helper and cytotoxic T-cells can be detected in fish (6).

4.2.1.1.2 Fish Immune Response

Immune cells recognize most pathogens and altered/infected cells through expressed pathogen-associated molecular patterns (PAMPS) or danger-associated molecular patterns (DAMPS), through nonself or MHC class I molecules and by presenting foreign nonself peptides of intracellular or extracellular origin. In order to destroy pathogens directly or by inhibiting their ability to replicate, specialized immune cells of the innate and adaptive responses were produced during evolution. The first line of defense is represented by the evolutionarily ancient macrophages and natural killer (NK) cells. These innate mechanisms are well developed in bony fish.

Adaptive cell-mediated cytotoxicity (CMC) requires key molecules expressed on cytotoxic T lymphocytes (CTLs) and target cells. CTLs kill host cells and attract intracellular pathogens by binding of their T-cell receptor (TCR) and its co-receptor CD8 to a complex of MHC class I and bound peptide molecule on the host infected cell. Alternatively, extracellular antigens are taken up by professional antigen-presenting cells such as macrophages, dendritic cells, and B-cells to process those antigens and present the resulting peptides in association with MHC class II to $CD4^+$ T-helper cells. Then, T-helper cells through activation of other immune cells eliminate the extracellular pathogens (29).

4.2.1.2 Vaccine Efficacy

Many factors affect the efficacy of vaccines. The factors include husbandry nature, handling method, fish age, smoltification, dominance hierarchies, pheromones, temperature, diet (vitamins and minerals), pollutants (metals and pesticides), seasonal variations, vaccine dose, vaccine route, nature of adjuvant usage, immunostimulants used, and antibiotics. Utmost care should be taken to achieve optimum levels of protective immunity when administering vaccines to minimize the effects of stressors (30).

4.2.1.3 Adverse Effects

Fish vaccines are no exception to the rule that adverse effects following vaccination are common problems. Vaccine producers considered adverse effects as a serious problem in recent years. The nature of delivery systems for oral or immersion administration would be an improvement to fish vaccination in the near future (4). Local injection site reactions remain an issue in aquaculture. Intraperitoneal inoculation may lead to local or diffuse peritonitis with adhesions. Multiple granulomas may also develop. The fish produce multiple autoantibodies and develop immune-complex glomerulonephritis, liver thrombosis, and spinal deformities. Both polyclonal IgM and antibodies to salmon red blood cells are elevated in vaccinated fish (6).

4.2.2 History of Vaccines

The history of fish vaccinology is a documentation of how the immune system of fish can be stimulated by vaccines to prevent accidental effects of pathogenic microorganisms (31). The first attempts to immunize fish started date back to the 1930s, but the economic incentive to commercialize vaccines was not until the mid- to late 1970s (30). Even though the fish vaccination process has a long history, only in the past few decades it has become a well-established protective measure against bacterial infections (32).

The first commercially available bacterial vaccines were against enteric red mouth disease and vibriosis, introduced in the USA in the late 1970s. Fish immersion vaccines were found to be effective against vibriosis in the USA. However, a new disease, furunculosis, appeared and injectable vaccines containing adjuvants were developed in the early 1990s.

The first viral vaccine was produced by Bioveta Company for fish in 1982. The vaccine was against a carp rhabdovirus, and was based on two inactivated strains of rhabdovirus emulsified in oil and administered by injection. Most available virus vaccines are based on inactivated virus or recombinant subunit proteins for aquaculture.

With the exception of the introduction of a recombinant virus vaccine in 1995, vaccination strategies in the fish farming industry have been more or less unchanged over the last 20 years. In the present scenario, trial and error has been the important strategy for the designing of fish vaccines.

Vaccination against the most serious bacterial diseases has been quite successful for the large-scale commercial farmed fish varieties with a few exceptions. The usage of modern sophisticated techniques is applied as a new trend for the development of new vaccines. The development of fish genomic studies and their information can be incorporated for the invention of new vaccines. Usage of DNA vaccines are safer than live attenuated vaccines but the legal issues related with the usage of DNA and genetically modified organism vaccines limited their availability. If only the method of administration, efficacy, and cost of production were considered, live attenuated vaccine would be chosen as an optimal type of vaccine. But a consumer safety measure prevents the usage of live attenuated vaccine type because of the reversal of infectivity features (4). To date, there are many vaccines used to control infections such as killed vaccines, live attenuated vaccines, subunit vaccines, and DNA vaccines (33).

The field of fish vaccinology has shown an amazing development recently. The comprehensive scientific research and valuable practical experience are responsible for the production of first generation fish vaccines, which will immensely contribute to social, environmental, and economical sustainability in aquaculture worldwide.

4.3 INFECTIOUS DISEASES OF FISH

Infectious diseases are adversely affecting aquaculture practices throughout the world including Asian countries. Indian aquaculture practices are facing serious problems due to a variety of infectious agents that are responsible for severe mortality and morbidity of all the cultured fish species leading to severe economic losses (34). As much as 10% of all cultured fish are lost due to infectious diseases which in turn amount to >10 billion USD in losses annually on a worldwide scale despite the improvements in the aquaculture industry (35).

Pathogens are of extracellular, facultative intracellular, and obligate intracellular origin. Most of the bacteria are of extracellular type; only few others may proliferate facultatively inside the infected cells. Viruses are usually obligate intracellular pathogens and their lifecycle depends upon the machinery of the host cells. Intracellular pathogens are more challenging for the immune system, because as long as they are not exposed extracellularly, the immune system is unable to recognize them through cellular interactions or humoral factors like antibodies. Thus, virus and intracellular bacteria are more dangerous than extracellular microbes (36). Fish mucosal organs are primary defense organs against diseases because they are directly exposed to aquatic environment and also suited for the colonization and growth of microorganisms. Thus, they play an important role in preventing pathogen entry and maintaining homeostasis (37).

Infectious causative agents are easily transmitted between fish where fish are cultured in high densities. Pathogens spread quickly within a population of cultured fish due to the easy transportation of pathogens in water and the large population of fishes used in large-scale commercial farming (4). Vaccination has had a significant impact on aquaculture industry for the prevention of specific bacterial diseases that are affecting commercially reared fish varieties (30).

4.3.1 Bacterial Diseases

Vaccines produced against many of the bacterial diseases that are causing major problems for the aquaculture industries are listed in Table 4.1 (6). The inactivated bacterin vaccine effectively controls vibriosis, but for other bacteria, it is difficult to control by vaccination process. A live attenuated *E. ictaluri* vaccine is efficacious through the process of immersion of fish (4).

4.3.2 Viral Diseases

Fishing industry suffers from serious setbacks due to viral diseases that can be rapidly spread throughout fish farms (38). Vaccines are available for the most common viral diseases with respect to salmonid farming. The commercially available viral fish vaccines are summarized in Table 4.2 (6). An inactivated viral vaccine against pancreas disease is available in Ireland and a vaccine against infectious salmon anemia (ISA) is available in Canada and the USA. Different recombinant subunit vaccines based on the infectious hematopoietic necrosis virus (IHNV) and viral

TABLE 4.1
Vaccines for Fish Bacterial Diseases

Sl. No.	Fish Type	Bacterial Diseases	Causative Organisms	Type of Vaccine	Delivery Method
1	Salmonids. Cod/halibut. Sea bass/bream. Amberjack/yellow tail, rainbow trout, turbot, and eel	Vibriosis	*Listonella anguillarum; Vibrio ordalii*	Inactivated Vibrio spp.	Intraperitoneal; Immersion
2	Atlantic salmon and rainbow trout	Coldwater vibriosis	*Aliivibrio salmonicida*	Killed bacterins	Immersion or injection
3	Salmonids	Furunculosis	*Aeromonas salmonicida*	Inactivated	Intraperitoneal; Immersion
4	Atlantic salmon and rainbow trout	Winter ulcer	*Moritella viscosa*	Inactivated	Injection
5	Salmonids	Yersiniosis	*Yersinia ruckeri*	Inactivated	Immersion or oral
6	Marine Fishes	Pasteurellosis	*Pasteurella piscicida*	Inactivated	Immersion or injection
7	Marine Fishes	Warm water vibriosis	*Vibrio alginolyticus, Vibrio parahaemolyticus,* and *Vibrio vulnificus*	Inactivated	Immersion or injection
8	Catfish	Edwardsiellosis	*Edwardsiella ictaluri*	Inactivated, live attenuated vaccine	Intraperitoneal Immersion
9	All freshwater finfish species, bream, bass, turbot, salmon	Flavobacteriosis	*Flavobacterium columnaris*	Inactivated, live attenuated vaccine	Immersion
10	Salmonids	Bacterial kidney disease	*Renibacterium salmoninarum*	Avirulent live; culture inactivated	Intraperitoneal; Immersion
11	Salmonids	Piscirickettsiosis	*Piscirickettsia salmonis*	Inactivated, live attenuated vaccine	Intraperitoneal; Immersion

TABLE 4.2
Vaccines for Fish Viral Diseases

Sl. No.	Fish Type	Viral Diseases	Causative Organisms	Type of Vaccine	Delivery Method
1	Salmonids	Infectious pancreatic necrosis	*Infectious pancreatic necrosis virus*	Subunit vaccines or inactivated	Intraperitoneal; Injection
2	Atlantic salmon	Infectious salmon anemia	*Infectious salmon anemia virus*	Subunit vaccines or inactivated	Intraperitoneal
3	Salmonids, sea bass, sea bream, turbot, Pacific cod	Infectious hematopoietic necrosis	*Infectious hematopoietic necrosis virus*	DNA-plasmid	Intramuscular injection
4	Salmonids	Salmon pancreas disease	Salmon pancreas disease virus	Inactivated	Intraperitoneal
5	Rainbow and brown trout, turbot, Japanese flounder	Viral hemorrhagic septicemia	Viral hemorrhagic septicemia virus	Inactivated; live attenuated; DNA	Immersion; oral
6	Catfish and trout	Spring viremia of carp virus	*Rhabdovirus carpio*	Subunit vaccines or inactivated	Intraperitoneal; oral
7	Carp	Koi herpesvirus disease	*Koi herpesvirus*	Live attenuated	Intraperitoneal; Immersion
8	Several marine fish species, e.g., sea bass, groupers, barramundi, halibut	Viral nervous necrosis	*Betanodavirus*	Inactivated	Intraperitoneal

hemorrhagic septicemia virus (VHSV) membrane glycoprotein are also available. DNA vaccines encoding the viral glycoproteins are remarkably efficacious. The lack of effective viral vaccines is one of the main problems facing fish vaccinology. In near future, new and improved virus vaccines will probably be developed for many viral diseases (4).

4.3.3 Protozoan Diseases

Amoebic gill disease (AGD) is caused by *Neoparamoeba perurans* and represents a significant threat to Atlantic salmon marine farming in several countries worldwide (39). *Ichthyophthirius multifiliis* is a ciliated protozoan that causes "itch" or "white spot disease." It is difficult to control it by conventional methods.

4.3.4 Types of Vaccines

Different types of vaccines are practiced in the fish farming industry. The variety of vaccines and their preparation mode along with the administration methods were depicted in Figure 4.1 (40). In this section, inactivated vaccines, live attenuated vaccines, subunit vaccines, virus-like particle (VLPs) vaccines, DNA vaccines, and RNA vaccines were discussed.

4.3.4.1 Inactivated Vaccines

Inactivated or killed vaccines are produced from virulent microbes, and their infectivity and pathogenic features are mutated. Without compromising the antigenicity of the microbes, these changes can be induced through physical, chemical, or radiation processes. Early vaccine trials focused on

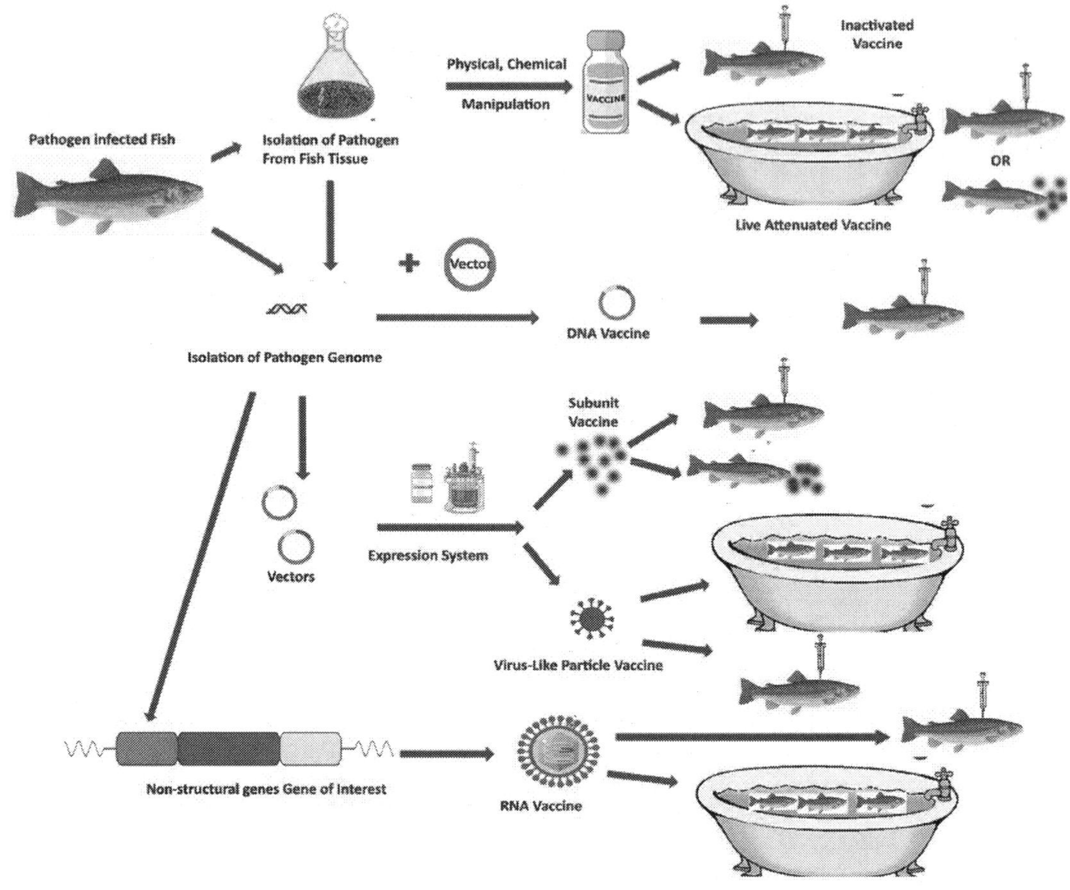

FIGURE 4.1 Vaccine types.

inactivated vaccines in aquaculture. *A. salmonicida* bacterin, an *A. salmonicida-Vibrio anguillarum-Vibrio ordalii-Vibrio salmonicida* bacterin, an ISA virus vaccine, and *Y. ruckeri* bacterin are the four of the eight licensed vaccines for aquaculture in the USA and all of these are killed vaccines. Killed vaccines are less efficient against diseases caused by intracellular bacterial and viral infections (40). Inactivated vaccines are administered through injection method to achieve high efficacy (41).

4.3.4.2 Live Attenuated Vaccines

Live attenuated vaccines are prepared by neutralizing the infectivity nature of the live microbes. The attenuated microbes are modified avirulent type. The major advantage of live attenuated vaccines is the ability to induce both humoral and cell-mediated immunity, which includes the induction of interferon, largely important in fighting viral infections. In addition to this, a smaller amount of attenuated vaccines is required to induce an effective immune response. Moreover, they can be administered through natural routes and also provide longer lasting protection, so that the need for a booster dose is not required. The licensing of live attenuated vaccines is difficult because of the safety concerns related to the reversion to a virulent form. Because of this, only few live attenuated vaccines are commercially available. However, with the usage of recent molecular methods and the identification of new and safer vectors, the introduction of more live vaccines is possible in the coming years. AQUAVAC-ESC is one of the commercially available vaccines for the catfish industry. AQUAVAC-ESC is produced from a strain of *E. ictaluri* (RE-33) and is attenuated by multiple passages in ascending concentrations of rifampin. Very few live attenuated viral vaccines are available currently. A koi herpesvirus mutant produced by serial cell culture transfer followed by irradiation to completely attenuate the virus and reduce the chances of reversion to a virulent form. A live attenuated vaccine against spring viremia of carp was also produced in China (42).

4.3.4.3 Subunit Vaccines

Subunit vaccines provide chances to target immune responses toward specific microbial determinants and they can be produced in a highly characterized state. They can be freeze-dried allowing for nonrefrigerated transport and storage (40). Subunit vaccines are produced using protein expression systems like bacteria, yeast, insect cells or tissues, protozoa, mammalian cell culture, and rarely even plants. One of the advantages of subunit vaccines is that they are safer than live attenuated vaccines since they are unable to invade the host genome or replicate within the host. But one of its main problems is the generation of incorrect or misfolded processed antigens that fail to induce a protective immune response. In recent years, the usage of *Tetrahymena thermophile* as an expression system reduces the problem of misfolding to a large extent. The commercially available subunit vaccines include hemagglutinin esterase/fusion proteins, infectious salmon anemia virus (ISAV), and the VP2 and VP3 capsid proteins against infectious pancreatic necrosis virus (IPNV). The VP28 envelope protein vaccine of white spot syndrome virus (WSSV) is also used to enhance survival of shrimp (42).

4.3.4.4 Virus-Like Particle Vaccines

The self-assembly of viral capsid proteins into virus-like particles is used as VLP vaccine because they mimic the natural structure of virus. Due to the lack of genomic materials, VLPs are unable to revert to their virulent form. Because of the presence of a particulate nature, VLPs can activate both innate and adaptive immune responses. Recently, various VLP vaccines for fish diseases have been developed. Nervous necrosis virus (NNV) VLP vaccines and two IHNV recombinant viruses displaying IPNV VP2 protein were generated against both IHNV and IPNV infection. Based on recent studies, VLPs have been shown to elicit strong immunogenicity and constitute a safe alternative to inactivated or attenuated vaccines (40).

4.3.4.5 DNA Vaccines

DNA encoding an antigenic protein is transformed in an expression plasmid vector under the control of a strong eukaryotic promoter applied in DNA vaccine production (42). DNA vaccines are produced by combining a specific gene that codes for an antigenic protein and an expression plasmid as vector. When DNA vaccines are administered, they expressed the antigenic protein in the host and are expected to elicit an effective immune response. They can strongly activate cellular and humoral immunity. The minimum requirement for the development of DNA vaccine is the identification of a protective antigen for the respective disease (40). The first DNA vaccine was tested in rainbow trout against IHNV in aquaculture (43). Another salmonid alphavirus subtype 3 DNA vaccine is also available in the European Union. DNA vaccines are considered safer than attenuated vaccines since they only express the antigenic protein and not the entire virulent organism (40).

4.3.4.6 RNA Vaccines

There are two types of RNA-based vaccines, namely, nonamplifying mRNA vaccine and self-amplifying mRNA vaccine. RNA-based vaccines are more advantageous because RNA is noninfectious and there is no potential risk of insertional mutagenesis. They are also potent stimulators of immunity. The alphavirus replicate functions in fish. Thus, the self-amplifying RNA vaccine produced by replacing the genes for the structural proteins of the alphavirus with a fish pathogenic antigen of interest could potentially protect against a number of important fish diseases. The untranslated regions of the salmonid alphavirus 3 (SAV3) genome might be used to construct an SAV3-based replicon. This replicon vaccine provides high protection against ISA. Hence, the SAV-based replicon represents a good vaccine candidate for aquaculture (40).

4.4 ADJUVANTS

Adjuvants are the helper substances that increase the magnitude of an adaptive response to a vaccine or its ability to prevent infection and death. The fish immune system is unresponsive to most foreign proteins that are injected in a soluble, deaggregated form, but when injected together with adjuvants, these foreign proteins can generate an immunity and memory to the antigen. The mechanisms by which adjuvants achieve these effects include the generation of antigen persistence, improved antigen presentation by activated antigen-presenting cells, and induction of co-stimulatory molecules to help direct the immune response. Thus, the vaccines are produced along with adjuvants that may induce robust immunity and vaccine efficacy (44). The main goal of a successful vaccine is the effective induction of long lost protective immunity without causing any side effects. Adjuvants induced increased immunity in several fish species against even with poor immunogenic antigens (45).

4.4.1 Adjuvants Usage

Adjuvants are grouped under two different categories, namely, signal 1 facilitators and signal 2 facilitators, during immunity induction. Both the presentation of an antigen (signal 1) and the additional secondary signals (signal 2) are required for the activation of specific adaptive immunity through T- and B-lymphocytes. The signal 1 facilitators increase the antigenicity by improving antigen availability, while signal 2 facilitators improve the co-stimulation signals during antigen recognition, which is responsible for the most adequate antigen-specific immune response. Many adjuvants induce immune response by promoting the detection of conserved molecular patterns, designated as pathogen-associated microbial patterns (PAMPs), of microbes through pathogen recognition receptors (PRRs), which include toll-like receptors (TLRs), NOD-like receptors (NLRs), dectin 1 or RIG-like helicases. These molecules are predominantly present on the immune cells (29). Adjuvanted nonviable immunogens result in long-term humoral immunity and the potential for long-lived memory, whereas soluble antigens tend to result in short or transient humoral immunity and the likely possibility is of no long-lived plasma cells (46).

4.4.2 Signal 1 Facilitators

4.4.2.1 Freund's Complete Adjuvant

Freund's complete adjuvant (FCA) is composed of heat-killed mycobacterial membrane combined with mineral oil and a surfactant. The antigen in an aqueous solution is mixed with the FCA producing a stable water-in-oil emulsion and then it is injected. Immunization with FCA with antigens results in strong T-cell responses via the MyD88 pathway (44).

4.4.2.2 Freund's Incomplete Adjuvant

Freund's incomplete adjuvant (FIA) that lacks the mycobacterial components of the emulsion is used as an alternative to FCA due to the latter's high toxicity issues. This adjuvant is showed as effective in vaccination with a significant reduction of toxicity for Atlantic cod (*Gadus morhua*) (29).

4.4.2.3 Montanide

Mineral oil adjuvants are registered under the trademark of montanide. It has been optimized in order to improve efficacy and stability of vaccine formulations. It also has reduced side effects. These adjuvants are based on either mineral oil, nonmineral oil, or a mixture of both, and may be used to manufacture different types of emulsions, W/O, O/W, or W/O/W, for the use in fish. Formalin-killed *P. plecoglossicida* bacterin was emulsified with montanide used for vaccination process (29).

4.4.2.4 Poly(lactide-co-glycolide) (PLGA) Particles

The encapsulation of vaccines in biodegradable PLGA polymers has been studied for over 20 years. Antigen is released from the nanoparticles by diffusion through pores and by matrix degradation. Biodegradation rates of these particles can be modulated by changing the polymer composition. Oral vaccines encapsulated in PLGA have been used in Japanese flounder, salmonids, and Atlantic salmon (44).

4.4.3 Signal 2 Facilitators

4.4.3.1 Alum

Aluminum phosphate and aluminum hydroxide are used as alum adjuvants. Alum adjuvants have been shown to induce cell-mediated immune response. A vaccine against *A. salmonicida*, mixed with alum as an adjuvant, was tested in Atlantic salmon, but the results were not statistically significant.

4.4.3.2 β-Glucans

β-Glucans are known to stimulate the nonspecific immune response of fish and the receptor dectin-1 is thought to be involved. To obtain protective effects against diseases, the glucan is injected intraperitoneally and there seems to be a short-lived protection.

4.4.3.3 Cytokines

Cytokines are small, cell-signaling protein molecules used in intercellular communication. Interferon regulatory factors 1 (IRF-1) have been shown to be involved in host defense against pathogens and in cytokine signaling. It has also been shown to induce an antiviral state in fish cells. In rainbow trout, vaccine with IL-8 was studied as an adjuvant against VHSV.

4.4.3.4 Polyinosinic Polycytidylic Acid (Poly I:C)

Poly I:C, a double-stranded polyribonucleotide, has been used as adjuvants in vaccine preparations. It is used to induce a type I interferon in fish. Poly I:C after binding to TLR3 and activation of intracellular signaling events induces a nonspecific antiviral state. The immunization process using Poly I:C was tested in rainbow trout infected with IHNV. The fish acquire a protective immunity against the corresponding viral disease when exposed to a virus.

4.4.3.5 Synthetic CpG Oligonucleotides

Synthetic oligodeoxynucleotides with CpG motifs trigger an immunostimulatory cascade that culminates in the maturation, differentiation, and proliferation of multiple immune cells, including B and T lymphocytes, NK cells, monocytes, macrophages, and dendritic cells. CpG motifs stimulate cells that express TLR9. Synthetic CpG oligonucleotides function as adjuvants when administered with vaccines and may both accelerate and magnify the immune response. Many studies have been carried out on the immunomodulatory effect of CpGs in salmonids, turbot, and the Japanese flounder.

4.4.3.6 Lipopeptides

Lipoproteins and lipopeptides have been found in a large number of microorganisms. Lipopeptides, which are most prominently present in mycobacteria and mycoplasms, have been found to exhibit both a strong innate (inflammatory) response and a long-lasting adaptive immune response. The adjuvant effect of polar glycopeptidolipids in vaccines against *A. salmonicida* was investigated (44).

4.5 NEW TECHNOLOGIES FOR VACCINE DEVELOPMENT

The different types of new technologies applied for the development of different types of vaccines are depicted in Figure 4.2 (47).

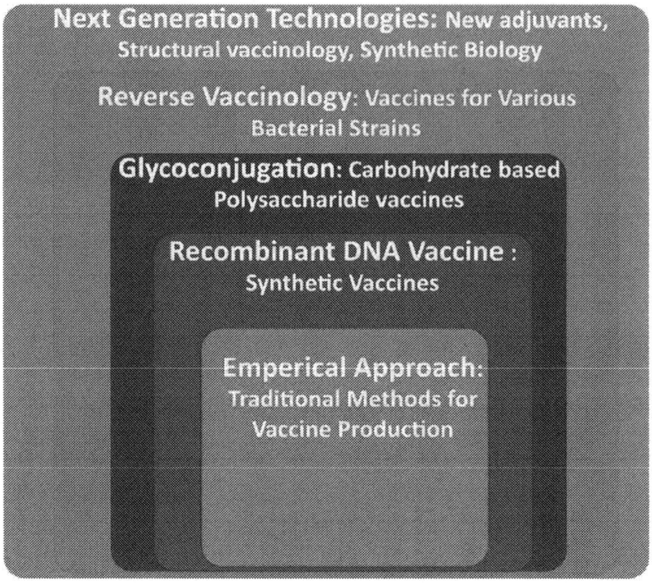

FIGURE 4.2 New technology approaches for vaccine development.

4.6 CONCLUSION

Disease prevention and control are essential in order to achieve a sustainable aquaculture, both economically and environmentally. Control measures based on stimulation of the immune system of the fish have been a vital measure for achieving this target. Fish vaccination has become a crucial part in the successful development of the aquaculture industry. Immunization protects fishes from diseases. Despite the outstanding success of vaccination against certain diseases, there are important challenges that need to be addressed. The knowledge gaps on antigenic variability and poor understanding of fish immunity have affected the design of good vaccines. Even though a considerable amount of work on fish vaccines has been performed by the pharmaceutical companies, only limited information is available through scientific publications (4). Availability and access to the vaccines are the major obstacles to utilize the available vaccines (7,47). A positive approach by scientists and constructive collaborations between government and industry will be necessary to move forward.

ACKNOWLEDGMENT

The authors express their immense gratitude to Dr. K.R. Venkatesan, Principal, and the Management of Sri Sankara Arts and Science College, Enathur, for providing permission to write this chapter on vaccines.

DISCLOSURE STATEMENT

The authors declare that they have no conflicts of interest.

REFERENCES

[1] A. Miccoli, P.R. Saraceni, and G. Scapigliati. Vaccines and immune protection of principal Mediterranean marine fish species. 94:800–809 ISSN 1050-4648. URL https://www.sciencedirect.com/science/article/pii/S1050464819309490.

[2] Yuta Matsuura, Sachiko Terashima, Tomokazu Takano, and Tomomasa Matsuyama. Current status of fish vaccines in Japan. 95:236–247 ISSN 1050-4648. URL https://www.sciencedirect.com/science/article/pii/S1050464819309155.

[3] Maxime Boutier, Yuan Gao, Owen Donohoe, and Alain Vanderplasschen. Current knowledge and future prospects of vaccines against cyprinid herpesvirus 3 (CyHV-3). 93:531–541. ISSN 1050-4648. URL https://www.sciencedirect.com/science/article/pii/S1050464819307892.

[4] Ingunn Sommerset, Bjørn Krossøy, Eirik Biering, and Petter Frost. Vaccines for fish in aquaculture. 4(1):89–101. ISSN 1476-0584. URL https://doi.org/10.1586/14760584.4.1.89. Publisher: Taylor & Francis.

[5] Po-Tsang Lee, Zhen-Hao Liao, Huai-Ting Huang, Chao-Yuan Chuang, and Fan-Hua Nan. β-glucan alleviates the immunosuppressive effects of oxytetracycline on the non-specific immune responses and resistance against vibrio alginolyticus infection in Epinephelus fuscoguttatus×Epinephelus lanceolatus hybrids. 100:467–475 ISSN 1050-4648. URL https://www.sciencedirect.com/science/article/pii/S105046482030214X.

[6] Ian R. Tizard. Chapter 21- fish vaccines. In Ian R. Tizard, editor, *Vaccines for Veterinarians*, pp 281–292.e1. Elsevier. ISBN 978-0-323-68299-2. URL https://www.sciencedirect.com/science/article/pii/B9780323682992000307.

[7] Jie Ma, Timothy J Bruce, Evan M. Jones, and Kenneth D. Cain. A review of fish vaccine development strategies: Conventional methods and modern biotechnological approaches. 7(11):569. ISSN 2076-2607. URL https://pubmed.ncbi.nlm.nih.gov/31744151. Publisher: MDPI.

[8] Chuchu Li, Zhicang Ye, Liangyou Wen, Ran Chen, Lihua Tian, Fukun Zhao, and Jianyi Pan. Identification of a novel vaccine candidate by immunogenic screening of vibrio parahaemolyticus outer membrane proteins. 32(46):6115–6121. ISSN 0264-410X. URL https://www.sciencedirect.com/science/article/pii/S0264410X14012419.

[9] Andrew J. Pollard and Else M. Bijker. A guide to vaccinology: From basic principles to new developments. 21(2):83–100. ISSN 1474-1741. URL https://doi.org/10.1038/s41577-020-00479-7.

[10] Jee Youn Hwang, Mun Gyeong Kwon, Jung Soo Seo, Seong Don Hwang, Ji Min Jeong, Ji Hoon Lee, Ah Reum Jeong, and Bo Young Jee. Current use and management of commercial fish vaccines in Korea. 102:20–27. ISSN 1050-4648. URL https://www.sciencedirect.com/science/article/pii/S1050464820302412.

[11] Takuya Yamaguchi, Edwige Quillet, Pierre Boudinot, and Uwe Fischer. What could be the mechanisms of immunological memory in fish? 85:3–8. ISSN 1050-4648. URL https://www.sciencedirect.com/science/article/pii/S1050464818300354.

[12] Takashi Aoki, Jun-ichi Hikima, Seong Don Hwang, and Tae Sung Jung. Innate immunity of finfish: Primordial conservation and function of viral RNA sensors in teleosts. 35(6):1689–1702. ISSN 1050-4648. URL https://www.sciencedirect.com/science/article/pii/S1050464813000661.

[13] Jie Xu, Yongyao Yu, Zhenyu Huang, Shuai Dong, Yanzhi Luo, Wei Yu, Yaxing Yin, Huili Li, Yangzhou Liu, Xiaoyun Zhou, and Zhen Xu. Immunoglobulin (ig) heavy chain gene locus and immune responses upon parasitic, bacterial and fungal infection in loach, Misgurnus anguillicaudatus. 86:1139–1150. ISSN 1050-4648. URL https://www.sciencedirect.com/science/article/pii/S1050464818308702.

[14] S. Picchietti, N. Nun˜ez-Ortiz, V. Stocchi, E. Randelli, F. Buonocore, L. Guerra, and G. Scapigliati. Evolution of lymphocytes. immunoglobulin t of the teleost sea bass (Dicentrarchus labrax): Quantitation of gene expressing and immunoreactive cells. 63:40–52. ISSN 1050-4648. URL https://www.sciencedirect.com/science/article/pii/S1050464817300657.

[15] Tracy S. Peterson. 3- Overview of mucosal structure and function in teleost fishes. In Benjamin H. Beck and Eric Peatman, editors, *Mucosal health in aquaculture*, pp 55–65. Academic Press. ISBN 978-0-12-417186-2. URL https://www.sciencedirect.com/science/article/pii/B9780124171862000030.

[16] Pedro Perdiguero, Alba Mart´ın-Mart´ın, Ottavia Benedicenti, Patricia D´ıaz-Rosales, Esther Morel, Estefan´ıa Mun˜oz-Atienza, M´onica Garc´ıa-Flores, Roc´ıo Sim´on, Irene Soleto, Andrea Cerutti, and Carolina Tafalla. Teleost IgD+IgM b cells mount clonally expanded and mildly mutated intestinal IgD responses in the absence of lymphoid follicles. 29(13):4223–4235.e5. ISSN 2211-1247. URL https://www.sciencedirect.com/science/article/pii/S2211124719316018.

[17] N. Nun˜ez Ortiz, M. Gerdol, V. Stocchi, C. Marozzi, E. Randelli, C. Bernini, F. Buonocore, S. Picchietti, C. Papeschi, N. Sood, A. Pallavicini, and G. Scapigliati. T cell transcripts and t cell activities in the gills of the teleost fish sea bass (Dicentrarchus labrax). 47(2):309–318. ISSN 0145-305X. URL https://www.sciencedirect.com/science/article/pii/S0145305X14001967.

[18] Tomonori Somamoto, Yuhei Miura, Teruyuki Nakanishi, and Miki Nakao. Local and systemic adaptive immune responses toward viral infection via gills in ginbuna crucian carp. 52(1):81–87. ISSN 0145-305X. URL https://www.sciencedirect.com/science/article/pii/S0145305X15000932.

[19] Nu Zhang, Xu-Jie Zhang, Dan-Dan Chen, J. Oriol Sunyer, and Yong-An Zhang. Molecular characterization and expression analysis of three subclasses of IgT in rainbow trout (Oncorhynchus mykiss). 70:94–105. ISSN 0145-305X. URL https://www.sciencedirect.com/science/article/pii/S0145305X16303962.

[20] Sukanta K. Nayak and Teruyuki Nakanishi. Direct antibacterial activity of CD8+/CD4+ t-cells in ginbuna crucian carp, Carassius auratus langsdorfii. 34(1):136–141. ISSN 1050-4648. URL https://www.sciencedirect.com/science/article/pii/S1050464812003816.

[21] Alf Seljenes Dalum, David James Griffiths, Elin Christine Valen, Karoline Skaar Amthor, Lars Austbø, Erling Olaf Koppang, Charles McLean Press, and Agnar Kvellestad. Morphological and functional development of the interbranchial lymphoid tissue (ILT) in Atlantic salmon (Salmo salar l). 58:153–164. ISSN 1050-4648. URL https://www.sciencedirect.com/science/article/pii/S1050464816305605.

[22] Maria Rosaria Coscia, Palma Simoniello, Stefano Giacomelli, Umberto Oreste, and Chiara Maria Motta. Investigation of immunoglobulins in skin of the antarctic teleost Trematomus bernacchii. 39(2):206–214. ISSN 1050-4648. URL https://www.sciencedirect.com/science/article/pii/S1050464814001351.

[23] Qingchao Wang, Wei Ji, and Zhen Xu. Current use and development of fish vaccines in china. 96:223–234. ISSN 1050-4648. URL https://www.sciencedirect.com/science/article/pii/S1050464819311386.

[24] Hung Bao Tran, Putri Nurhanida Rizky, Stanley Rob McDaniel Padgett, Yen- Hung Lee, Hso-Chi Chaung, and Ta-Chih Cheng. Molecular characterization of cobia (Rachycentron canadum) CD4 homologues revealed the first evidence of soluble CD4 in fish. 99:239–242. ISSN 1050-4648. URL https://www.sciencedirect.com/science/article/pii/S1050464820300954.

[25] Jorge Galindo-Villegas, Iv´an Mulero, Alicia Garc´ıa-Alcazar, Iciar Mun˜oz, Marcos Pen˜alver-Mellado, Sergio Streitenberger, Giuseppe Scapigliati, Jos´e Meseguer, and Victoriano Mulero. Recombinant TNF as oral vaccine adjuvant protects European sea bass against vibriosis: Insights into the role of the CCL25/CCR9 axis. 35(4):1260–1271. ISSN 1050-4648. URL https://www.sciencedirect.com/science/article/pii/S1050464813006980.

[26] Carmen W.E. Embregts and Maria Forlenza. Oral vaccination of fish: Lessons from humans and veterinary species. 64:118–137 ISSN 0145-305X. URL https://www.sciencedirect.com/science/article/pii/S0145305X16300982.

[27] Ida Bergva Aas, Lars Austbø, Knut Falk, Ivar Hordvik, and Erling Olaf Koppang. The interbranchial lymphoid tissue likely contributes to immune tolerance and defense in the gills of Atlantic salmon. 76:247–254. ISSN 0145-305X. URL https://www.sciencedirect.com/science/article/pii/S0145305X17302021.

[28] Rosario Castro and Carolina Tafalla. 2- overview of fish immunity. In Benjamin H. Beck and Eric Peatman, editors, *Mucosal health in aquaculture*, pp 3–54. Academic Press. ISBN 978-0-12-417186-2. URL https://www.sciencedirect.com/science/article/pii/B9780124171862000029.

[29] Carolina Tafalla, Jarl Bøgwald, and Roy A. Dalmo. Adjuvants and immunostimulants in fish vaccines: Current knowledge and future perspectives. 35(6):1740–1750. ISSN 1050-4648. URL https://www.sciencedirect.com/science/article/pii/S1050464813000909.

[30] Stephen G. Newman. Bacterial vaccines for fish. 3:145–185. ISSN 0959-8030. URL https://www.sciencedirect.com/science/article/pii/0959803093900338.

[31] Roar Gudding and Willem B. Van Muiswinkel. A history of fish vaccination: Science- based disease prevention in aquaculture. 35(6):1683–1688. ISSN 1050-4648. URL https://www.sciencedirect.com/science/article/pii/S1050464813007766.

[32] Marco Galeotti, Nicla Romano, Donatella Volpatti, Chiara Bulfon, Andrea Brunetti, Pietro Giorgio Tiscar, Francesco Mosca, Fabrizio Bertoni, Maria Gabriella Marchetti, and Luigi Abelli. Innovative vaccination protocol against vibriosis in Dicentrarchus labrax (l.) juveniles: Improvement of immune parameters and protection to challenge. 31(8):1224–1230. ISSN 0264-410X. URL https://www.sciencedirect.com/science/article/pii/S0264410X12018269.

[33] Hai-peng Zhang, Meng-yao Chen, Yu-xuan Xu, Guan-yi Xu, Jing-rui Chen, Yi- ming Wang, Yuan-huan Kang, Xiao-feng Shan, Ling-cong Kong, and Hong-xia Ma. An effective live attenuated vaccine against Aeromonas veronii infection in the loach (Misgurnus anguillicaudatus). 104:269–278. ISSN 1050-4648. URL https://www.sciencedirect.com/science/article/pii/S1050464820303533.

[34] Sukanta Kumar Nayak. Current prospects and challenges in fish vaccine development in India with special reference to Aeromonas hydrophila vaccine. 100:283–299. ISSN 1050-4648. .URL https://www.sciencedirect.com/science/article/pii/S1050464820301224.

[35] Alexandra Adams. Progress, challenges and opportunities in fish vaccine development. 90:210–214. ISSN 1050-4648. URL https://www.sciencedirect.com/science/article/pii/S1050464819303043.

[36] Takuya Yamaguchi, Fumio Takizawa, Mitsuru Furihata, Veronica Soto-Lampe, Johannes M. Dijkstra, and Uwe Fischer. Teleost cytotoxic t cells. 95:422–439. ISSN 1050-4648. URL https://www.sciencedirect.com/science/article/pii/S1050464819310101.

[37] Tomonori Somamoto and Teruyuki Nakanishi. Mucosal delivery of fish vaccines: Local and systemic immunity following mucosal immunisations. 99:199–207. ISSN 1050-4648. URL https://www.sciencedirect.com/science/article/pii/S105046482030005X.

[38] Daniel Gonzalez-Silvera, Francisco A. Guardiola, Crist´obal Espinosa, Elena Chaves- Pozo, M. Angeles Esteban, and Alberto Cuesta. Recombinant nodavirus vaccine produced in bacteria and administered without purification elicits humoral immunity and protects European sea bass against infection. 88:458–463. ISSN 1050-4648. URL https://www.sciencedirect.com/science/article/pii/S1050464819301615.

[39] Ylenia Pennacchi, Mark B. Adams, Barbara F. Nowak, and Andrew R. Bridle. Immune gene expression in the gills of Atlantic salmon (Salmo salar l.) following experimental reinfection with Neoparamoeba perurans. 464:410–419. ISSN 0044-8486. URL https://www.sciencedirect.com/science/article/pii/S004484861630374X.

[40] Jie Ma, Timothy J. Bruce, Evan M. Jones, and Kenneth D. Cain. A review of fish vaccine development strategies: Conventional methods and modern biotechnological approaches. 7(11):569. MDPI. ISSN 2076-2607. URL https://pubmed.ncbi.nlm.nih.gov/31744151.

[41] Hai Trong Nguyen, Thuy Thi Thu Nguyen, Ming-An Tsai, E. Ya-Zhen, Pei-Chyi Wang, and Shih-Chu Chen. A formalin-inactivated vaccine provides good protection against vibrio harveyi infection in orange-spotted grouper (Epinephelus coioides). 65:118–126. ISSN 1050-4648. URL https://www.sciencedirect.com/science/article/pii/S1050464817302152.

[42] Esteban Soto, Matt J. Griffin, and Jaime A. Tobar. 11- Mucosal vaccines. In Benjamin H. Beck and Eric Peatman, editors, *Mucosal health in aquaculture*, pp 297–323. Academic Press. ISBN 978-0-12-417186-2. URL https://www.sciencedirect.com/science/article/pii/B978012417186200011X.

[43] Chia-Jung Chang. Immune sensing of DNA and strategies for fish DNA vaccine development. 101:252–260. ISSN 1050-4648. URL https://www.sciencedirect.com/science/article/pii/S1050464820302321.

[44] J. Bøgwald and R.A. Dalmo. 10- Developments in adjuvants for fish vaccines. In Brian Austin, editor, *Infectious disease in aquaculture*, pp 244–274. Woodhead Publishing. ISBN 978-0-85709-016-4. URL https://www.sciencedirect.com/science/article/pii/B9780857090164500106.

[45] Jorge Galindo-Villegas, Alicia García-Alcazar, José Meseguer, and Victoriano Mulero. Aluminum adjuvant potentiates gilthead seabream immune responses but induces toxicity in splenic melanomacrophage centers. 85:31–43. ISSN 1050–4648. URL https://www.sciencedirect.com/science/article/pii/S1050464818301153.

[46] Jianmin Ye, Ilsa M. Kaattari, Cuiyan Ma, and Stephen Kaattari. The teleost humoral immune response. 35(6):1719–1728. ISSN 1050-4648. URL https://www.sciencedirect.com/science/article/pii/S1050464813008127.

[47] Rino Rappuoli, Christian W. Mandl, Steven Black, and Ennio De Gregorio. Vaccines for the twenty-first century society. 11(12):865–872. ISSN 1474-1741. URL https://doi.org/10.1038/nri3085.

5 Antigen Discovery

Nhat HM Truong and Nam Tri Vo
University of Science
Vietnam National University

Phuong Hong Vo
University of Science
Vietnam National University
Research Institute Aquaculture N02

Hoang Duc Nguyen
University of Science
Vietnam National University

CONTENTS

5.1 THE CHOICE OF VACCINE METHOD FOR FISH DISEASE VACCINE

5.1.1 The Current Situation of Fish Vaccine Development

Aquaculture is one of the highest-grossing food sectors worldwide, expected to surpass the total production of capture fisheries by the year 2025. Asian countries have been contributing 80% to global fish production, with the four biggest Asian producers being China, Indonesia,

DOI: 10.1201/9781003388548-7

India, and Vietnam [1]. However, the growing industry is being hampered by infectious diseases, which would cost farmers over 300 million Euros per year to handle the issue. Not only that, the development of resistant strains against chemotherapeutants has limited the use of antibiotics. This combined with the shortage of an effective vaccine is causing the costs of disease prevention to spike yearly [2].

Even though the development of aquaculture vaccines requires less strict evaluations compared to human vaccines, it is still lagging in terms of available vaccines. Several platforms of fish vaccines have been developed, ranging from traditional killed and attenuated vaccines to recombinant and vector technology [3]. DNA vaccines also got a boost from the recent technology used to develop SARS-CoV vaccines, but they still have some hurdles to overcome. With the rapid advancement in bioinformatics, reverse technology has been brought to attention as a complementary tool for vaccine design, especially for recombinant and vector vaccines. Starting from a repertoire of genome sequences of a pathogen, the computational analysis would be performed to narrow down peptides with immunogenic potential and have a wide coverage across the pathogen species. Later, immunoinformatic tools would be applied to identify antigens that have high reactivities with lymphocytes (B-cell and T-cell). The process would trim down a significant amount of expenditure and effort for the antigen screening process, reducing the vaccine production time from 5–10 years to under 2 years. Furthermore, adapting the bioinformatics approach offers the possibility of discovering antigens with high immunogenicity that are impossible to detect using practical approaches. In this chapter, we will discuss the application of bioinformatics techniques to discover new antigens, as well as the downstream validation steps using wet-lab methods (Table 5.1).

TABLE 5.1
Common Fish Vaccine Platforms [3]

Type of Vaccine	Advantage	Disadvantage
Killed or inactivated vaccine	Strong and long-lasting immune response Easy administration Safe to use	Too costly and less than satisfactory for viruses Adhesions associated with adjutants
Attenuated live vaccine	Can replicate, induce cellular and humoral immunity Does not require an adjuvant Mimics natural infection and immune response	Safety concerns both in terms of the vaccinated animals and in terms of environmental aspects Danger for reversion to virulence Not a good innate immunity stimulator
Recombinant protein	Ability to produce a large amount of vaccine Safe and low-cost method	Disturbance in glycosylation of the proteins and restoration of the tertiary structure
Vector technology	High level of heterologous antigen expression in the cytoplasm Low-level vector protein expression, induction of apoptosis in infected cells Biosafety	Lack of data regarding field performance
Genetically attenuated pathogen	Induces cell-mediated, humoral, and mucosal activity Low cost of production	Possible to retain its virulence Limits the use due to GMO
DNA vaccine	Induces humoral and cellular immunity	Some obstacles to DNA vaccine

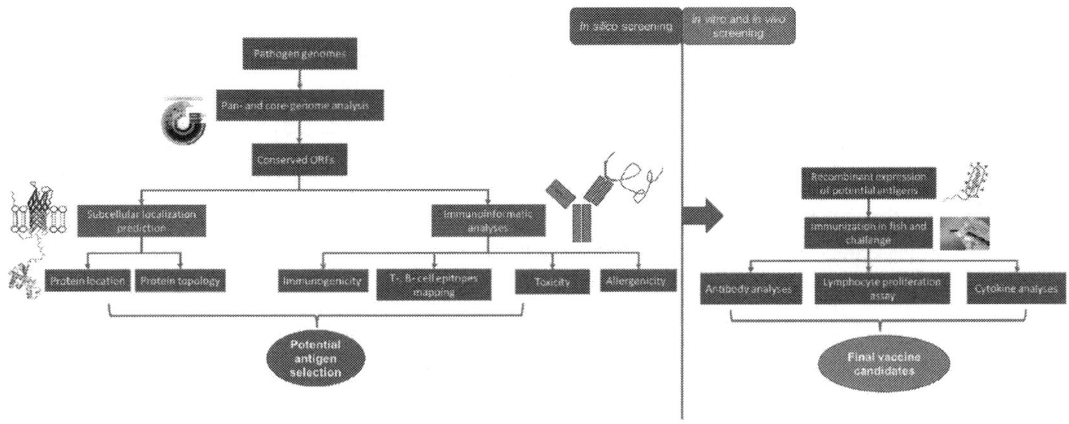

FIGURE 5.1 Schematic diagram of a combined antigen screening strategy.

5.1.2 A Proposed Workflow for Vaccine Development

This chapter is a comprehensive guide to methods to discover potential antigens from the pathogen's genomic data using the reverse vaccinology (RV) and immunoinformatic approach, as well as practical methods to determine the "true" antigenic protein from a list of potential antigens from the *in silico* prediction. For the sake of clarity, a visualization of our proposed pipeline was shown in Figure 5.1, which illustrates from the *in silico* antigen prediction steps to *in vitro* and *in vivo* examination steps.

5.2 SEQUENTIAL APPROACH IN SELECTING ANTIGEN

5.2.1 Reverse Vaccinology to Identify Antigens through Sequence Similarity

5.2.1.1 Conventional Reverse Vaccinology

The first application of bioinformatics could be dated back to the early 2000s, using RV to discover epitopes for MenB [4]. The idea of RV, which is defined as "the process of antigen discovery through the interrogation of an organism's complete antigenic repertoire, as coded in its genomic data," was spurred by the advancement in genomic sequencing in the late 90s. Currently, the availability of genomics and proteomics data, as well as the advancement in bioinformatics tools, has enabled high-throughput screening for vaccine candidates, which could save a considerable amount of efforts and expense for wet-lab screening steps. Overall, the strategy for RV would be different depending on the resources and target. However, a potential antigen would usually satisfy the following filters: suitable cellular localization, sequential conservation across strains, chemical properties compatible with the host's immune system, and the process of mass-producing antigens for vaccination [5].

The first step of RV starts with identifying the potential sequences as antigens. A satisfied sequence should be an ORF that is preferably conserved so that they could represent the whole strain and have a higher protective coverage against the said pathogen. Constructing phylogenetic trees is a common choice to cluster the diverse and enormous genomes of different strains based on their genotype, thus easing the comparative steps. There are several ways to build a genotypic tree, depending on how broad the vaccine's aim is. Multi-locus sequence typing (MLST) is the method of choice for many vaccine development projects due to its sensitivity in detecting different strains based on housekeeping genes' allelic profiling [4, 5]. This step could either be a confirmation step for knowledge-based or experimental antigen selection or preparation for the subsequent *in silico* selections. While the former was already shown with the case of MenB vaccine, the latter pipeline has emerged in this decade with the help of different bioinformatic tools.

5.2.1.2 The Pan-Genome and Comparative Genome Analysis

Based on the premise of RV, modifications have been made to enhance the accuracy in choosing antigenic proteins of this approach [6]. One common derivative, pan-genome RV, utilizes the availability of multiple genomes of the same strains. A pan-genome is the full gene library of different bacterial genomes of the same species, which comprises a core genome (common genes of all strains in the species), a dispensable genome of genes that are present in some strains only, and unique genes from each of the strain's genome. Based on alignment and clustering ORF of these strains, the pan-, core-, dispensable, and unique genes could be divided into groups, thus giving us a clearer picture of conserved antigens that could offer broadly protective vaccines [7]. There are several pipelines available for this method, which include Panseq [8], PGAP (pan-genomes analysis pipeline) [9], and PanRV (pan-genome-reverse vaccinology) [10].

With the aim of creating a vaccine that could target pathogenic strains of the bacteria, the comparative genome approach focuses on genes that are required for commensal strains to become virulent, which could either be original genes that are mutated or genes acquired through horizontal gene transfer. The result of this is vaccines that only target virulent strains and spare nonpathogenic bacteria [11].

5.2.2 Identify Antigens through Subcellular Localization

A large portion of immunogenic antigens come from the outer-membrane. Commonly, protective B-cell protein antigens are located in the outer-membrane protein (OMP) and extracellular environment (secreted proteins); hence, these predicted subcellular localizations were targets for selection [12]. Also, in some cases, OMP could interact with immune cells and trigger a protective immune response (dendritic cell, T- and B-cells) through the extrusion of outer-membrane vesicles (OMV) produced by both Gram-positive and Gram-negative bacteria during growth, which further justifies the use of OMP for vaccine development [13, 14].

There have been numerous *in silico* methods to detect the expression localization of antigens. One of the most used algorithms is PSORT (Protein Subcellular Localization Prediction Tool), which is a knowledge-based method capable of multi-category sorting. The method made use of different subprograms analyzing signal peptides, transmembrane structure, amino acid composition and structure, and so on [15]. Following this lead, other localization tools have also emerged, which are listed in Table 5.2.

Despite having high immunogenicity and intense attention from researchers, OMP vaccine still has its own obstacles to overcome. The fact that OMPs would usually contain at least one transmembrane hydrophobic domain means that the recombinant expression of these proteins will face a high risk of failure [4]. Furthermore, the variability in the surface structure of OMP among strains in

TABLE 5.2
Available Subcellular Localization Prediction Tools

Prediction Tools	Main Features	Site
PSORTb	Prediction of cellular sub-location of whole protein	https://www.psort.org/psortb/
CELLO2GO		http://cello.life.nctu.edu.tw/cello2go/
SignalP	Prediction of signal peptide's site on a protein	https://services.healthtech.dtu.dk/service.php?SignalP-5.0
TMHMM	Prediction of transmembrane topology in a protein	https://services.healthtech.dtu.dk/service.php?TMHMM-2.0
DeepTMHMM	Prediction of transmembrane topology and signal peptide	https://services.healthtech.dtu.dk/service.php?DeepTMHMM.
Phobius		https://phobius.sbc.su.se/cgi-bin/predict.pl

some species, which interact directly with B-cells, would be a challenge when choosing a conserved antigen for vaccine design. The appearance of hydrophobic structures in a membrane protein could lead to two scenarios: the protein was secreted to be buried in the periplasm or the outer-membrane, or the protein had a transmembrane domain. In the former case, the protein would not likely be immunogenic and will be discarded from the potential pool with the help of structural prediction tools (e.g., TMHMM for transmembrane helices prediction and PRED-TMBB for beta-barrel protein) [12, 16]. On the other hand, proteins with transmembrane domains could be trimmed off and expressed only in the outer-membrane domains, thus easing the high-throughput vaccine production steps. It should be noted that with the current prevalence of artificial neural network (ANN) algorithm, an enhanced transmembrane prediction tool from the previous hidden Markov model TMHMM was developed called DeepTMHMM, which now based upon an encoder-decoder deep learning model. Even though the tool still needs time to prove its effectiveness, it has shown superior performance to most traditional localization prediction tools [17].

5.3 IMMUNOINFORMATIC ASSISTANCE IN ANTIGEN DISCOVERY

5.3.1 Prediction of T-Cell and B-Cell Epitopes

While antigenicity could be assessed in the full antigen forms, sometimes the whole protein could not be easily produced due to the hydrophobic nature of some domains, like the transmembrane/cytosolic domains from OMPs. Thus, there is a need to "trim down" the peptide residues that could be expressed without losing the antigenic properties, and with this comes the demand for epitope prediction and mapping. Moreover, the determination of peptide fragments capable of being recognized by the immune system is one of the most crucial requirements of a vaccine candidate [18].

Epitopes are regions on the surface of the antigen that interact with B- and T-cells, the main cells that orchestrate the adaptive immune response. B-cell receptors could bind to the surface of a whole antigen, which is not limited to the continuous and discontinuous protein surface, polysaccharides, nucleic acids, and other organic molecules. T-cells, in contrast, are specifically bound to small peptide epitopes represented on the major histocompatibility complex (MHC). There are two classes of MHC corresponding to two T-cell-mediated pathways. MHC-I representing peptides transported by the transporter associated with antigen processing (TAP) to the endoplasmic reticulum (ER). These cytosolic peptides will trigger the activity of CD8+ T-cells (cell-mediated immune system), which would trigger infected cell death. On the other hand, CD+ T-helper triggered cells by the MHC-II class with the humoral immune pathway, which would promote the secretion of neutralizing antibodies from plasma cells [19].

Immunoinformatics is a subfield of bioinformatics that deals with algorithms for modeling the immune systems, subsequently enabling the mapping of B-cell and T-cell epitopes. From the basis of B-cell and T-cell recognition, there were four possible classes of epitopes that should be considered: MHC-I epitopes, MHC-II epitopes, B-cell continuous epitopes, and B-cell discontinuous epitopes. The prediction tools for each of the epitope classes are listed in Table 5.3.

It should be noted that epitopes should also be evaluated for other qualifications apart from potential binding affinities with immune receptors, namely, Kolaskar–Tongaonkar antigenicity, Emini surface accessibility prediction, Chou–Fasman beta-turn prediction, prediction of floppy-prediction, and Parker hydrophilicity prediction [20].

5.3.2 Molecular Docking

5.3.2.1 Obtaining Molecular Structures

To perform molecular docking experiments, receptor and ligand structures must be available. The Protein Data Bank (PDB) is the structural data database of biological macromolecules solved by X-ray crystallography, nuclear magnetic resonance (NMR) spectroscopy, cryo-electrical microscopy, and

TABLE 5.3
Antigen Mapping Software Classified by Epitope Types

Epitope Class	In Silico Tools
MHC-I epitope binding prediction	SVMHC
	TepiTope
	ProPred-I
	NetMHCpan
	CTLpred
	RANKPEP
	NetCTL
MHC-II epitope binding prediction	SVMHC
	TepiTope
	ProPred-I
	NetMHCpan
	CTLpred
	RANKPEP
	nHLAPred
	MHCpred
Discontinuous B-cell epitope prediction	Discotope
	ElliPro
	CBtope
Continuous B-cell epitope prediction	BCpred
	ABCpred
	Bepipred
	IEDB
	Ellipro

theoretical modeling. However, only a fraction of discovered proteins got their structures predicted with satisfying resolutions for docking. Thus, efforts have been made in the field of computational structural predictions, introducing several highly accurate and easy-to-operate prediction tools.

Homology modeling has been the best strategy for protein tertiary structure predictions, which was implemented in both online tools (e.g., SWISS-MODEL) and computer programs (e.g., Modeller and Deepview). However, the approach is limited by the requirement of protein template structures, which could affect the accuracy of the predicted structures. Fortunately, the rapid development of ANN algorithms has enabled an ANN-based model, Alphafold, to solve the dilemma that has existed for decades. The model is continued to be optimized and be easily accessed online through Google Colaboratory.

5.3.2.2 Molecular Docking Tools

Molecular docking was originally the most preferred structured-based method to design drug molecules for pharmaceutical research. The method simulates the receptor–ligand complex by predicting how different conformations of a molecule could bind to a "pocket" in the target protein [21]. Its applications to epitope mapping include predicting possible interaction and binding affinity between the epitopes and the cavities of antibodies, B- and T-cell receptors, and MHC molecules. Available tools for molecule docking, namely, Autodock Vina, ZDock, autodock 4, Patch dock, Molegro Virtual Docker, Mti auto dock, Cluspro 2.0, and PyRx were all applied in previous vaccine discovery studies. In addition, supplementary tools might be useful for analyzing molecular interactions (Schrodinger Academic Desmond MDS suite and/or PyMol and Jmol visualizing systems) [22, 23].

5.4 ANTIGEN VALIDATION BY EXPERIMENTAL APPROACHES

5.4.1 A Classic Pipeline of Antigen Discovery Validation

The traditional approach to discover a potential antigen would be to fractionate the pathogen and examine the fractions for immunoreactivity or be characterized by chromatography methods (LC-MS/MS) [24, 25]. With the current development of bioinformatics tools, the process could be semi-computational with the use of subcellular localization prediction tools, as well as other protein signal/protein secretion prediction tools, thus saving time and money [24]. However, it should be noted that the predictions of computational tools are not always aligned with practical applications, which could be attributed to the lack of data for training prediction tools. Hence, most often, the "candidate antigens" must go through other rounds of validation by different experimental steps.

In the classic example of vaccine screening for *N. meningitidis*, 570 potential antigen genes were cloned to *E. coli* for recombinant production of antigens, of which 350 were expressed successfully. After mouse immunization, FACS (fluorescence-activated cell sorting) analysis revealed that only 26% of the total injected vaccine showed the ability to induce humoral responses. Furthermore, by assessment of antibody titers for neutralizing antibodies against the target pathogens, only one-third of the candidates remained for the development of MenB vaccine [4]. The process has highlighted the importance of a rigorous screening for candidate antigens after *in silico* prediction steps.

5.4.2 Different Methods for Antigenic Evaluation of Predicted Antigens

5.4.2.1 Antibody Analysis

Generally, since the above approaches are mainly for subunit vaccines, the candidate protein must first be expressed in recombinant hosts. This also becomes the technical requirement for the antigens, as they should be preferably soluble when being produced in the host. Then, like other types of vaccines, these antigens would be tested for their immunostimulatory effect in animal models. The antisera from these models would go through numerous assays to confirm the presence of targeted antibodies, as well as bactericidal activity in the sera.

Active immunization is the state in which the host body can produce specific antibodies against a pathogen after exposure to antigens of the said pathogen, and this is also the major target of vaccines. Therefore, evaluating antibodies level against the vaccine is the first step in confirming the validity of chosen antigens. Enzyme-linked immunosorbent assay (ELISA) and/or Western Blot are standard methods applied to measure the quantity of antibody titer in the antisera and could even quantify the immunoreactivity of the antibodies [26].

5.4.2.2 Lymphocytes Analysis

Because the first stage in an immune response generates different classes of effector lymphocytes, one could predict that the immune pathway corresponded to an invading agent through the number of lymphocytes presenting in lymph nodes and organs [27]. Using lymphocyte proliferation assays, the rate of adaptive lymphocytes (T- and B-cells) differentiated from mononuclear cells at a specific site like the spleen could be enumerated, and if this rate is elevated in the challenge experiments, then the vaccine is proved to be effective in triggering long-term protection from lymphocytes, and could even have the potential to kick-start the cell-mediated immune response [28]. Apart from the former *in vitro* method, there are also *in situ* methods to count lymphoid cells, applying immunofluorescence and immunohistochemistry techniques, namely, microscopy and flow cytometry [29].

5.4.2.3 Cytokine Profile

Besides the ability to trigger the humoral immune response, the vaccine should also be able to induce T-cell responses. While the role of T-cells in assisting vaccine protection is still unclear, their importance in maintaining the memory response after the initial immunization has been observed.

TABLE 5.4
T-Helper Types and Their Associated Cytokines

T-Helper Subtype	Cytokine Involved (Stimulate and Being Secreted)
Th1	IL-1β (IL-1 F3), IL-2, IL-12, IL-23, IL-27, IFNγ, TNF, GM-CSF
Th17	IL-1β, IL-6, IL-12, IL-17A-F, IL-22, IL-23, IL-27, TGFβ
Th2	IL-1Ra (IL-1F3), IL-4, IL-5, IL-6, IL-10, IL-13, IL-25 (IL-17E)

T-cells are traditionally divided into cytotoxic CD8+ cells (CTL) and CD4+ T-helper (T_H) cells. While CTLs are effective in dealing with infected cells and tumor cells, Th cells are mainly in charge of modulating the "mode" of immune response to eliminate different pathogens efficiently. Therefore, CD4+ cells are further classified into Type 1 helper cell (Th1 cell), Type 2 helper cell (Th2 cell), and IL-17 secreting helper cell (Th17 cell). Particularly, Th1 cells are key players in monitoring defense against intracellular pathogens, including protozoa, bacteria, and viruses [30], while Th2 cells are responsible for host response against parasite infections, venom, and allergens. Finally, the third subtype Th17 cell will mount a defense against extracellular pathogens and fungi, ones that the Th1 and Th2 are not suited against.

Each type of T-helper cell can manifest a distinct pathway through the secretion of different cytokines set to deal, whereas the Th cells also proliferate under the regulation of cytokines and are guided by chemokines. Hence, cytokine and chemokine assays are also fundamental analyses of vaccine efficacy. Table 5.4 shows the main cytokines promoting Th1, Th2, and Th17 cells [30]. The assays could be carried out using high-performance liquid chromatography, bioassays, protein assays, ELISA, and other immunoassays [31].

5.5 CONCLUSION

With advances in RV, the process of antigen discovery could be operated *in silico*, starting from analyzing the pan-genome of the target pathogens. From the shortlist of genes providing high coverage, the subcellular localization of and the topology of these proteins will be considered so that only proteins exposed to the membrane surface will be chosen. Later, the potential antigens will be examined for antigenicity, as well as the ability to interact with T-cells, B-cells, MHC-I, and MHC-II. Finally, the satisfied proteins will be expressed in the recombinant host for challenge experiments, and subsequent immune response tests will be used.

ACKNOWLEDGEMENT

Research is supported by Vingroup Innovation Foundation (VINIF) in project code VINIF.2021.DA00034.

REFERENCES

[1] Towards Blue Transformation. *https://www.fao.org/state-of-fisheries-aquaculture/en/.* Accessed: June 30, 2022.

[2] Erkinharju, T., Dalmo, R.A., Hansen, M. and Seternes, T. (2021), "Cleaner fish in aquaculture: Review on diseases and vaccination", *Reviews in Aquaculture*, *13*(1), 189–237.

[3] Dadar, M., Dhama, K., Vakharia, V.N., Hoseinifar, S.H., Karthik, K., Tiwari, R., Khandia, R., Munjal, A., Salgado-Miranda, C. and Joshi, S.K. (2017), "Advances in aquaculture vaccines against fish pathogens: Global status and current trends", *Reviews in Fisheries Science & Aquaculture*, *25*(3), 184–217.

[4] Pizza, M. et al. (2000), "Identification of vaccine candidates against serogroup B meningococcus by whole-genome sequencing", *Science*, *287*(5459), 1816–1820.

[5] Del Tordello, E., Rappuoli, R. and Delany, I. (2017), *Human vaccines*, Academic Press, Boston, MA.
[6] Flower, D.R., Macdonald, I.K., Ramakrishnan, K., Davies, M.N. and Doytchinova, I.A. (2010), "Computer aided selection of candidate vaccine antigens", *Immunome Research*, *6*(Suppl 2), S1.
[7] Lefébure, T. and Stanhope, M.J. (2007), "Evolution of the core and pan-genome of Streptococcus: Positive selection, recombination, and genome composition", *Genome Biology*, *8*(5), R71.
[8] Laing, C., Buchanan, C., Taboada, E.N., Zhang, Y., Kropinski, A., Villegas, A., Thomas, J.E. and Gannon, V.P. (2010), "Pan-genome sequence analysis using Panseq: An online tool for the rapid analysis of core and accessory genomic regions", *BMC Bioinformatics*, *11*(1), 461.
[9] Zhao, Y., Wu, J., Yang, J., Sun, S., Xiao, J. and Yu, J. (2012), "PGAP: Pan-genomes analysis pipeline", *Bioinformatics*, *28*(3), 416–418.
[10] Naz, K., Naz, A., Ashraf, S.T., Rizwan, M., Ahmad, J., Baumbach, J. and Ali, A. (2019), "PanRV: Pangenome-reverse vaccinology approach for identifications of potential vaccine candidates in microbial pangenome", *BMC Bioinformatics*, *20*(1), 123.
[11] Kanampalliwar, A., Soni, R., Girdhar, A. and Tiwari, A. (2013), "Reverse vaccinology: Basics and applications", *The Open Vaccine Journal*, *4*, 1–5.
[12] Maiti, B., Dubey, S., Munang'andu, H.M., Karunasagar, I., Karunasagar, I. and Evensen, Ø. (2020), "Application of outer membrane protein-based vaccines against major bacterial fish pathogens in India", *Frontiers in Immunology*, 11.
[13] Alaniz, R.C., Deatherage, B.L., Lara, J.C. and Cookson, B.T. (2007), "Membrane vesicles are immunogenic facsimiles of *Salmonella typhimurium* that potently activate dendritic cells, prime B and T cell responses, and stimulate protective immunity in vivo", *The Journal of Immunology*, *179*(11), 7692–7701.
[14] Cai, W., Kesavan, D.K., Wan, J., Abdelaziz, M.H., Su, Z. and Xu, H. (2018), "Bacterial outer membrane vesicles, a potential vaccine candidate in interactions with host cells based", *Diagnostic Pathology*, *13*(1), 95.
[15] Gardy, J.L., Laird, M.R., Chen, F., Rey, S., Walsh, C.J., Ester, M. and Brinkman, F.S.L. (2005), "PSORTb v.2.0: Expanded prediction of bacterial protein subcellular localization and insights gained from comparative proteome analysis", *Bioinformatics*, *21*(5), 617–623.
[16] Solanki, V., Tiwari, M. and Tiwari, V. (2019), "Prioritization of potential vaccine targets using comparative proteomics and designing of the chimeric multi-epitope vaccine against Pseudomonas aeruginosa", *Scientific Reports*, *9*(1), 5240.
[17] Hallgren, J., Tsirigos, K.D., Pedersen, M.D., Armenteros, J.J.A., Marcatili, P., Nielsen, H., Krogh, A. and Winther, O. (2022), DeepTMHMM predicts alpha and beta transmembrane proteins using deep neural networks. bioRxiv.
[18] Purcell, A.W., McCluskey, J. and Rossjohn, J. (2007), "More than one reason to rethink the use of peptides in vaccine design", *Nature Reviews Drug Discovery*, *6*(5), 404–414.
[19] Lund, O., Nielsen, M., Lundegaard, C., Kesmir, C. and Brunak, S. (2005), *Immunological bioinformatics*, MIT Press, Cambridge, MA.
[20] Raoufi, E., Hemmati, M., Eftekhari, S., Khaksaran, K., Mahmodi, Z., Farajollahi, M.M. and Mohsenzadegan, M. (2020), "Epitope prediction by novel immunoinformatics approach: A state-of-the-art review", *International Journal of Peptide Research and Therapeutics*, *26*(2), 1155–1163.
[21] McConkey, B.J., Sobolev, V. and Edelman, M. (2002), "The performance of current methods in ligand–protein docking", *Current Science*, *83*(7), 845–856.
[22] Joshi, A., Pathak, D., Mannan, M.A. and Kaushik, V. (2021), "In-silico designing of epitope-based vaccine against the seven banded grouper nervous necrosis virus affecting fish species", *Network Modeling Analysis in Health Informatics and Bioinformatics*, *10*, 37.
[23] Schrodinger, LLC. "The PyMOL Molecular Graphics System, Version X.X.", *2010*.
[24] Li, W., Wang, H.-Q., He, R.-Z., Li, Y.-W., Su, Y.-L. and Li, A.-X. (2016), "Major surfome and secretome profile of *Streptococcus agalactiae* from Nile tilapia (Oreochromis niloticus): Insight into vaccine development", *Fish & Shellfish Immunology*, *55*, 737–746.
[25] Maji, S., Mali, P. and Joardar, S.N. (2006), "Immunoreactive antigens of the outer membrane protein of *Aeromonas hydrophila*, isolated from goldfish, *Carassius auratus* (Linn.)", *Fish & Shellfish Immunology*, *20*(4), 462–473.
[26] Sharma, M. and Dixit, A. (2015), "Identification and immunogenic potential of B cell epitopes of outer membrane protein OmpF of *Aeromonas hydrophila* in translational fusion with a carrier protein", *Applied Microbiology and Biotechnology*, *99*(15), 6277–6291.
[27] Desforges, J.-P., Sonne, C., Dietz, R. and Levin, M. (2018), *Marine mammal ecotoxicology*, Academic Press, Boston, MA.
[28] Sitz, K.V. and Birx, D.L. (1999), *HIV Protocols*, Humana Press, Totowa, NJ.

[29] Nikbakht, M., Pakbin, B. and Nikbakht Brujeni, G. (2019), "Evaluation of a new lymphocyte proliferation assay based on cyclic voltammetry; An alternative method", *Scientific Reports*, *9*(1), 4503.
[30] Butcher, M.J. and Zhu, J. (2021), "Recent advances in understanding the Th1/Th2 effector choice", *Faculty Reviews*, *10*, 30.
[31] Keustermans, G.C.E., Hoeks, S.B.E., Meerding, J.M., Prakken, B.J. and de Jager, W. (2013), "Cytokine assays: An assessment of the preparation and treatment of blood and tissue samples", *Methods*, *61*(1), 10–17.

6 Novel Advancements in Vaccine Delivery and Methods of Vaccination

Experimental Strategies

Abdullah Bin Abdul Nazar
Kerala University of Fisheries and Ocean Studies

Muhammed Salih Salim
Norwegian University of Science and Technology (NTNU)

Aswathi Ashok
Cochin University of Science and Technology (CUSAT)

Anuj Sharma, Sanchu Prakash, Malavika Biju Reni, Meril Mary Mathew, and Muzammil Siddiqui
Kerala University of Fisheries and Ocean Studies

Subramaniam Sivakumar
Sri Sankara Arts and Science College

Sreeja Lakshmi
Moredun Research Institute

Preetham Elumalai
Kerala University of Fisheries and Ocean Studies
Cochin University of Science and Technology (CUSAT)

CONTENTS

DOI: 10.1201/9781003388548-8

6.1 INTRODUCTION

Aquaculture is the world's most rapidly expanding and rising food sector. In the last few decades, aquaculture practices have undergone a revolution. Aquaculture has surpassed wild capture fisheries in terms of fish yield (FAO, 2014). People have shown a greater willingness to accept protein-rich foods from these sources. As the world's population grows, so will the need for fish and other aquaculture species. Existing aquaculture techniques, including pond culture, cage culture, and pen culture, have been put under strain as a result. As the demand for adequate protein sources develops in lockstep with the world population, farmers will have to expand their culture to meet the demand. Safe fish production and healthy aquaculture practices are of paramount significance for all. Infections caused by bacteria, viruses, parasites, and other pathogens will cause considerable losses in fish culture and pose biological risks to anyone who consumes raw fish products contaminated with zoonotic disease pathogens. If there is any complacency at any level involved, this condition can be realized as international trade grows. Infections are also on the rise due to the introduction of new species, increased hand-offs across phases and stages, biosecurity gaps, and other factors. In fish species, disease outbreaks were found to be 55% bacterial, 22% viral, and 20% parasitic (Dhar et al., 2014). Aquaculture management techniques that are not up to par result in annual losses of roughly $50 billion (Rather et al., 2011). In aquaculture, the probability of epidemics is unpredictably high. The best thing people can and should do is try to avoid such situations.

Despite numerous efforts, fish infections continue to surprise scientists and farmers, and diverse viruses continue to be a major cause of massive aquaculture losses around the world. Antibiotics have been used as a solution, although they have only been effective up to a certain point in many cases. However, while antibiotic use has improved our understanding of fish health, incorrect use has resulted in unfavorable outcomes such as food safety hazards. In this instance, drug resistance and safety concerns are common deterrents (Sneeringer et al., 2019). It is possible to manage deadly diseases by employing various preventative strategies. Vaccination is becoming more popular as a novel way to prevent infections, as it has a substantial impact on managing outbreaks in culture fishes (Adams and Thompson, 2006). The majority of fish vaccines have been designed to protect against bacterial diseases, with fewer being developed to protect against viral infections (Embregts and Forlenza, 2016). Improving vaccines for fish immunization is a never-ending procedure that requires additional progress. Because fish are poikilotherms, vaccine development considers parameters such as species, delivery methods, and routes (Dadar et al., 2017).

The first use of vaccines in fish was in 1938 when Snieszko and colleagues obtained immunity in fish by injecting dead bacteria into the fish. The procedure was thought to be too time-consuming, boring, and difficult to use in aquaculture operations (Dadar et al., 2017). In trout, the first oral fish immunization was done by feeding them a diet containing the chloroform-killed bacterium, *Aeromonas salmonicida*, which protected them from furunculosis (Duff, 1942). Vaccination has become more common since the 1990s, leading to a decrease in antibiotic use. During the twentieth century, fish immunization research developed, and knowledge grew dramatically in a fleeting period. For successful vaccination procedures, one requires a thorough grasp of the immune system of the species. Interferon-sensitive genes(IRGs) have been demonstrated to be important in fish innate immunity in previous investigations. They oversee the production of interferons, which are vital in the fight against viral infections (Langevin et al, 2013).

Adjuvant usage in vaccines has received a lot of attention recently and has got a key role in increasing the effectiveness of vaccine. Today, formulation of vaccines with powerful adjuvants have become one of the most appealing strategies for enhancing vaccination efficacy. Adjuvants work in a variety of ways and should be chosen based on the intended delivery method and the desired type of immune response (antibody, cell-mediated, or mucosal immunity) for a specific vaccination. The potential applications of nanoparticles have been expanded from vaccine delivery systems to adjuvants that can improve the immunogenicity of vaccine candidates due to their distinctive properties. Numerous types of nanoparticles have innate physical characteristics that can cause an immunological response. When compared to conventional vaccine methods, nano carrier-based delivery systems have a number of benefits, including greater adjuvant characteristics, superior stability, and increased protection against premature degradation. The utilization of a nanoparticle for coating or encapsulation of an immunogen can act to shield it against premature proteolytic destruction, thereby enabling investigation into other administration methods.

Modern molecular biology, biotechnology, vaccinology, and other sciences have made newer techniques to tackle these outbreaks available throughout time. Traditional vaccinations typically contain an antigen or a material capable of secreting the antigen. Live or attenuated creatures, DNA, RNA, nanoparticles, and subunits, among other things, are now included in novel vaccinations. In theory, these advancements are remarkably effective. Despite this, practical application has been limited due to the watery environment and rigorous cultivation methods (Ma et al., 2019).

6.2 NOVEL ADVANCEMENTS IN VACCINE DELIVERY

The discovery of new vaccine types has created the possibility of increasing the preventive potency against diseases. New vaccines are divided into two categories: genetic and subunit vaccines. In addition, new vaccines are becoming popular these days because of effective cargo protection, immunogenicity enhancement, targeted delivery, sustained antigen release, selective activation of humoral and/or cellular immune responses against specific antigens, and reduced adverse effects, due to their nature. As a result, vaccine delivery systems can have a big impact on the end result of genetic and other innovative vaccines, and they're crucial for their development. Natural split-product and subunit vaccines, recombinant subunit and protein or peptide vaccines are there among the primary inactivated/killed techniques to vaccination. It also includes modified live marker/differentiating vaccine from infected animal, live vector vaccines, and nucleic acid vaccines.

Vaccines based on recombinant proteins and DNA are not only less reactogenic but also less immunogenic than before. Hence, the situation necessitates new and improved vaccination adjuvants. Vaccine delivery methods and immunostimulatory adjuvants are two basic categories of adjuvants based on their main mechanisms of action. Vaccine delivery methods are typically particular (emulsions, microparticles, iscoms, and liposomes), and their primary role is to transfer linked antigens to antigen-receipting cells. Immunostimulatory adjuvants, on the other hand, are mostly produced from infections and frequently represent pathogen-associated molecular patterns (PAMPs) (e.g., lipopolysaccaride, monophosphoryl lipid A, and CpG DNA), which stimulate the innate immune system of the cells. Recent advances in innate immunity are beginning to shed light on how immune responses are initiated and how immunostimulatory adjuvants can aid this process. The development of preventative and therapeutic vaccinations against malignancies and chronic infectious diseases could be aided by the discovery of more effective adjuvants. In addition, novel adjuvants may make it possible to distribute vaccines through the mucosa.

Vaccine development is critical for the effective control of a number of devastating diseases. However, efficient preventative and therapeutic vaccinations for totally healing lethal diseases including cancer, malaria, HIV, and major microbial infections are yet to be created. As a result, vaccination candidates must be developed in order to elicit the desired immunological responses. Nanotechnology has been discovered to have a unique function in vaccine development, enhancing specificity and potency. Due to their beneficial advantages, such as improved antigen stability, targeted delivery, and long-term release, nano-scaled materials, such as virus-like particles, liposomes, polymeric nanoparticles (NPs), and protein NPs, have received considerable attention as potential carriers for the delivery of vaccine antigens and adjuvants over the last decade. Nanoparticle enables the programming of immune responses, overcoming many of the obstacles that come with vaccine development. Biomimetic NPs have evolved as novel natural imitating biosystems with applications in a variety of biomedical fields. This review addresses recent breakthroughs in biomimetic nanovaccines and their applications in antibacterial, anti-HIV, anti-malarial, anti-melittin, and antitumor immunity (Veena Vijayan et al., 2019).

6.2.1 A Brief Description on Adjuvants

One of the two components of vaccine formulation is the antigenic moiety, which is made up of antigenic components of microorganisms used in vaccine production and initiates the adaptive immune response. The other component is a non-specific moiety, which is made up of adjuvants that elicits the innate immune response. A formulation is chosen based on the nature of the immune response to be elicited, the avoidance of undesired side effects, and the vaccine's longevity. Vaccines are made up of proteins and sugars, and sometimes these components can fail to produce an adequate immune response. As a result, some vaccines contain "Adjuvant" chemicals that aid in the healing process. Adjuvants are substances that help the immune system respond to immunogens. They are substances that, when coupled with vaccine antigens, result in a more powerful and effective vaccine response than the vaccine alone. Adjuvants initiate early commencement of immunity, long persistence of effector responses such as antibody formation or cytotoxic T-cell activity, and avoidance of booster immunizations (Singh and O'Hagan, 2003). Adjuvants in vaccines also boost antigen immunogenicity, reducing the amount of antigen or number of vaccinations required for protective immunity, and improving vaccine efficacy in infants, the elderly, or people with compromised immune systems (Figure 6.1).

Adjuvants are widely utilized for enhancing the immune response to a vaccine and acting as an inducer to help defend against a wide range of infections by triggering non-specific defence

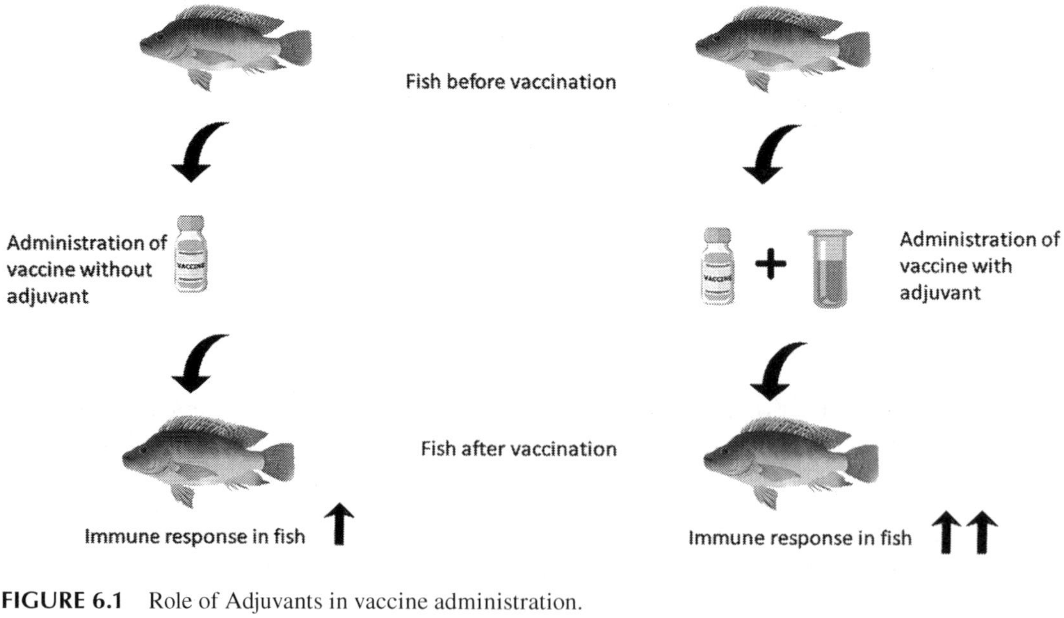

FIGURE 6.1 Role of Adjuvants in vaccine administration.

mechanism (Ellis, 1988). When coupled with bacterins, many substances enhance the formation of fish antibodies and exhibit adjuvant qualities such as light oils, incomplete adjuvants, and bacterial lipopolysaccharide (Dadar et al., 2017). Only certain compounds, such as mineral oil, glucans, and alum, are employed to create more efficient vaccine protection. Adjuvants can help with a variety of things, such as enhancing the immune response to the antigen while also extending the duration of immune responses; modulating the breadth, affinity, and specificity of antibody responses; stimulating the expression of potent cellular-mediated immune responses; increasing immunogenicity for weak antigens and seroconversion for low responders and promoting the induction of mucosal immunity (Tafalla et al., 2013).

There are two types of adjuvants: immunostimulatory adjuvants and vaccine delivery systems. Immunostimulatory adjuvants (immunopotentiators) are derived from pathogens and represent PAMP (e.g., lipopolysaccharides (LPS), monophosphoryl lipid (MPL), cytosine phosphokinase deoxynucleotides (CpG DNA)) that trigger innate immunity directly (via cytokines) or through pattern recognition receptors (PRRs). Delivery mechanisms, on the other hand, concentrate and display antigens, direct vaccination antigens to antigen-presenting cells (APCs) and aid in the co-localization of antigens and immunopotentiators (Puricelli et al., 2022). Antigen delivery systems can be utilized to direct the antigen processing route in the host. Intracellular antigens trigger a cellular immune response, whereas extracellular antigens are neutralized by humoral immune responses. Antigen fusion with bacterial toxins, emulsion suspensions, and biodegradable microparticles are among the delivery techniques being researched to route antigens for fish vaccines into intracellular or extracellular compartments.

The signal 1 concept includes the following features: (1) carrying antigens and immunomodulators; (2) allowing slow antigen release, reducing the need for frequent booster vaccination; (3) protecting the antigen during uptake; (4) being biodegradable; and (5) acting as an immunostimulant that attracts antigen-presenting sites to the site of antigen deposition. An emulsion is a system in which one immiscible liquid, referred to as the dispersed phase, is distributed in another liquid, referred to as the continuous phase. Water-in-oil adjuvants, in which the water phase transports the antigens in water globules and the oil phase serves as a stabilizer, are commonly used emulsion formulations for fish vaccinations. In Atlantic salmon, oil-adjuvant vaccinations against furunculosis and vibriosis/cold-water vibriosis result in the development of long-lasting and protective immunity (Tafalla et al., 2013). Oil-adjuvant vaccines have the disadvantage of causing visible injection-site sores that remain until harvest time (Midtlyng et al., 1996). Microparticles have been proposed as a vaccination carrier as an alternative to oil emulsions (Sinyakov et al., 2006).

The immunomodulator boosts signal 2 effects, resulting in a significant adaptive immune system priming. This includes activating appropriate cytokines and upregulating co-stimulatory molecules that attract various cell types, including as macrophages and dendritic cells (DCs), to antigen deposition sites for antigen uptake and processing before presentation to B- or T-cells. Signal 2 facilitating adjuvants is a term used to describe these modulators. Signal 2 adjuvants have designated conserved microbial molecular patterns as PAMPs by pathogen recognition receptors (PRRs) such as toll-like receptors (TLRs), Nucleotide oligomerization domain(NOD)-like receptors, dectin-1, or Retinoic acid-inducible gene I (RIG)-like helicases, which are predominantly present on the cells of the innate immune system.

Stabilizers and preservatives are chemicals that give vaccines a lengthy shelf life and keep their immunogenic properties after being absorbed by fish. Emulsifiers, like surfactants, which are chemical compounds with a hydrophobic and non-hydrophilic polar group made up of fatty acid chains, are used to stabilize the polarity of vaccines. Amino acids and sugars are other stabilizers in vaccine formulations that protect vaccines against harmful circumstances such as the freeze-drying process (Tafalla et al., 2013).

The adjuvants mixed with the vaccine during formulation also trigger regional inborn immune cells, along with APCs (Embregts and Forlenza, 2016). At present, adjuvants are mainly used in the salmon industry, and the oil adjuvanted vaccine is primarily used to inject in almost all salmon fishes for inducing an immune response. Oil in water (O/W), water in oil (W/O), oil in water in oil (O/W/O), and water in oil in water (W/O/W) are examples of oil-containing emulsions. The adjuvant commonly used for fish vaccine purposes is W/O for bringing about long-term protection (Aucouturier et al., 2001). Aqueous suspension of antigens integrated with oil to generate water in oil emulsion, which can provide long-lasting and extremely effective protection. In challenge and experimental studies against furunculosis, various oil-adjuvants like mineral oil have been noticed to provide a greater immune response and shielding instead of glucan and alum (Dadar et al., 2017). In commercial injectable bacterins, oil adjuvants are extensively used. Various substances are used as adjuvants such as levamisole, beta-1, 3 glucans, and chitosan.1, 3 glucans are highly used in oral fish vaccination formulation (Petit and Wiegertjes, 2016). It incorporates easily into the fish feed and protects against various pathogens and environmental stress (Dadar et al., 2017).

Some adjuvants consist of additional molecules such as heat shock protein and interleukins. The adjuvants liposomes and immune-stimulating complexes (ISCOMs) are effectively used in the aquaculture industry. It has been observed that carp vaccination using a liposome-encapsulated vaccine against Cyprinid Herpesvirus-3 (CyHV-3) showed a variable degree of protection that has been noticed after using liposome-encapsulated *A. hydrophila* (Miyazaki et al., 2008). Flagellin from *Salmonella spp.* and *Flagellin spp.* act as a good adjuvant. They can survive easily at the gastric pH and stimulate signals on the APCs through targeting membrane receptors on the leucocytes (Tafalla et al., 2013).

Over the last century, vaccine adjuvants have advanced from alum and emulsion-based adjuvants utilized in clinical trials from the 1920s to the 2000s to next-generation adjuvants that combines first-generation and molecular-specific adjuvants. Vaccine adjuvants are highly diverse so that there is no single integrated structure that can be used to define them. They are made up of a variety of naturally occurring or synthesized components that enhance the antigen's immunological impact. As a result, advantages such as reduced antigen dosage, fewer vaccinations, enhanced immune response quality, and, in some cases, increased vaccine product durability are possible.

The adjuvant is used for different purposes: (1) helping to improve the immunogenicity of purified and recombinant antigens; (2) providing protective immunity, which reduces the number of antigens; (3) improving the vaccine efficiency in new-born or immune-compromised person; and (4) facilitating the uptake of antigens by the mucosa and function as antigen delivery system (Petrovsky and Aguilar, 2004). The selection of adjuvant depends on various factors, which include antigens, species-specific going to be vaccinated, administration route, and side effects. The adjuvants, which are useful during vaccination formulation, should be durable with long shelf life, low cost, easy to produce, biodegradable, stimulating an appropriate immune response, and don't have the capability to produce immune responses against themselves (Edelman, 1980). The change in the efficiency of adjuvants has been observed, which depends on the administration route (Petrovsky and Aguilar, 2004).

6.2.2 Nanovaccine Delivery

The development of nanoparticle-based vaccines has opened up tremendous advancements in the field of biomedicine. Nanovaccines are characterized as adjuvants or dependable delivery mechanisms depending on how they work. With nanovaccines, antigens are protected from degradation, antigens are delivered to specific places, bioavailability is enhanced, and side effects are reduced (Vinay et al., 2018). Enveloping antigens with micro or nanoparticles has been tried to prevent antigen degradation due to reduced pH conditions inside the stomach. Because of their similar size, nanosized particles are chosen over microparticles for vaccine distribution. Nanosized particles such as chitosan and polylactide co-glycoside (PLGA) are commonly utilized (Dadar et al., 2017). Chitosan nanoparticles have a mucoadhesive feature that allows them to convey the antigen attached to them efficiently.

PLGA is being used more in the vaccination industry because it requires less effort during production and has a higher economic efficacy. Adjuvant nanoformulations activate a mechanism that promotes effective antigen absorption and use (Nagaraju, 2019). PLGA is a lactic acid and polyglycolic acid copolymer that is synthesized chemically. The use of PLGA in vaccine delivery is both environmentally benign and low in toxicity. On delivering PLGA at the location, it liberates glycolic and lactic acid after hydrolysis. These acids are later expelled from the body by tricarboxylic acid cycle (citric acid cycle). Largely explored nano-sized particle are environment-friendly polymers to utilize the benefits such as releasing of antigen in a controlled way, safety and gastrointestinal strength. Variety of nanoparticles used in research and delivery of vaccine comprises polymeric nanoparticles, nanoliposomes, virus-like particles, metal nano compounds and ISCOMs and nano emulsions (Vinay et al., 2018). Major types of nanoparticles and their benefits and drawbacks are as follows (Table 6.1).

6.2.2.1 Polymeric Nanoparticles

Polymeric nanoparticles stand out as an important technique for increasing drug bioavailability or precise distribution to the site of action. Polymers' adaptability makes them potentially excellent for meeting the needs of any individual medication delivery system.

TABLE 6.1
Types of Nanoparticles in Vaccine Delivery and Their Benefits and Drawbacks

Sl. No.	Types of Nanoparticles in Vaccine Delivery	Benefits	Drawbacks	Reference
1.	Polymeric nanoparticles	Improved immunogenicity can be achieved through simple surface modification, biodegradable antigen delivery, and targeted antigen delivery.	Water solubility is low that necessitate the use of organic solvents, antigen loading is low, premature antigen release, and insufficient antigen protection.	Vinay et al. (2018)
2.	Inorganic nanoparticles	It is simple to change, limited premature antigen release, and adsorbed antigens are well protected.	Low soluble in water and biodegradability.	Vinay et al. (2018)
3.	Nanoliposomes	When modified it has inherent adjuvant qualities, it can hold both hydrophilic and lipophilic antigens, and is generally stable in digestive juices.	Naked liposomes have poor gastrointestinal stability, little mucus penetration, and restricted antigen loading.	Vinay et al. (2018)
4.	Immune-stimulating complexes (ISCOMs)	Quill A has a built-in adjuvant characteristic that is simple to encapsulate.	Hydrophilic antigens are difficult to integrate and do not create a depot.	Vinay et al. (2018)
5.	Virus-like particles	It has self-adjuvant characteristics, it is virus-like, and has a high gastrointestinal stability.	Reproducibility issues.	Vinay et al. (2018)
6.	Nano emulsions	Self-adjuvant, encapsulating both hydrophilic and lipophilic antigens.	Antigens are released too soon, and the gastrointestinal system is unstable.	Vinay et al. (2018)

6.2.2.2 Inorganic Nanoparticles

As vaccine carriers, inorganic nanoparticles have distinct advantages. These nanoparticles, in particular, limit premature antigen release while extending antigen presentation, resulting in effective defence against infectious illnesses.

6.2.2.3 Nanoliposomes

Liposomes have been intensively researched as a carrier and immune-stimulating agent in the vaccine delivery sector. Liposomes are nanoparticles that mimic pathogen characteristics and can trigger humoral and cell-mediated immune responses.

6.2.2.4 ISCOMs

ISCOMs are self-assembled, spherical, hollow, cage-like particles (40–60 nm) made up of saponin Quill A adjuvant, antigen, phospholipids, and cholesterol. ISCOM matrix is a particle that does not contain antigen. Quill A has a cholesterol affinity, which promotes ISCOM matrix integrity.

6.2.2.5 Virus-like Particles

Although VLPs are most recognized for their immunogenic qualities, their adaptability allows for a wide range of uses. They have recently become the focus of research in vaccines, medication delivery, and gene therapy.

6.2.2.6 Nano Emulsions

An emulsifier, co-emulsifier, oil phase, and aqueous phase make up a nanoemulsion. This technology is effective as a vaccine adjuvant, and several nanoemulsions have been found to be safe.

Administration of DNA vaccine by oral method with a suitable carrier has also been discovered (Dadar et al., 2017).

Various inorganic nano compounds are also used for vaccine delivery research. Inorganic compounds of gold, aluminum, silver, carbon, calcium phosphate, etc, of which calcium phosphate and carbon nanotubes are most commonly used for the delivery of vaccine in aquatic organisms (Nagaraju, 2019).

The ultimate purpose of vaccination is to establish both safe and effective primary and secondary immune responses in the body. In a nutshell, when the body is first exposed to a disease or antigen, primary immune responses protect it from harm. Secondary immune responses, on the other hand, are induced in response to the immunological memory created when the initial exposure and serve to safeguard the body from future exposure with the same antigenic determinant. The use of nanoparticles to administer vaccines has shown to be an intriguing platform for boosting the body's primary and secondary immune responses. When nanoparticles are delivered, they elicit a range of immunological responses, but they are not immunogenic until they have been conjugated with an antigen. Activation of PRRs, induction of cytotoxic T lymphocytes, T-helper (Th) activation, cytokine synthesis in various ways, B-cell activation, and antibody creation have all been linked to the induction of immunological responses by nanoparticles. Depending on their size, nanoparticles are taken up by APCs by a variety of methods, including pinocytosis and phagocytosis (Vinay et al., 2018).

According to several studies, smaller particles provoke stronger immune responses than their larger counterparts (Vinay et al., 2018). Nanovaccines aid the enhancement of immune responses through a variety of mechanisms. Internalization, ability to co-deliver multiple antigens and adjuvants, inherent adjuvant property, rapid lymph node drainage, and efficient antigen presentation onDC's surface are all important mechanisms by which nanovaccines help increase the duration and magnitude of immune responses in immunized animals. Surprisingly, these nanoparticle characteristics have been heavily adjusted in the production of either preventive or therapeutic vaccinations in order to generate a desired immune response. Before constructing a safe and effective nanovaccine, it's vital to understand how the nanovaccine engages both innate and adaptive immunity in the body (Bhardwaj et al., 2020).

6.2.3 Nanovaccines Activate Innate Immunity

Therapeutic vaccinations (such as those used to treat cancer) are given intravenously, whereas prophylactic nanovaccines are frequently given subcutaneously, intramuscularly, or intranasally. Depending on the route of vaccination, immune cells such as neutrophils, macrophages, DCs, and natural killer (NK) cells quickly recognize the nanovaccine (foreign particle) based on PAMPs associated with it. PAMPs (e.g., bacterial LPS, teichoic acid, lectins, oligonucleotides, or carrier nanoparticle material) function as ligands for abundant PRRs (e.g., TLRs, C-type lectin receptors (CLRs), retinoic acid-inducible gene-I-like receptors (RLRs), etc). Macrophages preferentially endocytose larger particles (usually >500 nm) as a result of PAMP–PRR interaction, whereas DCs preferentially endocytose smaller particles. The slow breakdown of nanovaccine particles and encapsulated antigens occurs when nanovaccines are engulfed by macrophages. To avoid this, nanovaccine particles have been designed to withstand macrophage attack and deliver vaccines directly to APCs. Depending on the type of PAMPs, neutrophils and macrophages also produce a range of cytokines and chemokines that will activate APCs. In addition to PAMPs, there are a variety of undiscovered techniques for activating innate immunity, such as the activation of immunological responses (potential adjuvant effect) by the glycolic acid component of PLGA copolymer or cationic liposomes (Bhardwaj et al., 2020).

6.2.4 Nanovaccines Activate Adaptive Immunity

Adaptive immunity is divided into two parts: cell-mediated immunity and antibody-mediated immunity (AMI). Adaptive immune responses might last anywhere from a few days to many weeks. Both types of immunity are required for a long-lasting and protective immune response. However, in both preventive and therapeutic nanovaccines, their requirements are different. For example, powerful cell-mediated cytotoxic responses are favored in the case of a therapeutic nanovaccine in order to efficiently eradicate metastatic cancer cells (Bhardwaj et al., 2020).

6.2.5 Cell-Mediated Immunity Activation

Vaccination stimulates B cells and T cells to produce cell-mediated immunity (CMI), which neutralizes the pathogen/antigen while also creating immunological memory in the body. Migration of activated DCs to lymphatic organs (after uptake of nanovaccine particle) triggers CMI responses (spleen and lymph noted). CD8+ cytotoxic T lymphocytes (CTLs) receive antigens from activated DCs through the MHC class I receptor. Activated CTLs produce a lot of CMI and preferentially kill target cells (infected changed cells or malignant cells) by inducing programmed cell death. Multiple complement system proteins and co-stimulatory chemicals are involved in killing target cells, which is a sophisticated immunological mechanism to exvivo training of DCs through antigen adoptive transfer is the primary goal of therapeutic cancer nanovaccines. For example, Sipuleucel-T, a therapeutic vaccination, is authorized for the treatment of metastatic prostate cancer that is resistant to contraction. Activated DCs, on the other hand, transmit antigens to CD4+ T-helper cells (Th cells) through the MHC-II receptor. Effector Th cells are then activated and developed, causing them to secrete a variety of cytokine signals. Th populations are split into Th1 cells and Th2 cells subsets based on the kind of cytokine released. Many pro-inflammatory cytokines (including INF-alpha, TNF-gamma, and IL-1) are secreted by the Th1 subgroup, which promote CTL proliferation and boost CMI function. During antibody-mediated immune responses, Th2 cells release cytokines that promote B-cell proliferation (e.g., IL-4, IL-5, IL-6, IL-9, IL-10, and IL-13). A tight balance between Th1/Th2 responses determines the overall therapeutic or prophylactic vaccination potential of prospective nanovaccines. T-regulatory (Treg) cells, which normally regulate the activation and multiplication of effector T cells in the body, can also be reduced by nanovaccines (Bhardwaj et al., 2020).

6.2.6 Antibody-Mediated Immunity Activation

Nanovaccines excite naive B cells in the spleen and lymph nodes, which then bind to soluble antigen through B-cell receptors (BCRs). In the germinal core, the activated B-cell population proliferates. Furthermore, by somatic hyper mutation, B cells become specific to a certain antigen epitope, and only epitope specific B cells are reproduced. Activated B cells can either develop into antibody secreting plasma cells, which release soluble antibodies against the target antigen, or they can remain dormant as memory cells, waiting for the same antigen encounter (Bhardwaj et al., 2020).

6.2.7 Immunological Memory Is Boosted by Nanovaccines

Plasma cells have a short lifespan, and their numbers, like antibody tiers, steadily decline over time. In this case, memory cells (B cells and T cells) that have been created and stored in the lymphatic organs or bone marrow for several months come into play to protect the body from reinfection with the same antigen. In such situations, memory B cells proliferate rapidly and change into antibody secreting (mainly IgG) plasma cells to neutralize the antigen. By releasing more cytokine and chemokine signals, memory T cells (both CD4+ and CD8+) contribute to trigger stronger CMI and antibody-mediated responses. Memory cells, on the other hand, are unable to provide the necessary protection if the antigen (epitope) structure has been drastically altered by cleavage, aggregation, or refolding. Although first immunization can provide up to 90% protection in some instances, booster doses are scheduled to extend protection to 100% because even the remaining 10% can be lethal. Booster doses increase the creation of more and different types of memory cells. After the antibody titer from plasma cells has reduced sufficiently, booster doses are frequently given to let memory cells to compete more effectively with the injected antigen. Booster doses are typically provided after the antibody titer from plasma cells has decreased sufficiently, allowing memory cells to compete more efficiently with the injected antigen. Nanoparticles (e.g., PLGA-based nanovaccines) allow for the continuous release of intact antigen in the bloodstream, resulting in greater B-cell proliferation and hence more memory cells. Nanoparticles can also pass through the restricted lymphatic system to lymph nodes, causing additional memory cells to differentiate (Bhardwaj et al., 2020).

6.3 METHODS OF VACCINATION

The production of the aquaculture sector is increasing rapidly, and this increase has led to the spreading of different diseases. The likelihood of happening of disease is unforeseeable, which can cause substantial economic losses. To have sustainable development in aquaculture, the main hurdle faced by farmers is diseases caused. Unscientific usage of drugs and antibiotics may develop drug resistance to bacteria and pathogens. It can also induce antibiotic deposition, which further causes pollution.

Vaccines are administered in fish through different routes, and these channels determine the result of elicited immune response and level of protection against the pathogen of interest. The traditional method of vaccination of fish essentially included inactivated complete organisms. In contrast, a few live attenuated or subunit vaccine that are made of proteins were used for commercial purposes (Ma et al., 2019b). Bacteria killed using formalin were used in conventional vaccination of fishes. These vaccines were administered by either injection or immersions. This initiated adaptive immunity is manifested by the production of antibodies by B lymphocytes to some extent (Ma et al., 2019b). Intramuscular or intraperitoneal injection, oral administration (mixed with feed), and immersion by bath or dip vaccination are the three methods by which vaccine is commercially delivered to fish, and the method of delivery depends on the age and size of the fish (Table 6.2) (Palm et al., 1998) Anal administration and intramuscular injection are other routes carried out when required (Embregts and Forlenza, 2016). Intraperitoneal injections are traditionally used as a

TABLE 6.2
Vaccine Delivery Based on Nanoparticles

Nanoparticles	Species	Antigen	Pathogen	Vaccine Route	References
Chitosan	Danio rerio	rgpG	VHSV	IP	Kavaliauskis et al. (2016)
	Lates calcarifer	pFNCPE42	Nodavirus	Oral	Vimal et al. (2014)
		pVAOMP-DNA	Vibrio anguillarum	Oral	Vimal et al. (2014)
		pVAOMP38	Vibrio anguillarum	Oral	Kumar et al. (2008)
PLGA: Poly (Lactic-Co-Glycolic Acid)	Labeorohita	rOmpW	Aeromonas hydrophila	Oral	Dubey et al. (2016)
	Oncorhynchus mykiss	pCDNA-G	IHNV	Oral	Adomako et al. (2012)
Liposome	Cyprinus carpio	NKC03,IKC03	KHV	Oral	Yasumoto et al. (2006)
		Aeromonas salmonicida	KHV	Oral	Irie et al. (2005)
Carbon nanotubes	Ctenopharyngonidellus	rVP7	GCRV	Immersion	Zhu et al. (2014)
		pEGFP-vp5	GCRV	Immersion	Wang et al. (2015)
		pcDNA-vp7	GCRV	Immersion	Zhu et al. (2015)

GCRV, Grass carp reovirus; IHNV, infectious hematopoietic necrosis virus; KHV, Koi herpesvirus; VHSV, viral hemorrhagicsepticemia virus.

method to administer the oil-in-water type of vaccine. DNA plasmids are mostly administered by intramuscular injection. The systemic and mucosal vaccination leads to humoral along with cell-mediated immune responses in fish. A sequence of oral, immersion and injection vaccination have been utilized for various species to guarantee protection throughout the complete production cycle of fish. The specific delivery route, or perhaps numerous administrations using various modalities, may be required to achieve adequate and long-lasting immunity (Dadar et al., 2017). The significant parameters to examine during vaccine administration are the pathogen targeted, vaccine production methods, immunological memory status, infection pathway, fish species, and host life stage (Dadar et al., 2017). Vaccines are available for most fish species but are mostly against bacterial pathogens, but only a few are available against viruses (Embregts and Forlenza, 2016) (Figure 6.2).

6.3.1 Oral Method

The late 20th century viewed drastic research and study in the field of oral vaccines. Along with this timeline, physical and functional variations in the fish's gut were studied. It was observed that anal vaccination had more efficiency when compared with oral vaccination when introducing the vaccines to both the fish. The study was conducted in Salmon against *Vibrio anguillarum* and *Yersinia ruckeri*.

The requirement of antigen protection was validated by comparison of antigen absorption, induced immune responses, and efficiency after administration of the oral vaccine. Mucosal tolerance is related to the dose of antigen and the method of delivery of the vaccine. The tolerance by mucosal lining depends on the amount of antigen and the method of delivery. The excessive level of anti-inflammatory cytokinesis that assists the production and support of DCs and tolerogenic regulatory T cells is the driving force for mucosal tolerance. Vaccines that imitate the original pathogen infection and induce a suitable immune response as a defense to the pathogen show more efficacy.

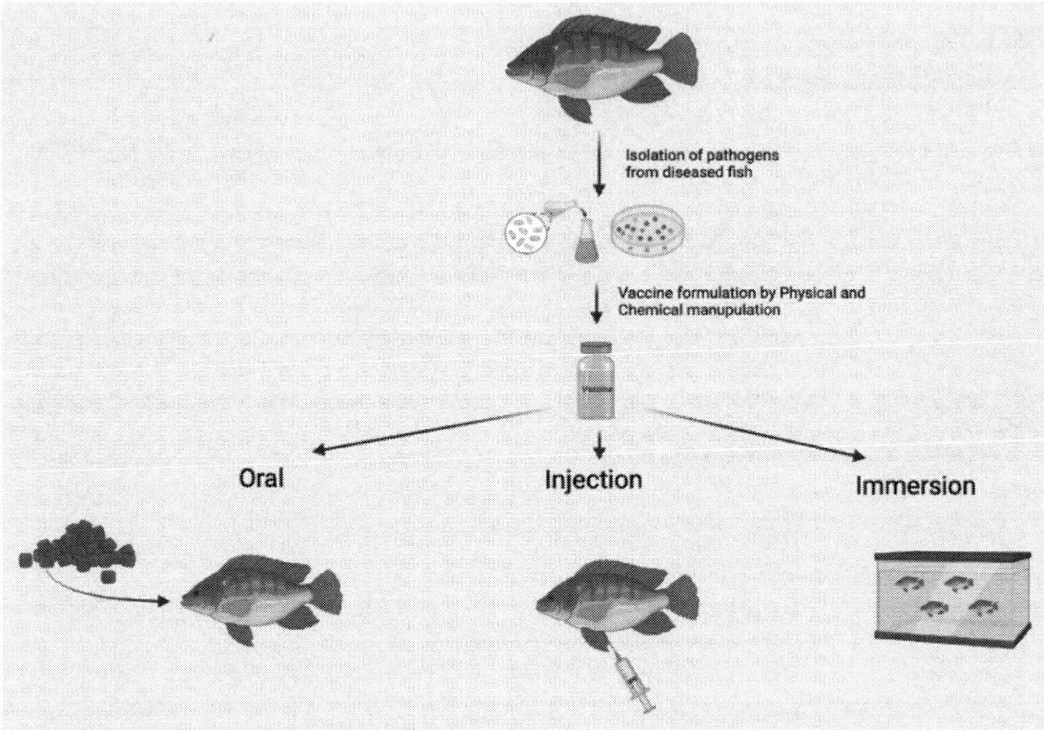

FIGURE 6.2 Different methods of vaccinations -oral, injection, and immersion methods.

When the vaccine is delivered orally, it has to overcome the same host immune response as that of commensals and enteric microbes. Orally administered vaccines will have to overcome conditions of the gut, striked by enzymes like nucleases and proteases, and will have to overcome a thick layer of mucus to enter the epithelial barrier. Because of this reason, when vaccines are delivered orally, it has to be given in large quantities as we cannot precisely know the amount of antigen administered to the epithelium of the gut (Embregts and Forlenza, 2016).

In view of animal welfare and handling costs, oral vaccination would be the best option among the mucosal routes. As feeding is practiced in fish farms, it is a convenient way to administer vaccines. The vaccine can be blended with the feed, could be sprayed onto feed particles, top dressed on the feed, and could even be bioencapsulated in oral vaccination. After prolonged feeding with chloroform-inactivated bacteria, the first successful oral vaccination on trout against a challenge with *Bacterium salmonicida* was observed in 1942. Vaccines of this kind are primarily involved with humoral immune response, then innate or cellular immunological activity. It's a reflection of overall immunity. This oral administration of antigens in fishes is being done using a large number of biological and synthetic materials to get a satisfactory immune action against possible pathogens. One example is *Chlamydomonas reinhardtii*, a microalga that may readily be used to transplant foreign genes into fish. As oral vaccinations for fish generally provide a few weeks or a limited length of protection, they are often used as a primary and booster immunization to establish protection from endemic diseases with an extended incubation period (Embregts and Forlenza, 2016 and Dadar et al., 2017). Although primary vaccination does not produce a strong immune response when given orally, a booster dose produces a strong secondary immunological response. Providing such antibody generating components through the oral way has many benefits. It makes production more cost-effective in most cases. This method can be used smoothly in fishes of all sizes or stages, and it is much less efficient. The injection vaccination comes with the disadvantage of handling fishes with stress, which can lead to mortality even though it can induce a higher immune response, so preferably, oral vaccination is adopted (Brudeseth et al., 2013). Oral vaccine efficacy is hampered by antigen degradation in the harsh digestive environment, as well as a severe tolerogenic gut environment and ineffective vaccine formulation. Antigen encapsulation is being done with the help of nano or microparticles because they can slow down or stop the breakdown of the antigen particle in the gut's acidic environment. The possibility of inducing immunity after giving an oral vaccine, particularly in immunologically immature juveniles, is thus a problem that fish immunologists address when creating mucosal vaccines (Embregts and Forlenza, 2016). To forecast the fate of antigens in orally administered vaccines, one can use transgenic fish lines with labeled antigens and the comparative donation of a particular type of cell in presentation, activation, and uptake of the immune response. Zebrafish can be used as transgenic model organisms as live vectors, or fluorescently labeled inactivated antigens can be effortlessly identified. Administration of encapsulated antigen could be observed in all stomach segments and preferably of different encapsulation methods such as stability or size.

Only very few numbers of oral vaccines are available for fish, similar to humans and other animals. Commercially there are five vaccines available for oral administration. The vaccines available on the market are only for a few pathogens and only for a limited number of species. When one considers the numerous species being cultured and the pathogen that affects them, the number of oral vaccines available on the market is not sufficient. The mixture of powerful mucosal adjuvant and weak oral antigens gives numerous combinations that are not yet completely studied or researched. Due to the significant genetic variability of teleost species, the fish mucosal immunology instills in its initial stage. In the last five to ten years, IgT and M-like sampling cells were found, which helped in the development of fish vaccine and immunology. The probability of exactly aiming M-like cells or APCs in fish's stomach is becoming a suitable choice for fish vaccine delivery. Purified recombinant proteins, DNA plasmids, and subunit vaccines also require some form of shielding to avoid disintegration of the stomach environment and make certain its uptake. The immune system of fish, unlike humans, doesn't have lymph nodes or payer patches and, like birds, doesn't have bursa of fabricus, but they have gut-associated lymphoid tissue (GALT).

Various elements in the development of vaccines and the amount of antigen influence the oral vaccine efficiency. To administer recombinant subunit vaccine, *Artemia nauplii* encapsulated with recombinant bacteria having the required antigen were given to fish. These fishes, which were administered with artemia, showed the release of encapsulated vaccine within the body. In *C. reinhardtii,* the protein renibacteriumsalmonicasalmon as whole or partial protein showed immunity through oral and immersion methods. The oral response resulted in systemic immunity, and immersion resulted in antibody response (Dadar et al., 2017).

6.3.2 Immersion: Dip and Bath Methods

Immersion immunization is one of the easiest ways of vaccination but is not appropriate for each aquaculture activity. Various dip or immersion vaccination methods incorporates spray, hyperosmotic infiltration (HI) and direct immersion (DI). In the process of HI, the fish is first dipped in sodium chloride or urea for a while; after that, it is immersed in the vaccine. In the direct method of vaccination, fishes are first transported to water having the vaccine for a definite time and then switched back to their holding tank. At first, HI was favored but later replaced because DI was more effective and fish were highly stressed. The method of spray vaccination was initially intended for conveying Vibrio vaccines and is most suitable for bigger fishes greater than 20 g. Still, it stresses the fish more than DI. Bath and flush immunization also evolved because, in spray immunization, fishes are caught by net, transported with a conveyer belt, and are to be sprayed with the vaccine. Other immunization methods such as bath Immunization are also practiced. In this method, the system's water level is reduced significantly, and the vaccine is diluted in the system. This offers the benefits of negligible handling stress, whereas it also comes with the limitation of usage of more quantity of vaccine for the process. A new evolving technique for vaccine delivery is also adopted, which uses ultrasound waves. The permeability of the cells in the fish's skin is significantly enhanced with the approximately 20 KHz high-frequency sound waves. Inactivated humpback grouper nervous necrosis virus was used for bath immunization of *Epinepheluscoioides* which showed how effectual is this way of immunization is. Zhou et al., assessed ultrasound to transport alginolyticus bacterium to grouper Ephinephalus aware, with the adequate frequency of the sound and suitable time period of ultrasound, it was observed that the ultrasound delivery method was providing defense as same as an intraperitoneal injection. The examination of bovine serum albumin absorption of *Carassius auratus* (goldfish) using ultrasound was tested along with HI. The study revealed that the ultrasound method used only one-fifth of the antigen used in bath immersion, and thus ultrasound method was found to be more effective (Plant and LaPatra) (Table 6.3).

Immersion vaccinations are made up of an emulsion of live attenuated bacteria and live bacterial and vector vaccines (Brudeseth et al., 2013). This practice is very convenient for fishes of smaller size, which weigh nearly 1–4 g. For immunizing fishes that cannot take much handling stress, this procedure is quick, effective, affordable, convenient, and less distressing. Antigens get directly exposed to immune cells of the fish skin and gills using this practice (short or long bath). In the skin mucosa, the immersion method elicited an antibody response. The immunity gained this way is effective for a short duration, such as between 3 months and 1 year, which cannot be sufficient for the culture of some fish species, thus, necessitating repeated immunization. As the immersion vaccination provides a diluted version of the vaccine, some of its components are also squandered while performing this process. Commercial immersion vaccinations are made up of formalin-inactivated and live bacterial vaccines suspended in suspension. In larger fishes, it is more challenging to choose the use of adjuvants, and even other immunity-inducing drugs, due to the high cost and limited time available. Ultrasound-mediated uptake, hyperosmotic dip, and multiple puncture instruments have all been shown to aid in the development of antigen uptake in immersion vaccinations. Antigens have also been delivered using a combination of immersion and puncture methods (Dadar et al., 2017).

Nakanishi et al. (2002) found a new immunization method by combining puncture and immersion methods. This method was not completely an immersion immunization method as the fishes were required to be punctured and handled. This way of immunization is similar in its action to intraperitoneal injection.

There are also methods of delivering foreign antigen through attenuated pathogen. In 1995, Noonan et al. did some experiments that utilized the attenuated strain of *A. salmonicida* to exhibit some glycoprotein. In order to get protection by immersion using this method, plasmids exhibiting viral hemorrhagic septicemia virus (VHSV) and infectious hematopoietic necrosis virus (IHNV) glycoprotein were transferred to *A. salmonicida*. This method is an effective way for immunization against multiple pathogens at a single time in an easy way (Plant and LaPatra).

6.3.3 Intraperitoneal (IP) Injection

Intraperitoneal injection is a method of vaccination where a small quantity of the required antigen is injected into the fish. This method of vaccination is more effective and gives protection for a more extended period. Intraperitoneal injections are mostly multivalent, having various bacteria or a mixture of bacteria and killed virus or viral proteins. Multivalent products available in the market tend to be profitable and can be a mixture of five vaccines. Injection vaccination with conventional bacteria is delivered through the intraperitoneal pathway, whereas the newly discovered DNA vaccines protecting against IHN and VHS viruses are more effective when delivered by intramuscular injection. Intraperitoneal injections are commonly used in *Salmo salar* (Atlantic salmon) production is economically practicable, as how much the fish can reach in size and various pathogens responsible for the numerous losses of the fishes at this bigger size (Plant and Lapatra, 2011).

Small fish are typically vaccinated orally or by immersion, although both methods have minimal efficiency and provide temporary protection. IP is the most effective injectable form of fish

TABLE 6.3
Various Fish Vaccines Used Globally for Aquaculture Purposes

Vaccine Type	Diseases	Antigens	Delivery Method	Major Fish Host	References
DNA	Infectious hematopoietic necrosis	G Glycoprotein	IM	Salmonids	Jie et al. (2019)
Inactivated	Infectious salmon anemia	Inactivated ISAV	IP	Atlantic salmon	Jie et al. (2019)
	Spring viremia of carp virus	Inactivated SVCV	IP	Carp	Dixon and Stone (2017)
	Furunculosis	Inactivated *A. salmonicida* spp.	IP or IMM	Salmonids	Jie et al. (2019)
Attenuated	Koi herpesvirus disease	Attenuated KHV	IMM or IP	Carp	Dixon and Stone (2017)
	Columnaris disease	Attenuated *F. columnare*	IMM	Freshwater finfish species, bream, bass, turbot, salmon	Shoemaker et al. (2011)
Vector technology		IHN	IP Injection	Salmon	Phenix et al. (2000)
		IPNV	IP Injection	Rainbow trout	Phenix et al. (2000)
Recombinant protein		IPNV/VP2	IP Injection	Salmon	Lecocq-Xhonneux et al. (1994)
		SVCV	IP Injection	Carp	Dadar et al. (2017)

IHN, infectious hematopoietic necrosis virus; IM, Intramuscular injection; IMM, Immersion; IP, Intraperitoneal injection; IPNV, infectious pancreatic necrosis virus; ISAV, Infectious salmon anemia virus; ISKNV, Infectious spleen and kidney necrosis virus; KHV, Koi herpesvirus; SVCV, Spring viremia of carp virus.

immunization; hence many of today's vaccines are delivered this way. This can be done manually with a needle or by using technologies like compressed air. The length of protection is longer in this method than in the immersion method. The vaccine is not diluted in this method. It is also supplemented with other substances, including adjuvants, carriers, bacterial cells, bacterial antigens, and others that aren't available through standard immunization methods (Dadar et al., 2017). Antigens for injectable vaccines can be conveniently kept at 4°C. Water-in-oil (w/o)-based injectable vaccinations are often delivered via intraperitoneal injection, whereas DNA plasmids are typically delivered via intramuscular injection (Embregts and Forlenza, 2016). Both humoral antibody and cellular antibody and cellular cytotoxic responses are highly efficient with injection.

An additional benefit of the injection vaccination methodology is that a multivalent vaccine can deliver numerous antigens from various infections simultaneously. But this method is not practiced commonly or recommended for fishes that are small in size or less than 5 g in weight as there are numerous risks in this way of administering the vaccine. As fishes would be very small, it would take much time to vaccinate enough biomass of fishes. This will be a highly complex process. Small fishes may also suspend their feeding activity temporarily. Adhesion formation and penetration of the intestines may also happen. Injury at the injecting point is also highly possible in very small-sized fishes, which can quickly worsen into another kind of infection. Moreover, w/o-based injecting vaccines have been caused by local side effects, which include many problems such as tissue inflammation. More issues such as necrosis and adhesion have been common occurrences in this process, decided by the type of adjuvant applied. Furthermore, many deaths have been reported because of ill-handling and stress on the body during the injection. One restriction associated with injectable vaccinations is that they can't be used several times during the fish production cycle. They cannot be used in early life stages because the immune system is still immature. During those stages, their body is delicate, and they can't take the handling stress during the injection of the vaccine. Frequently necessitates the use of sophisticated machinery or a big, experienced team.

Intraperitoneal injection can be administered both by manual and automatic methods. In the process of manual injection, the fish are anesthetized before injection. The injection is usually administered by an air-powered syringe. The vaccine can be administered to 1000–1200 fish in the manual method, and about 7000–9000 fish can be vaccinated using the automatic method. In the automatic method, fishes have to be of bigger and similar size for vaccine administration (Plant and Lapatra, 2011).

6.4 ADVANTAGES AND LIMITATIONS IN AQUACULTURE

Aquaculture is the fastest growing food sector globally. According to FAO, global production reached 73.8 million tons in the year 2014 (FAO, 2016). The human population has been growing steadily, though the growth rate may have shrunk by a small fraction. As a result, food demand obviously is going up, putting stress on production in all the sectors providing food. Food security has already been an issue in many of the underdeveloped and backward regions of the world. It is imperative to produce more and more food to continue to feed all people. Capture fisheries maybe started to show levels of reaching their peaks. As a result, the next step to get more results seems to be aquaculture. People started to take interest in cultural practices, and with the passage of time, this was also not proving to be enough. Newer measures and technologies were introduced to increase efficiency and production and minimize losses. Aquaculture entered places where it was not done earlier, and the older systems were intensified. Extensive culture systems were now changing into intensive culture systems.

Nearly around 600 aquatic organisms are being cultured in the whole world (FAO, 2010). They mean both, the primitive extensive culture species as well as the industrialized species which are produced on a large scale. Those species are cultivated at high density and the best care and simulation are given (Brudeseth et al., 2013). The intensification of culture systems and continued repeated cultivation have deteriorated the quality of various factors involved, and diseases are getting more and more common. As the density of culture species are very high, the loss incurred due to any infectious disease would be a large fraction for the aquaculture farmer. Such disease can also enter farms if the conditions and the species are in a healthy state and eliminate a large number of culture species. Intensifying culture means there is a constant looming break out of infectious disease by emerging and re-emerging pathogens even if all parameters have been maintained, so the occurrence of diseases cannot be predicted (Vinitnantharat et al., 1999). Mixed infections are also occurring (Hwang et al., 2020). US$ 50 billion are lost every year in aquaculture due to inadequate health management of fishes (Rather et al., 2011). Such diseases directly affect the livelihood of people and also the health of people consuming fish. For sustainable aquaculture practices to continue, they have to stay profitable for the farmer. Huge losses would be a significant deterrent for anyone who has invested in the venture.

Antibiotics provide useful means to fight against many bacterial diseases, but continued use of antibiotics has induced the development of bacteria resistant to them, thus making them ineffective. Accumulation of antibiotics also causes pollution (Karunasagar et al., 1994). The treatment thus becomes more costly and complex. Chemotherapy in fishes is not possible as diseased fish does not take feed properly (Hwang et al., 2020). Such a scenario has forced people to change strategy as to prevent the diseases altogether rather than treatment.

After many kinds of measures were tried, the fish diseases continue to surprise scientists and the farmer, and they continue to be a major reason for huge aquaculture losses throughout the globe. Drugs such as antibiotics have been applied as a solution, but they have been successful only to a threshold in many instances. Though the use of antibiotics has led to more understanding of fish health, improper usage has also led to undesirable results such as food safety hazards, drug resistance and safety issues are common deterrents in their case (Sneeringer et al., 2019). By using different prophylactic measures, managing fatal infections can be achieved. Vaccination had a major impact on controlling outbreaks in culture fishes. To date, most vaccines for fish have been developed against bacteria, and lesser numbers for viral infections (Embregts and Forlenza, 2016). Improvement of vaccines for fish is an ongoing process. There is an unknown number of diseases for which vaccines have to be developed. Fishes are poikilotherms, so vaccine development includes factors such as species, delivery measures, routes, etc. (Dadar et al., 2017).

Vaccination in fish is being practiced for over 50 years and provides stability and sustainability to aquaculture. A fish vaccine means something which has antigen or induces their production in the organism. An ideal fish vaccine would induce strong immunity for long periods, is not very costly to manufacture, has little or no side effects, and its method is not too complex (Ma et al., 2019). Vaccination is based on the principle of isolating, inactivating, and injecting, as stated by Louis Pasteur. Vaccines aim to provide safer food as well as maintain the environment of the surroundings. The advantages of immunization are manifold. With new techniques developing in microbiology and biotechnology, there is the rapid development of vaccines far from conventional ones. Sudden use of antibiotics in Norwegian Salmon after starting immunization has happened (Ma et al., 2019).

The first vaccination in fish was done in 1938 by Snieszko and co-workers in the carps against *Aeromonas punctata* by using killed bacteria, which induced immunity. But the process was too complicated to be replicated on the average aquaculture farms and people would discard the practice (Gudding and Van Muiswinkel, 2013). The first fish vaccine in the US was licensed in the 1970s and commercialized in the 1980s (Shao, 2001). Now, vaccines for 17 fish species are developed against 22 bacterial and six viral diseases. Large-scale vaccination was mostly done in Salmons in earlier times, but now other species also are taken. Vaccination is being done in more than 40 countries and is being given in several ways which include immersion, injection, and oral, among others (Brudeseth et al., 2013).

Immersion vaccination has been proven a convenient method. It also includes the dip method, bath method, and spray immersion. Fishes are simply immersed in a dilute vaccine for a short time and sent back to the culture area. Exposure duration can be between 30 and 60 seconds. But this has some limitations too, as, in cases of those fish that cannot be taken out of the culture area, this method can work when they are going to be put there. After that, they are not taken out. But, in large-sized fish, it helps in reducing the cost of the process. Smaller size fishes also can be immunized in large numbers at the same time. As it has a diluted version of the vaccine, some part of it would be lost, but it doesn't mean the efficacy is also reduced.

Intraperitoneal injection is the most effective method of vaccination. It allows the usage of adjuvants which will increase the time frame of its efficacy. Oil adjuvant is very commonly used in this case. Adjuvants are highly efficient and useful substances being put to use in vaccination. They form a part of the delivery mechanism of vaccines and make them many times more potent than earlier. They target specific cells or their organelles. They are abundantly taken up by mucosal surfaces and can easily transport through different barriers. Fishes are anesthetized and injected, then returned. Commercial operations use repeating injection guns that can inject 1000–2000 fish in just an hour. But this method is time taking and also requires skilled personnel, which is rare and also would increase the production costs by a number. Injecting smaller fish is not practical, as they may be too stressed due to handling. Wounds due to injection might not heal fully, leading to the loss of the fish (Vinitnantharat et al., 1999).

Oral vaccination means the fish can get the vaccine through their feed. This means extra costs of any kind can be avoided. No skilled labor is needed, no extra time would be spent on the same activity, and fishes of all sizes will be immunized. The vaccine is incorporated in the feed, so it needs some kind of protection also as it would be put in water thus leaving only some part of the vaccine available to the fish. Among the three most common methods, this method is the least potent due to this reason. The digestive system of the fish is a harsh and dynamic path through which this vaccine has to pass. So, preparing feed having adequate protection or covers make the method less smooth. In the intestines, the microfold cells are integral to the absorption of such vaccines to Peyer's patches and further into lymphoid follicles found in the mucosal layer. Microparticles are taken by the microfold cells spontaneously (Embregts and Forlenza, 2016).

The aim of vaccination is to induce immunity by activating the immune system. There are two components of immunity, innate and adaptive. Innate immunity is more reactive and acts spontaneously on sensing any infection. Surfaces such as skin, alimentary tract, gills, and mucous membranes form this arm of fish immunity. Growth inhibitors, for example, interferons, phagocytes, macrophages, and lysins, among numerous more constitute innate immunity, which is very rapid in its action. The adaptive immunity takes more time to get activated, but it gives specific memory, which is needed for the complete removal of the pathogenic substance. The adaptive immunity is made up of the humoral immunity, CMI, and immunological memory of the body. Humoral immunity, like other mammals, is due to the production of immunoglobulins, which have been found in three types, IgM, IgD, and IgT (Hu et al., 2010). CMI is due to T-cell involvement to combat pathogens whenever detected. The immunological memory is, obviously, the basis for identifying the antigen and making the required changes in the lymphoid cells (Secombes and Ellis, 2012). Immunization should result in activating both kinds, otherwise, it cannot be called successful by not having a lasting effect. For further enhancing the impact, adjuvants and nanoparticles are being used along with the traditional or conventional vaccines, which enhances the delivery system by preventing the degradation and loss of the integral structure and properties of the antigen. It has been seen and proven that without their application, the effect of vaccines is not satisfactory. Some studies have also shown that nanoparticles themselves act as an antigen, which is not desired and complicates the matter (Zaman et al., 2013).

6.4.1 Advantages

Vaccines for most diseases have not been developed, and many are still in progress. The commencement of practicing immunization has not completely eradicated infections, but it has significantly reduced mass mortality. Nanoparticles are now part of most vaccine formulations. They act by two mechanisms, either as an adjuvant or as becoming a part of the delivery system. Nanoparticles that act as the delivery system will deliver the antigen to the cell targeted. Adjuvant nanoparticles do not deliver but stimulate related pathways, which will enhance the absorption of these antigens, and thus, immunity is achieved (Zhao et al., 2014). There are various types of nanoparticles that are in use. They can be natural as well as synthetic, so lots of kinds are present as their usage has become more common in preparation. Nanoliposomes, immunostimulant complexes (ISCOMs), virus-like particles (VLPs), etc., are a few examples that are found useful.

DNA vaccines are now newly used for the same purpose. The definition given by the Norwegian Biotechnology Advisory Board is "the intentional transfer of genetic material (DNA or RNA) to somatic cells for the purpose of influencing the immune system." This definition differentiates it from gene therapy (Hølvold et al., 2014). There are significantly less number of DNA vaccines in fish, but examples such as vaccines against IHNV and VHSV have shown promise at experimental levels (Evensen and Leong, 2013). With time one can expect even more DNA vaccines as well as RNA vaccines. Polypeptides as vaccines are already in use widely, so there is an adequate chance that RNA vaccines will find success in the near future. The attribute of the DNA vaccine to induce all the arms of the immune system makes it very promising. Unlike conventional viral vaccines, DNA vaccines may conserve the structure and antigenicity of transgenic particles or proteins (Heppell et al., 1998). They show excellent efficiency when given in early life stages (Corbeil et al., 2000). They provide immunity in large ranges of temperatures, they are easy to produce, and processes for most are nearly the same (Lorenzen et al., 2009). As their replication is not a difficult task, modifications are not a tough task (Hansen et al., 1991). DNA plasmids cannot multiply in fishes but show great results and can be used at doses 100 times lower than in other animals (Embregts and Forlenza, 2016). Nucleic acids closely mimic antigens during intracellular pathogenic infection.

Bio-encapsulation in cases of oral vaccination for larva feed such as artemia and rotifers has been attempted, which has been successful. The amount of vaccine given is also in this way regulated as larvae do not consume a large amount of feed (Kawai et al., 1989). Microencapsulation of vaccines in polymers such as liposomes, PLGA, and alginates is very useful as they protect the antigen from the digestive tract (Tafalla et al., 2013).

In most cases, live vaccines are more effective as they mimic natural pathogenic infections. Modern ways include subunits, recombinant particles, and DNA or RNA particles as vaccine constituents. Immunization has enhanced economic potential for farmers in aquaculture. Antibiotics and therapeutic drugs have been used against diseases, but resistance to medication, overuse, etc., are the concerns. A sudden reduction in the use of antibiotics in Norwegian Salmon after starting immunization has been achieved because this method is sustainable and gives beneficial results (Ma et al., 2019). In Norway, the yearly usage of antibiotics has reduced from 47 tons to 1 ton (Lillehaug et al., 2003). Viral outbreaks such as koi herpesvirus disease (KHVD), infectious salmon anemia virus (ISAV), viral hemorrhagic septicemia (VHS), and bacterial outbreaks such as mycobacteriosis and disease caused by *Piscirickettsiasalmonis* have been identified as the primary reason for the destruction of cultures (Dadar et al., 2019).

Inactivated vaccines are produced by culturing pathogenic organisms, and formalin is used to kill microbes without affecting the protective immunity. The preparation cost of viral vaccines is very high, and the effect is weaker than anticipated. Some fish viruses do not grow well in the culture, so their preparation is tough (Dadar et al., 2019). Attenuated vaccines are being produced against diseases like KHV (Dhar et al., 2014). These undergo various scrutiny before being allowed for use on the farmer level. If the conventional ones do not produce good immunity, then-novel technologies create more potent vaccines. Genetically modified organisms have also been implied, though their release into the environment can be concerning. Synthetic peptide vaccines are subunit vaccines used for diseases like IHNV, rhabdovirus, birnavirus, etc. (Estepa et al., 1999). To enhance the performance of vaccines, various adjuvants are used to improve vaccine efficiency. Nanoparticles have acted as adjuvants as they increase the surface area and generate quantum size effects. Nanoparticles deliver antigens to cells, and adjuvants will activate specific pathways facilitating antigen uptake (Zhao et al., 2014). Virus-like particles do not contain infective nucleic acid but still have the structure for inducing immunity (Vinay et al., 2016).

6.4.2 Limitations

Immunization is a revolution in aquaculture, but improper methods have caused mishaps. Careful immunization after knowing wholly about the fish species' immune system is of primary importance. Vaccines are available only for considerable species, so people must choose carefully. In non-encapsulated oral fish vaccines, the protection was for a short duration which started to decrease after 2–3 months after vaccination (Embregts and Forlenza, 2016). In cases of oral vaccination, gut nature presents more challenges. Carnivorous fishes have more protease usage, and non-carnivorous fishes have more amylase activity to digest plant matter. Nanoparticles have given more options to prepare vaccines but come with some limitations. As they can cross the blood-brain barrier (BBB), their application is challenging (Yildirimer et al., 2011). They also have proven to be toxic, and the estimation of toxicity is difficult. Its result ranges from cell necrosis to reactive oxygen species (ROS), causing apoptosis (Elsaesser and Howard, 2012). This effect also can reach humans and harm them. Vaccines give protection and adjuvants also have some side effects in fish, such as inflammation, pigmentation, intra-abdominal adhesion, etc. Injecting fish means causing stress, and spinal deformities have been noticed as side effects (Midtlyng et al., 1996).

6.5 CONCLUSION

The vaccine development should be eco-friendly which should provide the desired result after administration. An ideal fish vaccine is one that is safe for the organsims and environment with which it is administered, as well as being cost-effective, simple to administer, stable, and multivalent. Adjuvants in vaccines have also helped to improve their efficacy, which has increased their acceptability in recent years. Adjuvants aid in the production of a strong immune response capable of protecting the organism from the disease being vaccinated against. The discovery of nanoparticle-based vaccines has also opened new avenues in vaccination. It is the best way to reduce the impact of antibiotics in the aquaculture sector which has a negative impact on human health also. The implication of vaccines should reach the farmer level which will help to solve the real problem of the fish culture and diseases related to it. A new vaccine should be developed for immunizing the fish against various pathogens. The vaccine development process should be environmentally friendly and it should be affordable to farmers.

Easy to administer, economically viable, cost-effective, and the capability of inducing strong long-term immunity with less side effects are the characteristics applicable for in an ideal vaccine. Vaccine development is still in its infancy in aquaculture because of the challenges it faces while addressing multicomponent affordable programs. On large-scale, low-cost vaccines are applied in the field even though the therapeutic effect provided by them are not the best. Moreover, oral vaccination by feed is more preferred by large-scale aquaculture farmers as the other methods are both time-consuming and highly expensive. Further studies must ensure the efficacy of mass vaccination through a reliable mode of administration.

REFERENCES

Adams, A., and Thompson, K. D. 2006. Biotechnology offers revolution to fish health management, *Trends in Biotechnology*, 24(5):201–205. https://doi.org/10.1016/j.tibtech.2006.03.004

Adomako, M., St-Hilaire, S., Zheng, Y., Eley, J., Marcum, R. D., Sealey, W., Donahower, B. C., Lapatra, S., and Sheridan, P. P. 2012. Oral DNA vaccination of rainbow trout, *Oncorhynchus mykiss* (Walbaum), against infectious haematopoietic necrosis virus using PLGA [Poly(D,L- Lactic-Co-Glycolic Acid)] nanoparticles. *Journal of Fish Diseases*, 35(3):203–214.

Aucouturier, J., Dupuis, L., and Ganne, V. 2001. Adjuvants designed for veterinary and human vaccines. *Vaccine*, 19(17):2666–2672.

Bhardwaj, P., Bhatia, E., Sharma, S., Ahamad, N., and Banerjee, R. 2020. Advancements in prophylactic and therapeutic nanovaccines. *Acta Biomaterialia*, 108:1–21. https://doi.org/10.1016/j.actbio.2020.03.020

Brudeseth, B. E., Wiulsrød, R., Fredriksen, B. N., Lindmo, K., Løkling, K.-E., Bordevik, M., Steine, N., Klevan, A., and Gravningen, K. 2013. Status and future perspectives of vaccines for industrialised fin-fish farming. *Fish & Shellfish Immunology*, 35(6):1759–1768.

Corbeil, S., LaPatra, S. E., Anderson, E. D., and Kurath, G. 2000. Nanogram quantities of a DNA vaccine protect rainbow trout fry against heterologous strains of infectious hematopoietic necrosis virus. *Vaccine*, 18(25):2817–2824.

Dadar, M., Alamian, S., Behrozikhah, A. M., Yazdani, F., Kalantari, A., Etemadi, A., and Whatmore, A. M. 2019. Molecular identification of Brucella species and biovars associated with animal and human infection in Iran. *Veterinary Research Forum*, 10(4):315–321. https://doi.org/10.30466/vrf.2018.89680.2171

Dadar, M., Dhama, K., Vakharia, V. N., Hoseinifar, S. H., Karthik, K., Tiwari, R., Khandia, R., Munjal, A., Salgado-Miranda, C., and Joshi, S. K. 2017. Advances in aquaculture vaccines against fish pathogens: Global status and current trends. *Reviews in Fisheries Science & Aquaculture*, 25(3):184–217. https://doi.org/10.1080/23308249.2016.1261277

Dhar, A. K., Manna, S. K., and Thomas Allnutt, F. C. 2014. Viral vaccines for farmed finfish. *Virusdisease*, 25(1):1–17.

Dixon, P. and Stone, D. 2017. Spring viraemia of carp. In Woo, P. T. K. and Cipriano, R. C., editors, *Fish Viruses and Bacteria: Pathobiology and Protection*, pp. 79–90. CABI, 1 edition.

Dubey, S., Avadhani, K., Mutalik, S., Sivadasan, S. M., Maiti, B., Paul, J., Girisha, S. K., Venugopal, M. N., Mutoloki, S., Evensen, Ø., Karunasagar, I., and Munang'andu, H. M. 2016. *Aeromonas hydrophila* OmpW PLGA nanoparticle oral vaccine shows a dose-dependent protective immunity in rohu (*Labeo rohita*). *Vaccine*, 4(2):21.

Duff, D. C. B. 1942. The oral immunization of trout against *Bacterium salmonicida*. *The Journal of Immunology*, 44(1):87–94.

Edelman, R. 1980. Vaccine adjuvants. *Reviews of Infectious Diseases*, 2(3):370–383.

Ellis, A. 1988. Current aspects of fish vaccination. *Diseases of Aquatic Organisms*, 4:159–164.

Elsaesser, A. and Howard, C. V. 2012. Toxicology of nanoparticles. *Advanced Drug Delivery Reviews*, 64(2):129–137.

Embregts, C. W. E. and Forlenza, M. 2016. Oral vaccination of fish: Lessons from humans and veterinary species. *Developmental & Comparative Immunology*, 64:118–137. https://doi.org/10.1016/j.dci.2016.03.024

Estepa, A., Fernandez-Alonso, M., and Coll, J. M. 1999. Structure, binding and neutralization of VHSV with synthetic peptides. *Virus Research*, 63(1):27–34.

Evensen, Ø. and Leong, J.-A. C. 2013. DNA vaccines against viral diseases of farmed fish. *Fish & Shellfish Immunology*, 35(6):1751–1758.

FAO. 2010. The state of world fisheries and aquaculture. Rome, FAO.

FAO. 2014. The state of world fisheries and aquaculture 2014, p. 223.

FAO. 2016. The state of world fisheries and aquaculture 2016. Number 2016 in the state of world fisheries and aquaculture, p. 190.

Gudding, R. and Van Muiswinkel, W. B. 2013. A history of fish vaccination: Science-based disease prevention in aquaculture. *Fish & Shellfish Immunology*, 35(6):1683–1688.

Hansen, E., Fernandes, K., Goldspink, G., Butterworth, P., Umeda, P. K., and Chang, K.C. 1991. Strong expression of foreign genes following direct injection into fish muscle. *FEBS Letters*, 290(1):73–76.

Heppell, J., Lorenzen, N., Armstrong, N. K., Wu, T., Lorenzen, E., Einer-jensen, K., Schorr, J., and Davis, H. L. 1998. Development of DNA vaccines for fish: Vector design, intramuscular injection and antigen expression using viral haemorrhagic septicaemia virus genes as model. *Fish & Shellfish Immunology*, 8(4):271–286.

Hølvold, L. B., Myhr, A. I., and Dalmo, R. A. 2014. Strategies and hurdles using DNA vaccines to fish. *Veterinary Research*, 45(1):21.

Hu, Y.-L., Xiang, L.-X., and Shao, J.-Z. 2010. Identification and characterization of a novel immunoglobulin Z isotype in zebrafish: Implications for a distinct B cell receptor in lower vertebrates. *Molecular Immunology*, 47(4):738–746.

Hwang, J. Y., Kwon, M. G., Seo, J. S., Hwang, S. D., Jeong, J. M., Lee, J. H., Jeong, A. R., and Jee, B. Y. 2020. Current use and management of commercial fish vaccines in Korea. *Fish & Shellfish Immunology*, 102:20–27.

Irie, T., Watarai, S., Iwasaki, T., and Kodama, H. 2005. Protection against experimental *Aeromonas salmonicida* infection in carp by oral immunisation with bacterial antigen entrapped liposomes. *Fish & Shellfish Immunology*, 18(3):235–242.

Karunasagar, I., Pai, R., Malathi, G. R., and Karunasagar, I. 1994. Mass mortality of *Penaeus monodon* larvae due to antibiotic-resistant *Vibrio harveyi* infection. *Aquaculture*, 128(3):203–209.

Kavaliauskis, A., Arnemo, M., Speth, M., Lagos, L., Rishovd, A.-L., Estepa, A., Griffiths, G., and Gjøen, T. 2016. Protective effect of a recombinant VHSV-G vaccine using poly(I:C) loaded nanoparticles as an adjuvant in zebrafish (*Danio rerio*) infection model. *Developmental & Comparative Immunology*, 61:248–257.

Kawai, K., Yamamoto, S., and Kusuda, R. 1989. Plankton-mediated oral delivery of *Vibrio anguillarum* vaccine to juvenile ayu. *Nippon Suisan Gakkaishi*, 55(1):35–40.

Kumar, S. R., Ahmed, V. P. I., Parameswaran, V., Sudhakaran, R., Babu, V. S., and Hameed, A. S. S. 2008. Potential use of chitosan nanoparticles for oral delivery of DNA vaccine in Asian sea bass (*Lates calcarifer*) to protect from *Vibrio (Listonella) anguillarum. Fish & Shellfish Immunology*, 25(1):47.

Langevin, C., Aleksejeva, E., Passoni, G., Palha, N., Levraud, J.-P., and Boudinot, P. 2013. The antiviral innate immune response in fish: Evolution and conservation of the IFN system. *Journal of Molecular Biology*, 425(24):4904–4920.

Lecocq-Xhonneux, F., Thiry, M., Dheur, I., Rossius, M., Vanderheijden, N., Martial, J., and de Kinkelin, P. 1994. A recombinant viral haemorrhagic septicaemia virus glycoprotein expressed in insect cells induces protective immunity in rainbow trout. *Journal of General Virology*, 75(Pt 7):1579–1587.

Lillehaug, A., Lunestad, B. T., and Grave, K. 2003. Epidemiology of bacterial diseases in Norwegian aquaculture a description based on antibiotic prescription data for the ten-year period 1991 to 2000. *Diseases of Aquatic Organisms*, 53(2):115–125.

Lorenzen, E., Einer-Jensen, K., Rasmussen, J. S., Kjær, T. E., Collet, B., Secombes, C. J., and Lorenzen, N. 2009. The protective mechanisms induced by a fish rhabdovirus DNA vaccine depend on temperature. *Vaccine*, 27(29):3870–3880.

Ma, J., Bruce, T. J., Jones, E. M., and Cain, K. D. 2019a. A review of fish vaccine development strategies: Conventional methods and modern biotechnological approaches. *Microorganisms*, 7(11):E569. https://doi.org/10.3390/microorganisms7110569

Ma, J., Bruce, T. J., Sudheesh, P. S., Knupp, C., Loch, T. P., Faisal, M., and Cain, K. D. 2019b. Assessment of cross-protection to heterologous strains of *Flavobacterium psychrophilum* following vaccination with a live-attenuated coldwater disease immersion vaccine. *Journal of Fish Diseases*, 42(1):75–84. https://doi.org/10.1111/jfd.12902

Midtlyng, P. J., Reitan, L. J., and Speilberg, L. 1996. Experimental studies on the efficacy and side-effects of intraperitoneal vaccination of Atlantic salmon (*Salmo salar*L.) against furunculosis. *Fish & Shellfish Immunology*, 6(5):335–350.

Miyazaki, T., Yasumoto, S., Kuzuya, Y., and Yoshimura, T. 2008. A primary study on oral vaccination with liposomes entrapping koi herpesvirus (KHV) antigens against KHV infection in carp. *Diseases in Asian Aquaculture*, 7:99–184.

Nagaraju, V. T. 2019. Nanovaccines in aquaculture. *Archives of Nanomedicine: Open Access Journal*, 2(1):153–159.

Nakanishi, T., Kiryu, I., and Ototake, M. 2002. Development of a new vaccine delivery method for fish: Percutaneous administration by immersion with application of a multiple puncture instrument. *Vaccine*, 20(31–32):3764–3769.

Palm, R. C., Jr, Landolt, M. L., and Busch, R. A. 1998. Route of vaccine administration: Effects on the specific humoral response in rainbow trout *Oncorhynchus mykiss. Diseases of Aquatic Organisms*, 33(3):157–166. https://doi.org/10.3354/dao033157

Petit, J. and Wiegertjes, G. F. 2016. Long-lived effects of administering beta-glucans: Indications for trained immunity in fish. *Developmental & Comparative Immunology*, 64:93–102.

Petrovsky, N. and Aguilar, J. C. 2004. Vaccine adjuvants: Current state and future trends. *Immunology and Cell Biology*, 82(5):488–496.

Phenix, K., McKenna, B., Fitzpatrick, R., Vaughan, L., Atkins, G., Liljestrom, P., and Todd, D. 2000. Cell culture evaluation of the Semliki Forest virus expression system as a novel approach for antigen delivery and expression in fish. *Marine Biotechnology*, 2(1):27–37.

Plant, K. P. and Lapatra, S. E. 2011. Advances in fish vaccine delivery. *Developmental & Comparative Immunology*, 35(12):1256–1262. https://doi.org/10.1016/j.dci.2011.03.007

Puricelli, C., Boggio, E., Gigliotti, C. L., Stoppa, I., Sutti, S., Rolla, R., and Dianzani, U. 2022. Cutting-edge delivery systems and adjuvants in tolerogenic vaccines: A review. *Pharmaceutics*, 14:1782. https://doi.org/10.3390/pharmaceutics14091782

Rather, M. A., Sharma, R., Aklakur, M., Ahmad, S., Kumar, N., Khan, M. and Ramya, V.L. 2011. Nanotechnology: A novel tool for aquaculture and fisheries development. A prospective mini-review. *Fisheries and Aquaculture Journal*, 16(1–5):3.

Secombes, C. and Ellis, A. 2012. The immunology of teleosts. In *Fish Pathology*, pp. 144–166. John Wiley &Sons, Ltd. https://onlinelibrary.wiley.com/doi/pdf/10.1002/9781118222942.ch4.

Shao, Z. J. 2001. Aquaculture pharmaceuticals and biologicals: Current perspectives and future possibilities. *Advanced Drug Delivery Reviews*, 50(3):229–243.

Shoemaker, C. A., Klesius, P. H., Drennan, J. D., and Evans, J. J. 2011. Efficacy of a modified live *Flavobacterium columnare* vaccine in fish. *Fish & Shellfish Immunology*, 30(1):304–308.

Singh, M. and O'Hagan, D. T. 2003. Recent advances in veterinary vaccine adjuvants. *International Journal for Parasitology*, 33(5–6):469–478. https://doi.org/10.1016/s0020-7519(03)00053-5

Sinyakov, M. S., Dror, M., Lublin-Tennenbaum, T., Salzberg, S., Margel, S., and Avtalion, R. R. 2006. Nano- and microparticles as adjuvants in vaccine design: Success and failure is related to host natural antibodies. *Vaccine*, 24(42–43):6534–6541. https://doi.org/10.1016/j.vaccine.2006.06.021

Sneeringer, S., Bowman, M., and Clancy, M. 2019. The U.S. and EU animal pharmaceutical industries in the age of antibiotic resistance (No. 1477-2019-2172).

Tafalla, C., Bøgwald, J., and Dalmo, R. A. 2013. Adjuvants and immunostimulants in fish vaccines: Current knowledge and future perspectives. *Fish & Shellfish Immunology*, 35(6):1740–1750.

Vijayan, V., Mohapatra, A., Uthaman, S., Park, I.-K. 2019. Recent advances in nanovaccines using biomimetic immunomodulatory materials. *Pharmaceutics*, 11, 534. https://doi.org/10.3390/pharmaceutics11100534

Vimal, S., Abdul Majeed, S., Nambi, K. S. N., Madan, N., Farook, M. A., Venkatesan, C., Taju, G., Venu, S., Subburaj, R., Thirunavukkarasu, A. R., and Sahul Hameed, A. S. 2014. Delivery of DNA vaccine using chitosan-tripolyphosphate (CS/TPP) nanoparticles in Asian sea bass, *Lates calcarifer* (Bloch, 1790) for protection against nodavirus infection. *Aquaculture*, 420–421:240–246.

Vinay, T. N., Bhat, S., Gon Choudhury, T., Paria, A., Jung, M.-H., Shivani Kallappa, G., and Jung, S.-J. 2018. Recent advances in application of nanoparticles in fish vaccine delivery. *Reviews in Fisheries Science & Aquaculture*, 26(1):29–41. https://doi.org/10.1080/23308249.2017.1334625

Vinay, T. N., Gon Choudhury, T., Paria, A., Gupta, S., and Sarkar, B. 2016. Nanovaccines: A possible solution for mass vaccination in aquaculture. *World Aquaculture*, 47(3):30–33.

Vinitnantharat, S., Gravningen, K., and Greger, E. 1999. Fish vaccines. *Advances in Veterinary Medicine*, 41:539–550.

Wang, Y., Liu, G.-L., Li, D.-L., Ling, F., Zhu, B., and Wang, G.-X. 2015. The protective immunity against grass carp reovirus in grass carp induced by a DNA vaccination using single-walled carbon nanotubes as delivery vehicles. *Fish & Shellfish Immunology*, 47(2):732–742.

Yasumoto, S., Kuzuya, Y., Yasuda, M., Yoshimura, T., and Miyazaki, T. 2006. Oral immunization of common carp with a liposome vaccine fusing koi herpesvirus antigen. *Fish Pathology*, 41(4):141–145.

Yildirimer, L., Thanh, N. T. K., Loizidou, M., and Seifalian, A. M. 2011. Toxicology and clinical potential of nanoparticles. *Nano Today*, 6(6):585–607.

Zaman, M., Good, M. F., and Toth, I. 2013. Nanovaccines and their mode of action. *Methods*, 60(3):226–231.

Zhao, L., Seth, A., Wibowo, N., Zhao, C.-X., Mitter, N., Yu, C., and Middelberg, A. P. J. 2014. Nanoparticle vaccines. *Vaccine*, 32(3):327–337.

Zhu, B., Liu, G.-L., Gong, Y.-X., Ling, F., Song, L.-S., and Wang, G.-X. 2014. Single-walled carbon nanotubes as candidate recombinant subunit vaccine carrier for immunization of grass carp against grass carp reovirus. *Fish & Shellfish Immunology*, 41(2):279–293.

Zhu, B., Liu, G.-L., Gong, Y.-X., Ling, F., and Wang, G.-X. 2015. Protective immunity of grass carp immunized with DNA vaccine encoding the vp7 gene of grass carp reovirus using carbon nanotubes as a carrier molecule. *Fish & Shellfish Immunology*, 42(2):325–334.

Section III

Adjuvants in Vaccination

An Underpinning

7 Role of Adjuvants in Vaccination Studies

Md Yasin Ina-Salwany, Tilusha Manchanayake, and Aslah Mohamad
Universiti Putra Malaysia

Md. Shirajum Monir
Bangladesh Fisheries Research Institute Freshwater Station

Mohammad Noor Amal Azmai, Salleh Annas, and Mohd Zamri-Saad
Universiti Putra Malaysia

CONTENTS

7.1 INTRODUCTION

Infectious diseases are a major cause of mortality and economic losses in the aquaculture industry. Today, vaccination is a key player in the development and sustainability of the aquaculture industry. It is a safe and cost-effective strategy to control many infectious diseases. Vaccines can be classified broadly as live-attenuated or inactivated vaccines. Live-attenuated vaccines made of pathogens are effective in developing immune responses because they stimulate the animal's immunity system similar to natural infection by replicating live organisms inside the host body. They are easy to produce, low cost, and generate a vigorous immune response, but they also pose a risk of reverting to virulence and causing disease in the immunocompromised animal. In contrast, inactivated vaccines that consist of killed or nonreplicating cellular components are safe but stimulate a weak and short-duration immunity because antigens alone are unable to stimulate the immune cells to achieve a high antibody titer. Thus, inactivated vaccines need to be incorporated with chemical substances

DOI: 10.1201/9781003388548-10

known as adjuvants, which act as pathogen-associated molecular patterns (PAMPs), to gain effective immunity through innate and adaptive mechanism (1). An adjuvant is a helper substance that is added to a vaccine to enhance the level and longevity of immune responses to antigens (2).

7.2 THE ORIGINS AND FUNCTIONS OF ADJUVANTS IN VACCINE PRODUCTION

7.2.1 History of the Discovery of Adjuvants

The word adjuvant refers to the Latin word *adjuvare* meaning "to help." The importance of adding substances to increase immune responses was discovered by chance like many other medical innovations. The history of the discovery of adjuvant dated back to the early 20th century to the works of Gaston Ramon and Alexander Thomas Glenny. Gaston Ramon, a French veterinarian, observed the higher antibody titers in horses when the animal developed inflammatory abscesses in Diphtheria vaccine injection sites. Ramon added different substances such as tapioca, bread crumbs, saponins, kaolin, calcium, and magnesium to the toxoid and observed the increased production of specific antibodies in injected animals. During the same time, Glenny, a British immunologist, used potassium aluminum sulfate to purify Diphtheria toxin to reduce irritability. He observed a single injection of washed toxoid precipitates was able to elicit higher immune responses with no irritating effects than the soluble toxin (3,4). Since then, aluminum was used in both human and animal vaccines. Another adjuvant experimented with was oil emulsions with mineral oil-in-water. However, vaccine adjuvant development has been a slow process owing to safety concerns resulting in alum being the only adjuvant licensed in the human vaccine for a few decades. In the 1990s, a few other substances such as MF59, AS01, AS03, AS04, and cytosine phosphoguanosin (CpG) 1018 were included in the license list as human vaccine adjuvants (5). Later in the 1990s, researchers discovered that the antigen-specific B- and T-cell responses can be activated by the activated dendritic cells (DCs) and the DC activation is governed by the activation of toll-like receptors (TLRs). The TLRs are expressed in antigen-presenting cells (APCs) such as DCs, which sense PAMPs in microbes. More pattern recognition receptors (PRRs), which can activate adaptive immune responses, were identified leading the adjuvant research to search for molecules that are able to bind with these receptors as possible vaccine adjuvants (5).

7.2.2 How Does Adjuvant Function in Vaccines?

A variety of compounds have been evaluated as vaccine adjuvants over the years including mineral salts, oil-in-water or water-in-oil emulsions, bacterial components or toxins, microparticles, saponins, cytokines, and liposomes (6). Vaccine adjuvant candidate is expected to fulfill properties such as being able to stimulate strong and long-lasting immunity, nontoxic, nonpyrogenic, not induce autoimmunity, and stable under a broad range of temperature, pH, and time (7).

In general, an adjuvanted vaccine is composed of two components, which act on different immune systems. The specific moiety (fraction), the antigenic component of the microorganism used, evokes a specific adaptive immune response. The adjuvant, which is the nonspecific moiety, induces an innate immune response (8). Adjuvants are helpful in increasing the immune responses to vaccine antigens through several mechanisms. The oldest and most commonly recognized mechanism of adjuvant action is the slow release of antigens at the site of injection. The constant presence of antigens ensures the continual stimulation of the immune system and the production of a high level of antibodies (9). Some adjuvants form a vaccine adjuvant depot at the injection site and sustained the release of antigens over time. For example, aluminum salt has a large surface area and it absorbs antigen through hydrophobic forces, electrostatic attractions, or ligand exchange. Absorbed antigens are released slowly upon exposure to interstitial fluid at the site of injection (10).

In a vaccine, the slow release of antigen can be achieved through the introduction of an emulsion as an adjuvant to the vaccine. In an emulsion, one liquid is dispersed in a second liquid. Vaccine antigen (water) and oil work as two phases in vaccine formulations. Surfactants are added to emulsions to stabilize them and the hydrophilic:lipophilic balance (HLB) of surfactant determines the type of emulsion it creates. Water/oil/water emulsion can be achieved using an emulsion with an intermediate value of HLB. In this emulsion, the aqueous phase (antigen) is trapped inside the oil droplets and has given long-term immune responses (11).

Adjuvants help in increasing the immunogenicity of weak antigens such as purified subunit antigens by forming antigen-adjuvant complexes for effective endocytosis (targeting). Once absorbed by the adjuvant, the soluble antigens convert to a particulate nature and facilitate the phagocytosis and uptake by APCs (12). Some adjuvants appear to create an immunocompetent environment at the injection site to enable the antigen to develop specific immune responses. MF59, an example of an oil-in-water emulsion adjuvant, stimulates phagocytosis of macrophages, monocytes, and granulocytes (10).

Antigens should reach the lymph nodes to initiate adaptive immune responses (13). Many adjuvants upregulate the secretion of chemical compounds (chemokines, cytokines) at the injection site to recruit immune cells (9). Recruited immune cells secrete more cytokines and chemokines to attract various immune cells to create a local pro-inflammatory environment at the injection site. A subclass of DCs that circulate in the peripheral tissues are thought to be uptaking antigens and delivering to lymph nodes. Then, the mature DCs process and present antigens to T-cells at the lymph nodes (9,14). The mechanisms of immune stimulation are described in detail in Section 7.5.

7.3 CLASSIFICATIONS OF DIFFERENT TYPES OF ADJUVANTS

Adjuvants are being grouped based on different criteria to compare rationally. Origin, chemical nature, physicochemical properties, and mechanism of action can be used to classify adjuvants. But related compounds may have different immunomodulatory properties (7).

Based on the mechanism of action of adjuvants, they can be broadly classified into three groups: delivery systems, immunomodulators/immunopotentiators, and mucosal adjuvants. The mucosal adjuvant group consists of compounds that can fit into both the above classes (Table 7.1) (6).

Adjuvants differ in their cellular receptors, and thereby the mode of action. Based on this, another classification of adjuvants can be done according to the capacity to stimulate either innate or adaptive immunity. Innate immunity adjuvants consist of LPS (endotoxin), β-Glucan (fungal polysaccharide), bacterial CpG-DNA, and imidazoquinolines. They bind to TLRs or PRRs on APCs. They activate the My88D system and stimulate the production of inflammatory cytokines, B7 ligand production, and adaptive immune responses. Adaptive immunity adjuvants are endogenous cytokines and acylated quillaja saponins and semi-synthetic derivatives. These adjuvants bind to highly specific receptors on T-cells to activate co-stimulatory signals and induce cytokine-related genes (7).

Alternatively, adjuvants can be classified according to the immune responses they elicit, Signal 1 facilitators (antigen presentation) and Signal 2 facilitators (secondary signals) and they are summarized in Table 7.2 (11).

7.4 REQUIREMENTS FOR USING ADJUVANTS

Traditionally, adjuvants, in the context of vaccines, are the components capable of increasing the antigen-specific immune responses. However, the role of adjuvants in guiding the adaptive immune response type to achieve more specific immunity to vaccine pathogens has become equally interesting.

In clinical terms, the adjuvant is added to increase the vaccine response in a given population with increasing antibody titers and increased protectivity from the infection. Inactivated vaccines are an ideal choice for some fish diseases considering the safety but may need a high dose of antigens to elicit the desired immunoprotection. Adjuvants used in such vaccines decrease the dose of antigens needed (dose sparing) (15). The addition of adjuvants to the vaccine enables a reduction of the number of doses of the vaccine to achieve the target antibody titers (16,17).

TABLE 7.1
Type of Adjuvant Groups, Adjuvant, and Mode of Action (Broad)

Type	Adjuvant	Mode of Action (Broad)
Delivery systems		Act as antigen carriers
Mineral salts	Alum	Create local inflammatory reactions
	Calcium phosphate	Recruit innate immune cells
	Freund's incomplete adjuvant (FIA)	
Lipid particles	MF59	
	Cochleates	
	Virus-like particles	
Microparticles	Virosomes	
	Polylactic acid (PLA)	
	Poly(lactide-co-glycolide) (PLG)	
Immune potentiators	Poly (I:C)	Activate innate immune system directly or through PRRs
	Flagellin	
	CpG oligodeoxynucleotides (ODN)	
	Monophosphoryl lipid A (MPL)	
	LPS	
	Imidazoquinolines	
	Muramyl dipeptide (MDP)	
	Saponins	
Mucosal adjuvants	Cholera toxin	Induction of local mucosal immunity
	Heat-labile enterotoxin	
	Chitosan	

Source: Adapted from de Souza et al. 2016 (6).

TABLE 7.2
Signal 1 Facilitators and Signal 2 Facilitators, Their Immunomodulatory Compounds, Receptors They Act on and Principal Immune Responses They Elicit

Adjuvant	Immunostimulatory Component/s	Receptor/Process	Principal Immune Response
	Signal 1 facilitators		
Freund's complete adjuvant (FCA)	Heat-killed mycobacteria Mineral oil	NLR, TLR, MyD88 pathway	Ab, Th1, T17
Incomplete Freund's adjuvant (FIA)	Mineral oil/paraffin oil + surfactant	Not defined	Ab, Th1 + Th2
Montanide	Mineral oil		Ab
PLGA particles	Biocompatible and biodegradable Poly-(lactic-co-glycolide)		Ab (combined with FIA)
	Signal 2 facilitators		
Alum	Aluminum salts	NLRP3	Ab, Th2
β-Glucans			Ab
Cytokines	IRF (interferon regulatory factor)		Ab
Poly I:C	Synthetic dsRNA	TLR3	Ab, Th1, CD8+ cells
Synthetic CpG oligonucleotides	Synthetic phosphorothioate-linked DNA Oligonucleotides with optimized CpG motifs	TLR9	Ab, Th1, CD8+ cells (when conjugated)

The other importance of incorporating adjuvants in vaccines is to gain an effective immunity, which cannot be achieved by antigen alone. A vaccine made from live-attenuated or killed microorganisms produces strong immune responses owing to naturally occurring adjuvants. Advances in biotechnology enable the production of highly purified microbial components as vaccine antigens. They are safe antigens but the immunogenicity is low. Hence, the adjuvants are added to enhance the efficacy of the immune responses of such weak antigens.

The effective vaccine should trigger potent antibody response as well as cell-mediated immunity including Th and cytotoxic T-cells. Therefore, the careful selection of appropriate adjuvants will be aimed at antigen-specific innate and/or adaptive immunity (18). The selection of adjuvant depends on the desired type of immune response desired. For example, Th 1 versus Th2 or cytotoxic CD8+ versus helper CD4+ cells. Adjuvants help in increasing the generation of T-cell memory and the initial response to an infection (16). In human vaccine developments, adjuvants have shown cross-reactivity and broadened antibody response profile against pathogens showing a high variation of strains or antigenic drifts (19, 20).

7.5 MECHANISM OF IMMUNE STIMULATION

Immune response to a vaccine (or to an infectious organism) can be divided into two main phases: innate and adaptive immune responses. When the antigens were exposed, innate immune cells such as macrophages, neutrophils, monocytes, and DCs recognize them through PRRs. PPRs bind to the PAPMs of antigens and identify them as threats (1). Once a vaccine with an adjuvant is administered, cascade of responses occurs in the animal and it is hard to distinguish the specific mechanism of action of adjuvants.

Recognition of pathogens by PPRs of immune cells, such as TLRs, NOD-like receptors, retinoic acid-inducible gene (RIG)-like helicases, or dectin-1, leads to a complex series of events including phagocytosis, activation and maturation of APCs, upregulating the expression of major histocompatibility complex (MHC) and co-stimulatory molecules and cytokine gene expression, antigen processing and presenting, secreting of cytokines and chemokines to recruiting more immune cells to the injection site, trafficking activated APCs to drainage lymph nodes, interacting with antigen-specific B-cells (antibody production), and activating effector CD8+ T-cells (7,9,11). Some natural or synthetic adjuvants such as bacterial cell wall components or yeast polysaccharide glucan are closely related with PAMPs structurally and are able to act or mimic as PAMPs in order to be identified as a threat to induce innate immune responses (14). When PAMPs of antigens or adjuvants bind to TLR receptors on APCs, an intracellular signaling pathway is activated and nuclear factor-κB is induced resulting in expressing cytokine genes to induce inflammation and activation of co-stimulatory ligand B7-1 and B7-2, which in turn activate the adaptive immunity (7) (Figure 7.1).

The processing and presenting of endogenous (intracellular) and exogenous (extracellular) antigen are different (21). Intracellular antigens such as virus antigens that are delivered in to the cytoplasm are degraded into small peptide particles. Then, the MHC I-peptide complexes are formed and presented at the cell surface to bind with CD8+ T-cells. Extracellular antigens such as bacteria are phagocytosed and degraded into peptides in the lysosomes. These peptides bind to MHC II molecules and MHC II-peptide complexes are presented to the CD4+ T-cells at the cell membrane (7). T-cell activation is governed by two signals. One signal is derived when the MHC-antigen peptide complex interacts with T-cell receptor (TCR). The other signal is a co-stimulatory signal that is derived when the bond between B7-1 and B7-2 ligand of APCs and CD28 receptor on T-cells is formed. Some adjuvants can substitute the B7-1 or B7-2 ligands and provide the co-stimulatory signal (7,22).

Nowadays, research on adjuvants has highlighted the fact that the type of immune response can be influenced by the adjuvant. According to the two-signal model, both signal 1 (antigen) and signal 2 (additional secondary signal) are important to activate antigen-specific B and T lymphocytes. Adjuvants that facilitate signal 1 have an effect on the immune availability of antigen in terms of

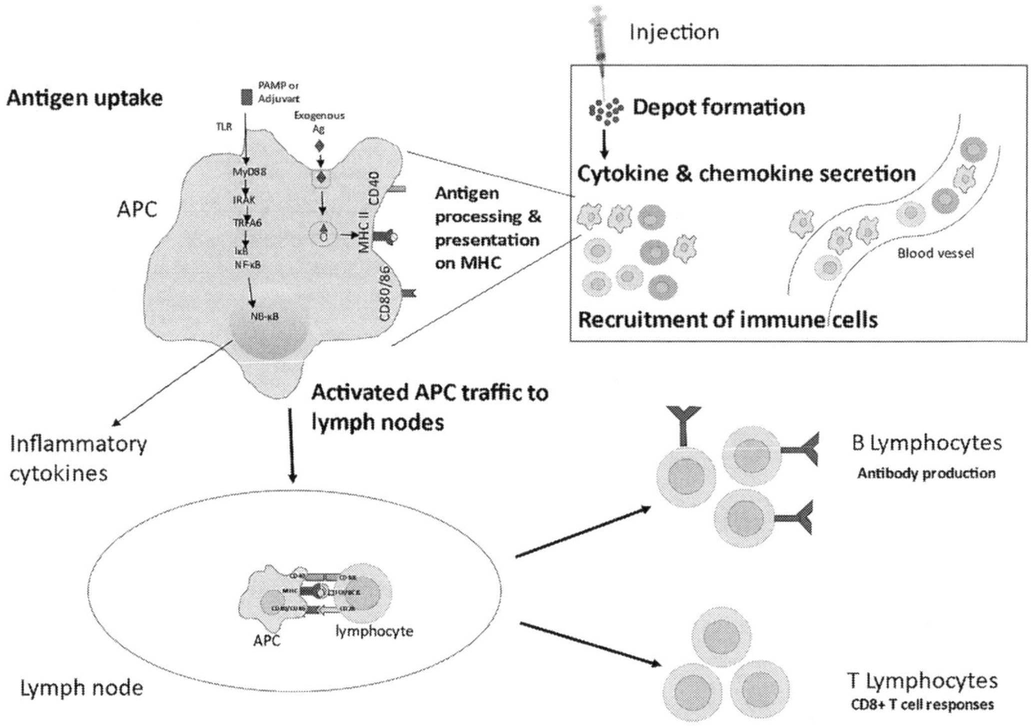

FIGURE 7.1 Mechanism of action of adjuvants. At the site of injection, some adjuvants form a depot of vaccine and sustain the release of antigen. Adjuvants activate the secretion of cytokines as chemokine to attract immune cells to the site of infection. APCs express MHC and co-stimulatory molecules and process antigens. Mature APCs migrate to lymph nodes and interact with antigen-specific B or T lymphocytic cells. (Modified from Marciani et al. (7) and Awate et al. (9).)

place, time, and amount. Signal 2 facilitators supply a co-stimulation to antigen recognition and thereby enhance the vaccine efficacy (11). Immunomodulatory compounds, binding receptors, and main immune responses they elicit are summarized in Table 7.2 (11,17).

7.6 OPTIMAL CONDITIONS IN THE ADJUVANT FORMULATION

Safety of adjuvant in the vaccine preparation is an utmost important criterion in the adjuvant formulation. It should not be reactogenic to the recipient. The reactions are manifested locally (swelling, redness, pain) or systematically (fatigue, fever). Freund's complete adjuvant, an excellently potent adjuvant, causes inflammation and granulation at the injection site. Another safety issue is the stimulation of unwanted immune reactions due to the immunogenic properties of the adjuvant (23). The ideal immunopotentiators should be delivered to the specific target cells at the correct time. Therefore, it should be formulated in such a manner that optimal molecules will be delivered and target the appropriate cells to induce only the desired immune responses.

When a vaccine is formulated with an adjuvant, the nature and reactivity of the adjuvant surface are important factors. The physical properties of adjuvants determine their interaction with antigens. For example, electrostatic forces, ligand exchange, hydrophobic interactions, hydrogen bonds, and van der Waals forces are needed to absorb antigen to aluminum adjuvant force (24). Aluminum hydroxide surface has an electrically positively charge while aluminum phosphate has a negative charge and therefore at physiological pH, aluminum hydroxide will absorb antigens having acidic isoelectric points and aluminum phosphate will attract the antigens with basic isoelectric points (24).

Excipients such as sodium chloride, buffers, and stabilizers in the medium can also affect absorption. Sodium chloride in the preparation will decrease the adsorption because the electrical charges are covered by the salt (25).

The mixture of adjuvants with different mechanisms of action can be included in a single vaccine formulation to facilitate a robust immune response and increase the potency of the vaccine (26). Adjuvant system 01 (AS01) consists of monophosphoryl lipid A (MPL), saponin, and liposome, which improves the antigen presentation, the CD8 T-cell responses, and antibody titer. Adjuvant system 02 (AS02) consists of MPL, saponin, and emulsion, which induces Th1-type responses and high antibody responses (18).

7.7 CHARACTERISTICS OF AN IDEAL ADJUVANT

Ideal adjuvant candidates should have a long and stable shelf-life. It should be nonexpensive to produce. The raw materials need to be cheap and readily available in mass production. Furthermore, an ideal adjuvant has a simple synthetic pathway with repeatability and reproducibility. It should be biodegradable without leaving long-term residues. The adjuvant should be safe for the animal and should not be associated with long-term adverse effects. It should not develop serious local reactions at the site of administration. An ideal adjuvant should not produce immune responses against itself. It will only promote the desired immune response of the antigen. Moreover, a successful adjuvant should be compatible with different kinds of antigens and be absent from any adverse impact on the antigen. In addition, an ideal adjuvant should be capable of delivering together with antigens and other formulation components (27).

7.8 ADJUVANTED VACCINE FOR FISH

One of the popular adjuvants in the laboratory is Freund's adjuvant. It is a traditional oil adjuvant that generates a water-in-oil emulsion with the antigen with the help of an emulsifying agent (28). Freund's adjuvant in its complete form (FCA) comprises killed, dried mycobacteria, mainly *Mycobacterium tuberculosis*, which is not present in the incomplete form (FIA). Freund's adjuvant is one of the most powerful adjuvants, inducing significant immune responses at both the humoral and cellular levels that have yet to be matched by any other adjuvant (29).

Another often used adjuvant in experimental studies is Montanide, a well-known brand of vaccine adjuvant that has already been employed in a range of terrestrial farm animal models (30). Montanide IMS 1312 VG, for example, consists of water-dispersed liquid nanoparticles as a vehicle stimulating vaccine delivery and is suitable for mass vaccination via the immersion route (30, 31). Montanide ISA 760VG, on the other hand, is the traditional water in polymer emulsion adjuvant (32) and consists of mineral oil with a refined and nonionic surfactant from an ester of the sugar mannitol and purified oleic acid (33). It is a metabolizable oil adjuvant that has been used in both mammalian and fish vaccines (34). The adjuvant was reported to induce more balanced cell-mediated and humoral immune responses in higher vertebrates (35) and to be quite effective in stimulating a protective immune response in different fish species (36).

Li et al. (2020) reported using Freund's incomplete adjuvant incorporated with *V. anguillarum* serotype O1 and O2 bivalent inactivated vaccines against *V. anguillarum* in turbot (*Scophthalmus maximus*) (37). Their study found that the fish vaccinated with the inactivated bivalent vaccine together with Freund's incomplete adjuvant gave a high relative percentage survival (97.56%) after challenge with pathogenic wild-type *V. anguillarum.* Another study by Xu et al. (2019) showed that the formalin-killed cells of *V. harveyi* with commercial adjuvant Montanide™ ISA 763 A VG gave a high RPS of more than 75% after intramuscularly challenged with 5.0×10^6 CFU/fish of *V. harveyi* 4 weeks post-vaccination (38). Moreover, the vaccinated fish had an enhanced antibody titer, lysozyme activity, total serum protein, and antibacterial properties in their serum. Table 7.3 shows some of the studies on the adjuvanted vaccine tested using different kinds of antigens.

TABLE 7.3
Vaccination Trial Study for Vaccine Efficacy Post-Challenge

Study	Vaccine Composition	Efficacy (Relative Percentage Survival Post-Challenge)	Reference
1	*Yersinia ruckeri* bacterin+Montanide™ IMS 1312 VG	100	(30)
2	*Edwardsiella tarda* TX1 bacterin+aluminum hydroxide	53	(29).
	Edwardsiella tarda TX1 bacterin+aluminum phosphate	69	
	Edwardsiella tarda TX1 bacterin+Freund's incomplete adjuvant	81	
3	*Streptococcus agalactiae* strain (serotype Ib) bacterin+Freund's incomplete adjuvant	77.8	(28)
	Streptococcus agalactiae strain (serotype Ib) bacterin+aluminum hydroxide	59.3	
4	*Nocardia seriolae* 024013 strain+bacterin Montanide™ ISA763	26.	(39)
	Nocardia seriolae 024013 strain+bacterin Montanide™ ISA763+glycolipids	52	
5	Inactivated viral hemorrhagic septicemia (VHS) virus strain FP-VHS2010–1+Montanide™ IMS 1312 VG	89	(31)
6	*Streptococcus agalactiae* bacterin+palm oil	70	(40)
	Streptococcus agalactiae bacterin+Freund's incomplete adjuvant	45	
7	*Mycobacterium marinum* bacterin+Montanide™ ISA 760 VG	91	(41)
8	*Aphanomyces invadans* strain INM20101 bacterin+adjuvant Montanide™ ISA 763 AVG	66.7	(36)
9	The recombinant protein (CP-S5E) from Nervous necrosis virus (NNV)+Montanide™ ISA 763 AVG	72.2	(42)

7.9 CHALLENGES IN FISH VACCINE ADJUVANT DEVELOPMENT

The development of effective adjuvanted vaccines has been hampered by a lack of efficient adjuvants and fundamental knowledge of the immune response (43). The cost of a commercial adjuvant may also be prohibitive in some countries leading to a higher vaccine price, thus making the farmers use therapeutic measures such as using antibiotics rather than vaccination strategy (44). These issues must be addressed as the aquaculture industry continues to grow globally.

Besides, commercially available killed vaccines are usually developed from mineral oil as an adjuvant, and the mineral oil administered intraperitoneally can cause internal adhesions and granulomatous lesions at the site of injection, skeletal deformation, growth inhibition, and even increased mortality (45, 46). It may also have an adverse effect, acting as a carcinogen on consumers (44). Sun et al. (2020) reported that adjuvant residue was observed in the peritoneal cavity of vaccinated turbot until 16 weeks post-vaccination when fish were vaccinated with a commercial adjuvant (47). Therefore, many researchers have been searching for other types of adjuvants that can be used as an alternative to mineral oil and other harmful adjuvants (40).

7.10 CONCLUSION

Research on fish vaccines has increased throughout the centuries. Many researchers have developed the use of adjuvants and immunostimulants in fish vaccines, together with delivery methods. Current vaccine applications for large-scale fish farming operations include inactivated, live-attenuated, and DNA vaccines with adjuvants. More research is currently focusing on alternative methods other than using harmful emulsion oil to be formulated with the vaccine. A new generation of adjuvants is employed to give better efficacies. The underlying mechanisms involved in the adjuvant effects of modern adjuvants need to be further explored. As an effective vaccine requires both a suitable antigen and a suitable adjuvant, an ideal adjuvant with a long and stable shelf-life, nonexpensive to produce, safe to the animal with no long-term adverse effects, and compatible with different kinds of antigens should be found to produce a highly effective vaccine. With various significant diseases in aquaculture, finding the best adjuvant to be used with a proper antigen is essential with the expanding demand for aquaculture products that promote the intensive culture strategy.

DISCLOSURE STATEMENT

The authors declare that they have no conflicts of interest.

REFERENCES

[1] Di Pasquale, A., Preiss, S., Da Silva, F.T., and Garçon, N. (2015) Vaccine adjuvants: From 1920 to 2015 and beyond. *Vaccines. 3*:320–343. ISSN 2076-393X. https://doi.org/10.3390/vaccines3020320

[2] Guy, B. (2007) The perfect mix: Recent progress in adjuvant research. *Nature Reviews Microbiology. 5*:505–517. ISSN 1740-1526. https://doi.org/10.1038/nrmicro1681

[3] Burakova, Y., Madera, R., McVey, S., Schlup, J.R., and Shi, J. (2018) Adjuvants for animal vaccines. *Viral Immunology. 31*(1):11–22. ISSN 0882-8245. https://doi.org/10.1089/vim.2017.0049

[4] Relyveld, E.H. (2011) Chapter 8 - A history of toxoid. In S.A. Plotkin, editor, *History of vaccine development*, pp 57–64.el. Springer, New York, NY. e-ISBN 978-1-4419-1339-5. https://doi.org/10.1007/978-1-4419-1339-5_8

[5] Pulendran, B., Arunachalam, P.S., and O'Hagan, D.T. (2021) Emerging concepts in the science of vaccine adjuvants. *Nature Reviews Drug Discovery. 20*:454–475. ISSN 1474-1776. https://doi.org/10.1038/s41573-021-00163-y

[6] de Souza Apostólico, J., Santos Lunardelli, V.A., Coirada, F.C., Boscardin, S.B., and Rosa, D.S. (2016) Adjuvants: Classification, Modus Operandi, and licensing. *Journal of Immunology Research. 2016*:1–16. ISSN 2314-8861. http://doi.org/10.1155/2016/1459394

[7] Marciani, D.J. (2003) Vaccine adjuvants: Role and mechanisms of action in vaccine immunogenicity. *Drug Discovery Today. 8*(20):934–943. ISSN 1359-6446. https://doi.org/10.1016/s1359-6446(03)02864-2

[8] R. Gudding, A. Lillehaug, and Ø. Evensen. (1999) Recent developments in fish vaccinology. *Veterinary Immunology and Immunopathology. 72*:203–212. ISSN 0165-2427. https://doi.org/10.1016/S0165-2427(99)00133-6

[9] Awate, S., Babiuk, L.A., and Mutwiri, G. (2013) Mechanisms of action of adjuvants. *Frontiers in Immunology. 4*:114. ISSN 1664-3224. https://doi.org/10.3389/fimmu.2013.00114

[10] De Gregorio, E., Caproni, E., and Ulmer, J.B. (2013) Vaccine adjuvants: Mode of action. *Frontiers in Immunology. 4*(214):1–6. ISSN 1664-3224. https://doi.org/10.3389/fimmu.2013.00214

[11] Tafalla, C., Bøgwald, J., and Dalmo, R.A. (2013) Adjuvants and immunostimulants in fish vaccines: Current knowledge and future perspectives. *Fish & Shellfish Immunology. 35*:1740–1750. ISSN 1095-9947. http://doi.org/10.1016/j.fsi.2013.02.029

[12] HogenEsch, H. (2013) Mechanism of immunopotentiation and safety of aluminium adjuvants. *Frontiers in Immunology. 3*(406):1–13. ISSN 1664-3224. https://doi.org/10.3389/fimmu.2012.00406

[13] Zinkernagel, R.M., Ehl, S., Aichele, P., Oehen, S., Kundig, T., and Hengartner, H. (2006) Antigen localisation regulates immune responses in a dose and time-dependent fashion: A geographical view of immune reactivity. *Immunological Reviews. 156*:199–209. ISSN 1600-065X. https://doi.org/10.1111/j.1600-065x.1997.tb00969.x

[14] Singh, M. and O'Hagan, D.T. (2003) Recent advances in veterinary vaccine adjuvants. *International Journal for Parasitology. 33*:469–478. ISSN 0020-7519. https://doi.org/10.1016/S0020-7519(03)00053-5

[15] Sun, X., Jin, P., Liu, Q., Wang, Q., Zhang, Y., and Liu, X. (2020) A CpG-riched plasmid as vaccine adjuvant reduce antigen dose of an inactivated *Vibrio anguillarum* vaccine in turbot (*Scophthalmus maximus* L.). *Fish & Shellfish Immunology. 98*:312–317. ISSN 1095-9947. https://doi.org/10.1016/j.fsi.2020.01.031

[16] Coffman, R.L., Sher, A., and Seder, R.A. (2010) Vaccine adjuvants: Putting innate immunity to work. *Immunity. 33*(4):492–503. ISSN 1074-7613. https://doi.org/10.1016/j.immuni.2010.10.002

[17] Reed, S.G., Orr, M.T., and Fox, C.B. (2013) Key roles of adjuvants in modern vaccines. *Nature Medicine. 19*(12):1597–1608. ISSN 1078-8956. https://doi.org/10.1038/nm.3409

[18] Lee, S. and Nguyen, M.T. (2015) Recent advances of vaccine adjuvants for infectious diseases. *Immune Network. 15*(2):51–57. ISSN 1598-2629. http://doi.org/10.4110/in.2015.15.2.51. pISSN 1598-2629 eISSN 2092-6685

[19] Galli, G., Hancock, K., Hoschler, K., et al. (2009) Fast rise of broadly cross-reactive antibodies after boosting long-lived human memory B cell primed by an MF59 adjuvanted prepandemic vaccine. *Frontiers in Immunology. 106*:7962–7967. ISSN 1091-6490. https://doi.org/10.1073/pnas.0903181106

[20] Akira, S., Takeda, K., and Kaisho, T. (2001) Toll-like receptors: Critical proteins linking innate and acquired immunity. *Nature Immunology. 2*: 675–680. ISSN 1529-2908. https://doi.org/10.1038/90609

[21] Parkin, J. and Cohen, B. (2001) An overview of the immune system. *Immunology. 357*(9270):1777–1789. ISSN 0140-6736. https://doi.org/10.1016/S0140-6736(00)04904-7

[22] Wang, W. and Singh, M. (2011) Selection of adjuvants for enhanced vaccine potency. *World Journal of Vaccines. 1*:33–78. ISSN 2160-5815. http://doi.org/10.4236/wjv.2011.12007

[23] Guerrero Manriquez, G.G. and Tuero, I. (2021) Adjuvants: Friends in vaccine formulations against infectious diseases. *Human Vaccines & Immunotherapeutics. 17*(10):3539–3550. ISSN 2164-5515. http://doi.org/10.1080/21645515.2021.1934354

[24] Morefield, G.L. (2011) A rational systematic approach for the development of vaccine formulations. *Journal of the American Association of Pharmaceutical Scientists. 13*:2. ISSN 1550-7416. http://doi.org/10.1208/s12248-011-9261-1

[25] HogenEsch, H., O'Hagan, D.T., and Fox, C.B. (2018) Optimizing the utilization of aluminum adjuvants in vaccines: You might just get what you want. *NPJ Vaccines 3*, 51. ISSN 2059-0105. https://doi.org/10.1038/s41541-018-0089-x

[26] Petrovsky, N. and Aguilar, J.C. (2004) Vaccine adjuvants: Current state and future trends. *Immunology and Cell Biology. 82*:488–496. ISSN 0818-9641. http://doi.org/10.1111/j.1440-1711.2004.01272.x

[27] O'Hagan, D.T. and De Gregorio, E. (2009) The path to a successful vaccine adjuvant - 'The long and winding road'. *Drug Discovery Today*. ISSN 1359-6446. http://doi.org/10.1111/j.0818-9641.2004.01272.x

[28] Ramos-Espinoza, F.C., Cueva-quiroz, V.A., Yunis-aguinaga, J., and Alvarez-Rubio, N.C., Paganoti de Mello, N., Engrácia de Moraes, J.R., Paganoti, N., Mello, D., Paganoti de Mello, N., Engrácia de Moraes, J.R., 2020. Efficacy of two adjuvants administrated with a novel hydrogen peroxide-inactivated vaccine against *Streptococcus agalactiae* in Nile tilapia fingerlings. *Fish & Shellfish Immunology. 105*, 350–358. https://doi.org/10.1016/j.fsi.2020.07.051

[29] Jiao, X.-d., Cheng, S., Hu, Y.-h, and Sun, L., 2010. Comparative study of the effects of aluminum adjuvants and Freund's incomplete adjuvant on the immune response to an *Edwardsiella tarda* major antigen. *Vaccine. 28*, 1832–1837. https://doi.org/10.1016/j.vaccine.2009.11.083

[30] Soltani, M., Shafiei, S., Yosefi, P., Mosavi, S., and Mokhtari, A., 2014. Effect of Montanide™ IMS 1312 VG adjuvant on efficacy of *Yersinia ruckeri* vaccine in rainbow trout (*Oncorhynchus mykiss*). *Fish & Shellfish Immunology. 37*, 60–65. https://doi.org/10.1016/j.fsi.2013.12.027

[31] Hwang, J.Y., Kwon, M.G., Kim, Y.J., Jung, S.H., Park, M.A., and Son, M.H., 2017. Montanide IMS 1312 VG adjuvant enhances the efficacy of immersion vaccine of inactivated viral hemorrhagic septicemia virus (VHSV) in olive flounder, *Paralichthys olivaceus*. *Fish & Shellfish Immunology. 60*, 420–425. https://doi.org/10.1016/j.fsi.2016.12.011

[32] Hoare, R., Jung, S.-J.J., Ngo, T.P.H., Bartie, K., Bailey, J., Thompson, K.D., and Adams, A., 2019. Efficacy and safety of a non-mineral oil adjuvanted injectable vaccine for the protection of Atlantic salmon (*Salmo salar* L.) against *Flavobacterium psychrophilum*. *Fish & Shellfish Immunology. 85*, 44–51. https://doi.org/10.1016/j.fsi.2017.10.005

[33] Zhang, Z., Yu, A., Lan, J., Zhang, H., Hu, M., Cheng, J., Zhao, L., Lin, L., and Wei, S., 2017. GapA, a potential vaccine candidate antigen against *Streptococcus agalactiae* in Nile tilapia (*Oreochromis niloticus*). *Fish & Shellfish Immunology. 63*, 255–260. https://doi.org/10.1016/j.fsi.2017.02.019

[34] Bøgwald, J. and Dalmo, R., 2012. Developments in adjuvants for fish vaccines, in: *Infectious disease in aquaculture*, pp. 244–274. Woodhead Publishing Series in Food Science, Technology and Nutrition. https://doi.org/10.1533/9780857095732.2.244

[35] Veenstra, K.A., Wang, T., Alnabulsi, A., Douglas, A., Russell, K.S., Tubbs, L., Ben, A.J., and Secombes, C.J., 2017. Analysis of adipose tissue immune gene expression after vaccination of rainbow trout with adjuvanted bacterins reveals an association with side effects. *Molecular Immunology. 88*, 89–98. https://doi.org/10.1016/j.molimm.2017.05.026

[36] Pradhan, P.K., Sood, N., Yadav, M.K., Arya, P., Chaudhary, D.K., Kumar, U., Kumar, C.B., Swaminathan, T.R., and Rathore, G., 2018. Effect of immunization of rohu *Labeo rohita* with inactivated germinated zoospores in providing protection against *Aphanomyces invadans*. *Fish & Shellfish Immunology*. *78*, 195–201. https://doi.org/10.1016/j.fsi.2018.04.041

[37] Li, J., Tang, L., Li, S., Li, G., and Mo, Z., 2020. The efficacy and side-effects of oil-based adjuvants emulsified *Vibrio anguillarum* bivalent inactivated vaccine in turbot (*Scophthalmus maximus*) under production mode. *Aquaculture*. *524*, 735259. https://doi.org/10.1016/j.aquaculture.2020.735259

[38] Xu, W., Jiao, C., Bao, P., Liu, Q., Wang, P., Zhang, R., Liu, X., and Zhang, Y., 2019. Efficacy of Montanide™ ISA 763 A VG as aquatic adjuvant administrated with an inactivated *Vibrio harveyi* vaccine in turbot (*Scophthalmus maximus* L.). *Fish & Shellfish Immunology*. *84*, 56–61. https://doi.org/10.1016/j.fsi.2018.09.024

[39] Matsumoto, M., Araki, K., Nishimura, S., Kuriyama, H., Nakanishi, T., Shiozaki, K., Takeuchi, Y., and Yamamoto, A., 2018. Adjuvant effect in aquaculture fish of cell-wall glycolipids isolated from acid-fast bacteria. *Developmental and Comparative Immunology*. *85*, 142–149. https://doi.org/10.1016/j.dci.2018.04.012

[40] Aminudin, A., Kamal, F.M., Zamri-saad, M., Abdullah, S., Ridzuan, M.S., Yusoff, H.M., Hashim, S., Sudirwan, F., Salihin, I., and Sulaiman, S., 2018. Effect of incorporating different concentrations of palm oil as adjuvant in fish vaccine. *International Journal of Biosciences*. *12*, 35–41. https://doi.org/10.12692/ijb/12.1.35-41

[41] Ravid-Peretz, S., Colorni, A., Sharon, G., and Ucko, M., 2019. Vaccination of European sea bass Dicentrarchus labrax with avirulent *Mycobacterium marinum* (iipA::kan mutant). *Fish & Shellfish Immunology*. *90*, 317–327. https://doi.org/10.1016/j.fsi.2019.04.057

[42] Lin, C.F., Jiang, H.K., Chen, N.C., Wang, T.Y., and Chen, T.Y., 2018. Novel subunit vaccine with linear array epitope protect giant grouper against nervous necrosis virus infection. *Fish & Shellfish Immunology*. *74*, 551–558. https://doi.org/10.1016/j.fsi.2018.01.029

[43] Adams, A., 2019. Progress, challenges and opportunities in fish vaccine development. *Fish & Shellfish Immunology*. *90*, 210–214. https://doi.org/10.1016/j.fsi.2019.04.066

[44] Roslindawani, M.N., Syafiqah, A.S., Jesse, F.F.A., Effendy, A.W., and Zamri-Saad, M., 2016. Recombinant caseous lymphadenitis vaccine with palm oil as adjuvant enhances the humoral and cell-mediated immune responses in rat model. *Journal of Animal Health Production*. *4*, 22–25. https://doi.org/10.14737/journal.jahp/2016/4.1.22.25

[45] Noia, M., Domínguez, B., Leiro, J., Blanco-Méndez, J., Luzardo-Álvarez, A., and Lamas, J., 2014. Inflammatory responses and side effects generated by several adjuvant-containing vaccines in turbot. *Fish & Shellfish Immunology*. *38*, 244–254. https://doi.org/10.1016/j.fsi.2014.03.020

[46] Rivas-Aravena, A., Fuentes, Y., Cartagena, J., Brito, T., Poggio, V., La Torre, J., Mendoza, H., Gonzalez-Nilo, F., Sandino, A.M., and Spencer, E., 2015. Development of a nanoparticle-based oral vaccine for Atlantic salmon against ISAV using an alphavirus replicon as adjuvant. *Fish & Shellfish Immunology*. *45*, 157–166. https://doi.org/10.1016/j.fsi.2015.03.033

[47] Sun, X., Jin, P., Liu, Q., Wang, Q., Zhang, Y., and Liu, X., 2020. A CpG-riched plasmid as vaccine adjuvant reduce antigen dose of an inactivated *Vibrio anguillarum* vaccine in turbot (*Scophthalmus maximus* L.). *Fish & Shellfish Immunology*. *98*, 312–317. https://doi.org/10.1016/j.fsi.2020.01.031

8 Development of Mucosal Adjuvants for Fish Vaccination

Ina Salwany Md Yasin, Aslah Mohamad, Mohammad Noor Amal Azmai, Annas Salleh, and Mohd Zamri Saad
Universiti Putra Malaysia

CONTENTS

8.1 INTRODUCTION

Vaccination has proven to improve the fish's immune response, and it is frequently employed in aquaculture to protect the fish from diseases (1,2). However, few vaccination cases have demonstrated adequate levels of protection (3). A powerful adjuvant is another crucial component for a successful vaccine, in addition to a highly specific and immunogenic antigen (4,5). Adjuvants are a group of structurally heterogeneous compounds capable of slowing down and modulating an antigen's intrinsic immunogenicity (6,7). The word "adjuvants" came from the Latin word *adjuvare*, which means "to help" (6). Adjuvants have traditionally been known as helper substances added to a vaccine to stimulate and enhance the immune response's magnitude and durability, thereby improving defense against the intended disease (8). In some fish vaccines, adjuvants provide vital substances or compounds for the vaccine efficacy (9). Even so, several of the adjuvants used in aquaculture can have negative consequences, including myelin buildup, tissue adhesion, granulomatous peritonitis, growth suppression, increased mortality, skeletal deformity, and even autoimmunity in the host (3), while some may cause an adverse effect and act as a carcinogen on consumers (10). Thus, this chapter discusses the necessity of adding adjuvants in vaccination, particularly on vaccines used in the aquaculture industry.

8.2 ROLES OF ADJUVANT IN VACCINATION

Adjuvants primarily work by stimulating the aggregation and activation of antigen-presenting cells (APC) at the site of antigen exposure to boost T- and B-cell activation (11). As shown in Figure 8.1, adjuvants such as emulsions provide a delivery system and immunomodulatory properties that enhance the antigen uptake. Adjuvants primarily work by stimulating the aggregation and activation of APC at the site of antigen exposure to boost T- and B-cell activation (2). The role of adjuvants in fish is crucial for the advancement in immunology and vaccinology. Adjuvants like aluminum hydroxide and Freund's adjuvant increased the quantity of immunoglobulin significantly (12).

DOI: 10.1201/9781003388548-11

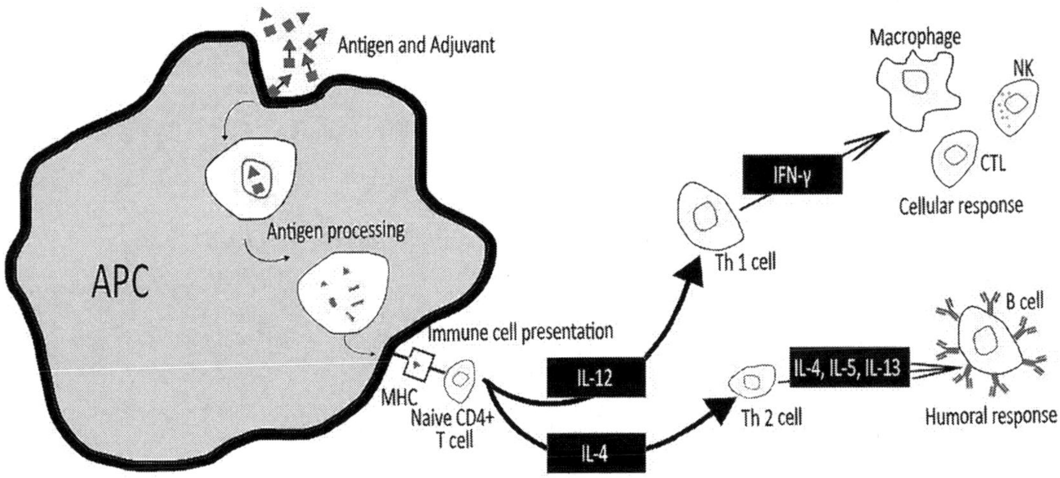

FIGURE 8.1 Putative mechanisms of action of adjuvant via influencing antigen presentation by MHC.

These substances can stimulate the immune response so that antigens become very effective by magnifying the immune pathways, and also will be helpful in increasing the actions of very small quantities or doses of antigen to increase the immune responses to levels that can protect diseases in fish culture (9). Currently, commercially available inactivated vaccines for fish are usually used with mineral oil adjuvant (e.g., Montanide™ ISA 763 A VG adjuvant). Oil-based adjuvants work by either recruiting effector cells to the injection site or building a depot of antigen that increases antigen delivery to APC (13).

8.3 CLASSIFICATION OF ADJUVANT

A brief overview of adjuvant classifications is found in Table 8.1. Adjuvants are categorized as Signal 1 and 2 facilitators according to how they affect the immune system's adaptive immunological response (1,7,14,15). Oily emulsions are most frequently employed in Signal 1 facilitators (16). These adjuvants, which have been demonstrated to increase vaccine effectiveness when used in fish vaccines, include commonly used commercial adjuvants such as Freund's complete adjuvant (FCA), Freund's incomplete adjuvant (FIA), and Montanide (1). Contrarily, during the antigen recognition stage, Signal 2 facilitators activate cytokines and control costimulatory molecules, enhancing the immune response during vaccination (15,16). Aluminum compounds like aluminum hydroxide and aluminum phosphate are some of the Signal 2 facilitators' adjuvants (16,17). This two-signal paradigm states that in order for certain T- and B-cells, which make up the adaptive immune system, to become activated, both the presence of an antigen (Signal 1) and the additional secondary signals (Signal 2) are necessary (1,7,15). While Signal 2 facilitators give costimulatory signals during antigen recognition, Signal 1 facilitators control the course of the vaccination antigen in terms of timing, place, and concentration, ultimately enhancing its immunological availability (14,18).

8.4 THE USE OF ADJUVANTS IN FISH VACCINATION

Several commercial adjuvants are commonly used and mixed with different fish vaccines. In the laboratory, Freund's adjuvant, which keeps high antibody levels following vaccination, is the most often used commercial adjuvant (5). It is one of the most potent adjuvants and has generally not been topped by any other adjuvant in terms of effectiveness in inducing robust immune responses at the humoral and cellular levels (17). Freund's adjuvant in its complete form (FCA)

TABLE 8.1
Class of Adjuvants Based on the Induced Immunological Events

Class of Adjuvants	Description	Types of Adjuvants	Example	Reference
Signal 1	Influences the fate of the vaccine antigen in time, place, and concentration, ultimately improving its immune availability	Oil emulsions	• Freund's complete adjuvant • Freund's incomplete adjuvant • Montanide • AJ-oil (Alpha Ject 5200)	(19–21)
		Nano-/microparticles	• Poly lactic-co-glycolic acid (PLGA) particles • Immunostimulating complex (ISCOMs)	(9,22)
Signal 2	Provides the co-stimulation signals during antigen recognition that will provide an adequate environment for an adequate antigen-specific immune response	Aluminum containing adjuvants	• Matrix-Q (Isco- nova, Sweden) • Imject™ aluminum hydroxide (Thermo Sci, USA) • Aluminum hydroxide gel (Alhydrogel, Sigma, USA)	(23,24)
		β-Glucans	• Yeast glucan • Vitastim-Taito • Lentinan • Laminaran • Chitosan • Raffinose	(7,25)
		Saponins	• Quil-A • Ginseng Saponins • Quillaja Saponaria Saponin (QSS)	(2,7,9,26)
		Polyinosinic polycytidylic acid (Poly I:C)	• Poly-IC12U (Ampligen) • Poly-ICLC (Hiltonol)	(11)
		Lipopeptides	• Polar glycopeptidolipids (pGPL-Mc) • LP1–30 • LP1–34 • LP2-2 • LP2–3	(27,28)
		Flagellin	• FlgD • FliC • FlaA • FlaB • FlaE	(27,29,30)
		CpG oligodeoxynucleotide (ODN)	• CpG-A ODN • CpG-B ODN	(31,32)
		Cytokines	• Inter- leukin (IL)-2 • IL-6 • IL-12 • IL-1β • Interferon (IFN)-γ • Granulocyte–macrophage colony-stimulating factor (GM-CSF)	(22)

comprises killed, dried mycobacteria, mainly *Mycobacterium tuberculosis*, which is not present in the incomplete form (FIA) (17). Li et al. (2020) reported using FIA incorporated with *Vibrio anguillarum* serotype O1 and O2 bivalent inactivated vaccines against *Vibrio anguillarum* in turbot (*Scophthalmus maximus*) (19). Their study found that the fish vaccinated with the inactivated bivalent vaccine and FIA gave a high relative percentage survival (97.56%) after the challenge with pathogenic wild-type *V. anguillarum*.

Another study by Xu et al. (2019) showed that the formalin-killed cells of *V. harveyi* with commercial adjuvant Montanide™ ISA 763 A VG gave a high relative percentage survival (RPS) of more than 75% after intramuscularly challenged with 5.0×10^6 CFU/fish of *V. harveyi* at 4 weeks post-vaccination (33). Moreover, the vaccinated fish had an enhanced antibody titer, lysozyme activity, total serum protein, and antibacterial property in their serum. An inactivated FCA added vaccine was tested for the first time successfully against *Aeromonas hydrophila* for tilapia in 1986 (34), while in 2008, an FIA incorporated vaccine was produced against *A. hydrophila* for tilapia, and it protected tilapia against the homologous virulent isolates (35). Moreover, Jiang et al. (2015) showed that subunit vaccine incorporated with FIA adjuvant provided 85% of RPS against *Streptococcus iniae* infection in channel catfish (*Ictalurus punctatus*) (36).

Another often used adjuvant in experimental studies is Montanide, a well-known vaccine adjuvant that has already been employed in a range of terrestrial farm animal models (37). It is an ISA adjuvant, a type of recently developed water-in-oil emulsion (5). It is a traditional oil adjuvant that generates a water-in-oil emulsion with the antigen with the help of an emulsifying agent (16). A variety of Montanide adjuvant, Montanide IMS 1312 VG, for example, consists of water-dispersed liquid nanoparticles as a vehicle stimulating vaccine delivery and is suitable for mass vaccination via the immersion route (37,38). On the other hand, another variety of Montanide adjuvant, Montanide ISA 760VG, is the traditional water in polymer emulsion adjuvant (39) and consists of mineral oil with a refined and nonionic surfactant from an ester of the sugar mannitol and purified oleic acid (5). It is a metabolizable oil adjuvant used in mammalian and fish vaccines (27). Wang et al. (2014) developed a recombinant vaccine with non mineral oil adjuvant Montanide ISA 763, showing high protection in tilapia against *S. agalactiae* (40). The adjuvant was reported to induce balanced cell-mediated and humoral immune responses in higher vertebrates (20) and be quite effective in stimulating a protective immune response in different fish species (41). Table 8.2 shows some examples of the different types of adjuvants, their application, and their efficacy post-vaccination in different types of fish.

8.5 FISH ADJUVANT

Commercial fish vaccines are usually in the form of inactivated antigens. Although typically providing only moderate immunoprotection, the killed vaccines are excellent choices for achieving the best protection while retaining safety and high applicability in aquaculture. Consequently, certain adjuvants, minerals, or nonmineral oil are typically used in the formulation of commercial inactivated vaccines (50). They do, however, have a number of harmful side effects. For instance, granuloma at the injection site might be brought on by FCA (6), and incomplete Freund's adjuvant (IFA) can cause peritonitis in the Atlantic cod. Besides, adding FCA to the inactivated vaccine did not significantly improve the protective effect against *Pasteurella piscicida* infection in yellowtail (3,51). In addition, when fish were vaccinated with a commercial adjuvant, the adjuvant residue was observed in the peritoneal cavity of vaccinated turbot until 16 weeks post-vaccination (50).

On the other hand, some vaccines were prepared without adding any adjuvants but still work well and have high efficacy. For example, when *Escherichia coli*-derived NNV VLPs were employed as the vaccine antigen in the grouper, *Epinephelus lanceolatus*, a study found that an adjuvant was not

TABLE 8.2
Some Examples of the Different Types of Adjuvants, Their Application, and Efficacy Post-Vaccination in Different Types of Fish

Adjuvant	Antigen	Vaccine Type-Delivery	Efficacy	Reference
Freund's incomplete adjuvant (FIA)	*Yersinia ruckeri* serotype O1 biotype 1 and 2, *Vibrio anguillarum* serotype O1 and O2a and *Aeromonas salmonicida* subsp. *salmonicida*.	Inactivated-Injection	96.8%, 100%, 100%, 96.97%, and 72.7% relative percentage survival (RPS) in vaccinated rainbow trout (*Oncorhynchus mykiss*) post-challenge with *Y. ruckeri* serotype O1 biotype 1 and 2 and *V. anguillarum* serotype O1 and O2a	(42)
	A. salmonicida subsp. *salmonicida* (040617-1/1A)	Inactivated-Injection	87% RPS in vaccinated rainbow trout (*Oncorhynchus mykiss*) post-challenge with *A. salmonicida* (040617-1/1A)	(43)
	A. hydrophila B11, *Edwardsiella anguillarum* B79 and *V. vulnificus* B88	Recombinant-Injection	83.33%, 55.56%, and 55.56% RPS in vaccinated European eel (*Angullia angullia*) post-challenge with *A. hydrophila* B11, *E. anguillarum* B79, and *V. vulnificus* B88, respectively	(44)
	Streptococcus agalactiae strain (serotype Ib)	Inactivated-Injection	77.8% RPS in vaccinated Nile tilapia (*Oreochromis niloticus*) post-challenge with a homologous *S. agalactiae* strain	(16)
Freund's complete adjuvant (FCA)	*Pasteurella* sp.	Inactivated-Injection	44.83% RPS in vaccinated lumpsucker (*Cyclopterus lumpus*) post-challenge with *Pasteurella* sp.	(45)
	Vibrio harveyi	Recombinant-Injection	76% RPS in vaccinated humphead snapper (*Lutjanus sanguineus*) post- challenge with *V. harveyi*	(46)
	Neoparamoeba perurans	Recombinant-Injection	75% RPS in vaccinated Atlantic salmon (*Salmo salar*) post-challenge with *N. perurans*	(47)
Montanide	Viral hemorrhagic septicemia virus (VHSV) strain FP-VHS2010–1	Inactivated-Immersion	67% RPS in vaccinated olive flounder, *Paralichthys olivaceus* subjected to VHSV	(38)
	V. harveyi	Inactivated-Injection	83.87% RPS in vaccinated turbot (*Scophthalmus maximus*) groups post-challenge with *V. harveyi*	(33)
	Aphanomyces invadans strain INM20101	Inactivated-Injection	66.7% RPS in immunized groups of *Labeo rohita* following challenge with *Aphanomyces invadans*	(41)
CpG oligodeoxy-nucleotides (ODNs)	*V. harveyi* strain VH4	Inactivated Injection	96.2% RPS in vaccinated orange-spotted grouper (*Epinephelus coioides*) subjected to *V. harveyi* strain AOD99026-2sk	(48)
	V. anguillarum	Subunit-injection	60% RPS in vaccinated tilapia (*O. niloticus* subjected to *V. anguillarum*)	(49)
Aluminum hydroxide	*Flavobacterium psychrophilum*	Inactivated-Injection	85.71% RPS in vaccinated Atlantic salmon (*Salmo salar*) against *F. psychrophilum*	(39)
	Edwardsiella tarda TX1	Recombinant-Injection	81% RPS in vaccinated Japanese flounder (*Paralichthys olivaceus*) subjected to *E. tarda* TX1	(17)

TABLE 8.3
Some Examples of Nonadjuvanted Fish Vaccines

Antigen	Vaccine Type-Delivery	Efficacy	Reference
Flavobacterium psychrophilum	Inactivated- Injection	85.71% RPS in vaccinated Atlantic salmon (*Salmo salar*) post-challenge with *F. psychrophilum*	(39)
Yersinia ruckeri biotype 1 BA19	Inactivated- Injection	94% RPS in vaccinated rainbow trout (*Oncorhynchus mykiss*) post-challenge with *Y. ruckeri* biotype 1	(54)
Yersinia ruckeri O1 biotype 2 (100415-1/4)	Inactivated- Injection	87% RPS in vaccinated rainbow trout (*Oncorhynchus mykiss*) post-challenge with *Y. ruckeri*	(55)
Streptococcus parauberis 2019TSP	Inactivated- Injection	70% RPS in vaccinated turbot (*Scophthalmus maximus*) post-challenge with *S. parauberis* 2019TSP	(56)
Vibrio mimicus strain 04–14	Inactivated- Injection	76.67% RPS in vaccinated grass carps (*Ctenopharyngodon idella*) post-challenge with *V. mimicus* strain 04–14	(57)
Streptococcus parauberis SNUFPC-050803	Inactivated- Injection	80% RPS in vaccinated olive flounder (*Paralichthys olivaceus*) post-challenge with *S. parauberis*	(58)
Pseudomonas fluorescens (ATCC 13525)	Recombinant-immersion	65% RPS in vaccinated Indian major carp (*Cirrhinus mrigala*) post-challenge with *P. fluorescens*	(59)
Nocardia seriolae (Strain TC 258)	Recombinant-injection	75% RPS in vaccinated ginbuna crucian carp (*Carassius auratus langsdorfii*) post-challenge with *Nocardia seriolae*	(60)
V. harveyi (strain ZJ0603)	DNA-Injection	77.5% RPS in vaccinated orange-spotted grouper (*Epinephelus coioides*) challenged with *V. harveyi*	(61)
Lactococcus garvieae	Inactivated- Injection	70.83% RPS in vaccinated rainbow trout (*Oncorhynchus mykiss*) challenged with *L. garvieae*	(62)

required to boost the efficacy of a VLP vaccine and that a VLP dosage of 1 μg/g fish body weight was sufficient to generate a successful immune response (52). Interestingly, in South Korea, 29 commercial vaccines for 10 fish pathogens have been approved after their safety and efficacy were confirmed following the Pharmaceutical Affairs Act. These vaccines include single and combination vaccines containing more than two inactivated pathogens. All the 29 vaccines are manufactured as formalin-inactivated vaccines. However, only one is an adjuvanted vaccine, while another 28 are non adjuvanted vaccines (53). Thus, certain vaccines work well in some cases without adding any adjuvants. Table 8.3 shows some examples of fish vaccines using different antigens without including any adjuvants in their formulations.

8.6 NECESSITY OF THE USE OF ADJUVANT IN VACCINATION

As both vaccines, with adjuvants and without adjuvants, are commercially available in the market and were reported to be highly effective, is adjuvant necessary in fish vaccination? To answer this question, we need to understand several key points. The first point is whether the adjuvant is a crucial component of the vaccine's formulation. The second point is the enhancement of the vaccine's efficacy after adding adjuvants, and the third is the safety of the adjuvants to the host. Thus, answering these questions will provide sufficient information on whether adjuvants are necessary for vaccinations.

A. *Adjuvant requirements in vaccine formulation*
Numerous studies recommended that when conducting immunization trials, a variety of fish vaccine formulation variables be considered. These include the dose of the vaccine (antigen), the duration of the vaccination, the antigen uptake during the immunization of fish, the performance of the adjuvant, the environment, fish size (age), the booster, how the experimental pathogen challenges are conducted, general management procedures related to aquaculture production conditions, and others (2,63–65). Thus, in general, the adjuvant can be a crucial component in vaccine formulations. Therefore, a good vaccine requires both a suitable antigen and adjuvant (2). Besides, some studies reported that vaccines without the presence of adjuvant could be less immunogenic (1,16), not effective (66), induce only a weak immune response, especially in subunit vaccines (67,68), and not usually able to confer protection for a sufficiently long time (69,70).

B. *Effectiveness*
Although there were some studies that reported that the addition of the adjuvant in the fish vaccine does not significantly improve the efficacy of the vaccine (51) and certain adjuvants (e.g., rTNFα) may counteract poor vaccine efficacy (63), yet, there were numerous reported studies that proved that the adjuvants have been indispensable for the improvement of vaccine efficacy (16,22,71,72). Thus, we believe that adding certain adjuvants might significantly improve the effectiveness of an adjuvant, but a proper adjuvant selection is needed to ensure that the adjuvant works well in improving the vaccine's efficacy. Galindo-Villegas et al. (23) listed some of the critical factors for adjuvants' selection, which include (1) proving to be effective and safe, (2) inducing a long-lasting protective immunity, (3) complying with the human food safety regulation, (4) feasible for scale-up production, and (5) cost-effective. Table 8.4 shows some examples of the comparative results between the adjuvanted and nonadjuvanted vaccines in the same experimental design.

C. *Safety issues*
Finding the appropriate adjuvant is one of the major challenges in animal vaccine development (23). Using the right vaccine adjuvants reduces the number of immunizations or the amount of antigen needed for immunization (39). Unfortunately, several studies report side effects after their use in farmed fish, including inflammation, granulomas, pigmentation at the site of infection, and connective tissue in internal organs. Thus, there is a need for new and improved adjuvants that can stimulate the immune system in order to prevent diseases while causing fewer or no side effects (77). One of the alternatives is using cytokine adjuvants that were reported to be safer and more effective than the traditional adjuvants (78). Adjuvants enable dose sparing, which lowers the quantity of vaccine that must be made and administered and may make otherwise unfeasible vaccine candidates functional (11). In addition, substitute adjuvants like CpG patterns and the simultaneous administration of immunostimulants like glucans or heat shock proteins may offer safer options in the future (1).

Therefore, from the points given, in some cases, an antigen can stand alone without adding adjuvants with an acceptable range of vaccine efficacy. Thus, adjuvants are actually not compulsory ingredients in vaccine formulation. However, adding a certain amount of adjuvant in a vaccine might improve the vaccine's efficacy and the antigen delivery to the host. Although some studies have been reported on some traditional adjuvants, which gave a negative side effect to the host, an alternative adjuvant should be considered to prevent the incident while retaining the high improvement of immune response and a better protective efficacy to the host.

TABLE 8.4
Comparative Study between the Adjuvanted and Non Adjuvanted Vaccine

Study	Composition: With Adjuvant	Composition: Without Adjuvant	Efficacy (Relative Percentage Survival Post-Challenge): With Adjuvant	Efficacy (Relative Percentage Survival Post-Challenge): Without Adjuvant	Reference
1	*Yersinia ruckeri* bacterin+Montanide™ IMS 1312 VG	*Yersinia ruckeri* bacterin	100%	82.2%	(37)
2	*Edwardsiella tarda* TX1 bacterin+aluminum hydroxide	*Edwardsiella tarda* TX1 bacterin	53%	34%	(17)
	Edwardsiella tarda TX1 bacterin+aluminum phosphate		69%		
	Edwardsiella tarda TX1 bacterin+Freund's incomplete adjuvant		81%		
3	*Streptococcus agalactiae* strain (serotype Ib) bacterin+Freund's incomplete adjuvant	*Streptococcus agalactiae* strain (serotype Ib) bacterin	77.8%	40.7%	(16)
	Streptococcus agalactiae strain (serotype Ib) bacterin+aluminum hydroxide		59.3%		
4	*Nocardia seriolae* 024013 strain+bacterin Montanide™ ISA763	*Nocardia seriolae* 024013 strain bacterin	26%	26%	(73)
	Nocardia seriolae 024013 strain+bacterin Montanide™ ISA763+glycolipids		52%		
5	Inactivated viral hemorrhagic septicemia (VHS) virus strain FP-VHS2010-1+Montanide™ IMS 1312 VG	Inactivated viral hemorrhagic septicemia (VHS) virus strain FP-VHS2010-1	89%	30%	(38)
6	*Streptococcus agalactiae* bacterin+palm oil	*Streptococcus agalactiae* bacterin	70%	25%	(74)
	Streptococcus agalactiae bacterin+Freund's incomplete adjuvant		45%		
7	*Mycobacterium marinum* bacterin+Montanide™ ISA 760 VG	*Mycobacterium marinum* bacterin	91%	85%	(75)
8	*Aphanomyces invadans* strain INM20101 bacterin+adjuvant Montanide™ ISA 763 AVG	*Aphanomyces invadans* strain INM20101 bacterin	66.7%	0%	(41)
9	The recombinant protein (CP-S5E) from Nervous Necrosis Virus (NNV)+Montanide™ ISA 763 AVG	The recombinant protein (CP-S5E) from Nervous Necrosis Virus (NNV)	72.2%	61.1%	(76)

8.7 CONCLUSION

For an effective and safer design of contemporary vaccination adjuvants, it is necessary to keep improving our understanding of the mechanisms of action of the current adjuvants. Many adjuvants fail during vaccine development for various reasons, including low efficacy, unstable nature (the same adjuvant can boost, inhibit, or have no impact), unacceptable tolerance levels, and safety issues. For an ideal vaccine, the kind and dosage of the immunogen, the stability of the adjuvant formulation, the timing and route of vaccination, and the animal species and strain used in the study are some of the significant drivers of immunogenicity to be the focus. In addition to immunologic enhancement with no undesirable side effects and adequate protection against challenged organisms, selecting an adjuvant for a vaccination trial may depend upon cost and commercial availability. Because the choice of an adjuvant often depends upon expensive trial and error, and due to the continuing concerns about adjuvant safety, future vaccine development should focus more on unique synthetic antigen concepts in the hope of evading the need to administer extraneous chemical or biological adjuvants and to shorten the time of the vaccine development.

DISCLOSURE STATEMENT

The authors declare that they have no conflicts of interest.

REFERENCES

[1] Wang, Q., Ji, W. & Xu, Z. Current use and development of fish vaccines in China. *Fish & Shellfish Immunology. 96*, 223–234 (2020).

[2] Soltani, M., Lymbery, A., Song, S. K. & Hosseini Shekarabi, P. Adjuvant effects of medicinal herbs and probiotics for fish vaccines. *Reviews in Aquaculture. 11*, 1325–1341 (2019).

[3] Rivas-Aravena, A. et al. Development of a nanoparticle-based oral vaccine for Atlantic salmon against ISAV using an alphavirus replicon as adjuvant. *Fish & Shellfish Immunology. 45*, 157–166 (2015).

[4] Nguyen, H. T., Nguyen, T. T. T., Wang, Y. T., Wang, P. C. & Chen, S. C. Effectiveness of formalin-killed vaccines containing CpG oligodeoxynucleotide 1668 adjuvants against Vibrio harveyi in orange-spotted grouper. *Fish & Shellfish Immunology. 68*, 124–131 (2017).

[5] Zhang, Z. et al. GapA, a potential vaccine candidate antigen against *Streptococcus agalactiae* in Nile tilapia (*Oreochromis niloticus*). *Fish & Shellfish Immunology. 63*, 255–260 (2017).

[6] Liu, G. et al. Development of *Streptococcus agalactiae* vaccines for tilapia. *Diseases of Aquatic Organisms. 122*, 163–170 (2016).

[7] Tafalla, C., Bøgwald, J. & Dalmo, R. A. Adjuvants and immunostimulants in fish vaccines: Current knowledge and future perspectives. *Fish & Shellfish Immunology. 35*, 1740–1750 (2013).

[8] Pulendran, B., S. Arunachalam, P. & O'Hagan, D. T. Emerging concepts in the science of vaccine adjuvants. *Nature Reviews. Drug Discovery. 20*, 454–475 (2021).

[9] Tafalla, C., Bøgwald, J., Dalmo, R. A., Munang'andu, H. M. & Evensen, Ø. Adjuvants in fish vaccines. in *Fish vaccination*, vol. 9780470674, 68–84 (John Wiley & Sons, Ltd, 2014).

[10] Wanasawaeng, W., Tawatsin, A., Sasipreeyajan, J., Poomvises, P. & Chansiripornchai, N. Development of inactivated Newcastle disease vaccine using palm oil as an adjuvant. *Thai Journal of Veterinary Medicine. 39*, 9–16 (2009).

[11] Martins, K. A. O., Bavari, S. & Salazar, A. M. Vaccine adjuvant uses of poly-IC and derivatives. *Expert Review of Vaccines. 14*, 447–459 (2015).

[12] Gudding, R. & Van Muiswinkel, W. B. A history of fish vaccination: Science-based disease prevention in aquaculture. *Fish & Shellfish Immunology. 35*, 1683–1688 (2013).

[13] Thim, H. L. et al. Vaccine adjuvants in fish vaccines make a difference: Comparing three adjuvants (Montanide ISA763A oil, CpG/poly I:C combo and VHSV glycoprotein) alone or in combination formulated with an inactivated whole salmonid alphavirus antigen. *Vaccines. 2*, 228–251 (2014).

[14] Guo, M., Tang, X., Sheng, X., Xing, J. & Zhan, W. The effects of IL-1β, IL-8, G-CSF and TNF-α as molecular adjuvant on the immune response to an *E. tarda* subunit vaccine in flounder (*Paralichthys olivaceus*). *Fish & Shellfish Immunology. 77*, 374–384 (2018).

[15] Adams, A. Progress, challenges and opportunities in fish vaccine development. *Fish & Shellfish Immunology. 90*, 210–214 (2019).
[16] Ramos-Espinoza, F. C. et al. Efficacy of two adjuvants administrated with a novel hydrogen peroxide-inactivated vaccine against Streptococcus agalactiae in Nile tilapia fingerlings. *Fish & Shellfish Immunology. 105*, 350–358 (2020).
[17] Jiao, X. dong, Cheng, S., Hu, Y. hua & Sun, L. Comparative study of the effects of aluminum adjuvants and Freund's incomplete adjuvant on the immune response to an *Edwardsiella tarda* major antigen. *Vaccine. 28*, 1832–1837 (2010).
[18] Wangkahart, E., Secombes, C. J. & Wang, T. Studies on the use of flagellin as an immunostimulant and vaccine adjuvant in fish aquaculture. *Frontiers in Immunology. 10*, 1–22 (2019).
[19] Li, J., Tang, L., Li, S., Li, G. & Mo, Z. The efficacy and side-effects of oil-based adjuvants emulsified *Vibrio anguillarum* bivalent inactivated vaccine in turbot (*Scophthalmus maximus*) under production mode. *Aquaculture. 524*, 735259 (2020).
[20] Veenstra, K. A. et al. Analysis of adipose tissue immune gene expression after vaccination of rainbow trout with adjuvanted bacterins reveals an association with side effects. *Molecular Immunology. 88*, 89–98 (2017).
[21] Embregts, C. W. E. & Forlenza, M. Oral vaccination of fish: Lessons from humans and veterinary species. *Developmental and Comparative Immunology. 64*, 118–137 (2016).
[22] Guo, M. & Li, C. An overview of cytokine used as adjuvants in fish: Current state and future trends. *Reviews in Aquaculture. 13*, 996–1014 (2021).
[23] Galindo-Villegas, J., García-Alcazar, A., Meseguer, J. & Mulero, V. Aluminum adjuvant potentiates gilthead seabream immune responses but induces toxicity in splenic melanomacrophage centers. *Fish & Shellfish Immunology. 85*, 31–43 (2019).
[24] Noia, M. et al. Inflammatory responses and side effects generated by several adjuvant-containing vaccines in turbot. *Fish & Shellfish Immunology. 38*, 244–254 (2014).
[25] Petit, J. & Wiegertjes, G. F. Long-lived effects of administering β-glucans: Indications for trained immunity in fish. *Developmental and Comparative Immunology. 64*, 93–102 (2016).
[26] Yuan, L. et al. Sunflower seed oil containing ginseng stem–leaf saponins (E515-D) is a safe adjuvant for Newcastle disease vaccine. *Poultry Science. 99*, 4795–4803 (2020).
[27] Bøgwald, J. & Dalmo, R. Developments in adjuvants for fish vaccines. in *Infectious disease in aquaculture*, 244–274 (Woodhead Publishing Limited, 2012). doi:10.1533/9780857095732.2.244.
[28] Mao, L. et al. Novel synthetic lipopeptides as potential mucosal adjuvants enhanced SARS-CoV-2 rRBD-induced immune response. *Frontiers in Immunology. 13*, 1–16 (2022).
[29] Liu, X. et al. Flagellin enhances the immunoprotection of formalin-inactivated *Edwardsiella tarda* vaccine in turbot. *Vaccine. 35*, 369–374 (2017).
[30] Jia, P. pan et al. Comparative study of four flagellins of *Vibrio anguillarum*: Vaccine potential and adjuvanticity. *Fish & Shellfish Immunology. 34*, 514–520 (2013).
[31] Chen, S. P., Peng, R. H. & Chiou, P. P. Modulatory effect of CpG oligodeoxynucleotide on a DNA vaccine against nervous necrosis virus in orange-spotted grouper (*Epinephelus coioides*). *Fish & Shellfish Immunology. 45*, 919–926 (2015).
[32] Krieg, A. M. CpG motifs in bacterial DNA and their immune effects. *Annual Review of Immunology. 20*, 709–760 (2002).
[33] Xu, W. et al. Efficacy of MontanideTM ISA 763 A VG as aquatic adjuvant administrated with an inactivated Vibrio harveyi vaccine in turbot (*Scophthalmus maximus* L.). *Fish & Shellfish Immunology. 84*, 56–61 (2019).
[34] Caipang, C. M. A., Lucanas, J. B. & Lay-Yag, C. M. Updates on the vaccination against bacterial diseases in tilapia, *Oreochromis* spp. and Asian Seabass, Lates Calcarifer. *AACL Bioflux. 7*, 184–193 (2014).
[35] Shirajum Monir, M., Yusoff, S. M., Mohamad, A. & Ina-Salwany, M. Y. Vaccination of Tilapia against Motile Aeromonas Septicemia: A review. *Journal of Aquatic Animal Health. 32*, 65–76 (2020).
[36] Jiang, Y. et al. Isolation and expression of grass carp toll-like receptor 5a (CiTLR5a) and 5b (CiTLR5b) gene involved in the response to flagellin stimulation and grass carp reovirus infection. *Fish & Shellfish Immunology. 44*, 88–99 (2015).
[37] Soltani, M., Shafiei, S., Yosefi, P., Mosavi, S. & Mokhtari, A. Effect of MontanideTM IMS 1312 VG adjuvant on efficacy of *Yersinia ruckeri* vaccine in rainbow trout (Oncorhynchus mykiss). *Fish & Shellfish Immunology. 37*, 60–65 (2014).
[38] Hwang, J. Y. et al. Montanide IMS 1312 VG adjuvant enhances the efficacy of immersion vaccine of inactivated viral hemorrhagic septicemia virus (VHSV) in olive flounder, Paralichthys olivaceus. *Fish & Shellfish Immunology. 60*, 420–425 (2017).

[39] Hoare, R. et al. Efficacy and safety of a non-mineral oil adjuvanted injectable vaccine for the protection of Atlantic salmon (*Salmo salar* L.) against *Flavobacterium psychrophilum*. *Fish & Shellfish Immunology*. *85*, 44–51 (2019).

[40] Wang, J., Zou, L. L. & Li, A. X. Construction of a *Streptococcus iniae* sortase A mutant and evaluation of its potential as an attenuated modified live vaccine in Nile tilapia (*Oreochromis niloticus*). *Fish & Shellfish Immunology*. *40*, 392–398 (2014).

[41] Pradhan, P. K. et al. Effect of immunization of rohu *Labeo rohita* with inactivated germinated zoospores in providing protection against *Aphanomyces invadans*. *Fish & Shellfish Immunology*. *78*, 195–201 (2018).

[42] Marana, M. H. et al. A pentavalent vaccine for rainbow trout in Danish aquaculture. *Fish & Shellfish Immunology*. *88*, 344–351 (2019).

[43] Rømer Villumsen, K., Koppang, E. O. & Raida, M. K. Adverse and long-term protective effects following oil-adjuvanted vaccination against *Aeromonas salmonicida* in rainbow trout. *Fish & Shellfish Immunology*. *42*, 193–203 (2015).

[44] He, L. et al. Immunization of a novel outer membrane protein from Aeromonas hydrophila simultaneously resisting A. hydrophila and Edwardsiella anguillarum infection in European eels (Angullia angullia). *Fish & Shellfish Immunology*. *97*, 300–312 (2020).

[45] Ellul, R. M. et al. Protection and antibody reactivity in lumpsucker (*Cyclopterus lumpus* L.) following vaccination against *Pasteurella* sp. *Fish & Shellfish Immunology*. *95*, 650–658 (2019).

[46] Huang, P. et al. An IL-6 gene in humphead snapper (*Lutjanus sanguineus*): Identification, expression analysis and its adjuvant effects on Vibrio harveyi OmpW DNA vaccine. *Fish & Shellfish Immunology*. *95*, 546–555 (2019).

[47] Valdenegro-Vega, V. A., Cook, M., Crosbie, P., Bridle, A. R. & Nowak, B. F. Vaccination with recombinant protein (r22C03), a putative attachment factor of *Neoparamoeba perurans*, against AGD in Atlantic salmon (*Salmo salar*) and implications of a co-infection with *Yersinia ruckeri*. *Fish & Shellfish Immunology*. *44*, 592–602 (2015).

[48] Nguyen, H. T. et al. A formalin-inactivated vaccine provides good protection against Vibrio harveyi infection in orange-spotted grouper (*Epinephelus coioides*). *Fish & Shellfish Immunology*. *65*, 118–126 (2017).

[49] Kwon, H. C. & Kang, Y. J. Effects of a subunit vaccine (FlaA) and immunostimulant (CpG-ODN 1668) against *Vibrio anguillarum* in tilapia (*Oreochromis niloticus*). *Aquaculture*. *454*, 125–129 (2016).

[50] Sun, X. et al. A CpG-riched plasmid as vaccine adjuvant reduce antigen dose of an inactivated *Vibrio anguillarum* vaccine in turbot (*Scophthalmus maximus* L.). *Fish & Shellfish Immunology*. *98*, 312–317 (2020).

[51] Ji, Q., Wang, S., Ma, J. & Liu, Q. A review: Progress in the development of fish *Vibrio* spp. vaccines. *Immunology Letters*. *226*, 46–54 (2020).

[52] Jeong, K. H., Kim, H. J. & Kim, H. J. Current status and future directions of fish vaccines employing virus-like particles. *Fish & Shellfish Immunology*. *100*, 49–57 (2020).

[53] Hwang, J. Y. et al. Current use and management of commercial fish vaccines in Korea. *Fish & Shellfish Immunology*. *102*, 20–27 (2020).

[54] Scott, C. J. W., Austin, B., Austin, D. A. & Morris, P. C. Non-adjuvanted flagellin elicits a non-specific protective immune response in rainbow trout (*Oncorhynchus mykiss*, Walbaum) towards bacterial infections. *Vaccine*. *31*, 3262–3267 (2013).

[55] Jaafar, R. M. et al. Effects of adjuvant MontanideTM ISA 763 A VG in rainbow trout injection vaccinated against Yersinia ruckeri. *Fish & Shellfish Immunology*. *47*, 797–806 (2015).

[56] Gao, Y. et al. Isolation, identification and vaccine development of serotype III Streptococcus *parauberis* in turbot (*Scophthalmus maximus*) in China. *Aquaculture*. *538*, (2021).

[57] Cao, J., Huang, A. L., Zhu, X. C., Li, L. & Li, J. N. Construction of *Vibrio mimicus* ghosts as a novel inactivated vaccine candidate and its protective efficacy against ascites disease in grass carps (*Ctenopharyngodon idella*). *Aquaculture*. *485*, 147–153 (2018).

[58] Woo, J. et al. Superiority of PLGA microparticle-encapsulated formalin-killed cell vaccine in protecting olive flounder against *Streptococcus parauberis*. *Aquaculture*. *509*, 67–71 (2019).

[59] Rajalakshmi, P. & G.B. Brindhadevi., D. Effect of using bivalent vaccines (whole cell and outer membrane protein) against the infections of *Edwardsiella tarda* and *Pseudomonas fluorescens* in *Cirrhinus mrigala*. *International Journal of Current Research in Chemistry and Pharmaceutical Sciences*. *3*, 75–82 (2016).

[60] Nayak, S. K., Shibasaki, Y. & Nakanishi, T. Immune responses to live and inactivated Nocardia seriolae and protective effect of recombinant interferon gamma (rIFN γ) against nocardiosis in ginbuna crucian carp, *Carassius auratus langsdorfii*. *Fish & Shellfish Immunology*. *39*, 354–364 (2014).

[61] Wang, H. et al. Construction of glutathione peroxidase (GPx) DNA vaccine and its protective efficiency on the orange-spotted grouper (*Epinephelus coioides*) challenged with *Vibrio harveyi*. *Fish & Shellfish Immunology. 60*, 529–536 (2017).

[62] Lin, K. et al. Immunity to nervous necrosis virus infections of orange-spotted grouper (*Epinephelus coioides*) by vaccination with virus-like particles. *Fish & Shellfish Immunology. 56*, 136–143 (2016).

[63] Bøgwald, J. & Dalmo, R. A. Review on immersion vaccines for fish: An update 2019. *Microorganisms. 7*, 1–28 (2019).

[64] Ke, X. et al. The immune efficacy of a *Streptococcus agalactiae* immersion vaccine for different sizes of young tilapia. *Aquaculture. 534*, 736289 (2021).

[65] Sudheesh, P. S., Zimmerman, J. K. & Cain, K. D. Dietary effects on immunity, stress, and efficacy of two live attenuated *Flavobacterium psychrophilum* vaccine formulations. *Aquaculture. 454*, 35–43 (2016).

[66] Mottier, A., Serpentini, A., Dallas, L., James, A., Lebel, J. M., & Costil, K. In vitro effects of glyphosate-based herbicides and related adjuvants on primary culture of hemocytes from *Haliotis tuberculata*. *Fish and Shellfish Immunology. 100*(October 2019), 1–8 (2020). https://doi.org/10.1016/j.fsi.2020.02.058

[67] Beck, B. R., Lee, S. H., Kim, D., Park, J. H., Lee, H. K., Kwon, S. S., … Song, S. K. A Lactococcus lactis BFE920 feed vaccine expressing a fusion protein composed of the OmpA and FlgD antigens from *Edwardsiella tarda* was significantly better at protecting olive flounder (*Paralichthys olivaceus*) from edwardsiellosis than single antigen vac. *Fish and Shellfish Immunology, 68*, 19–28 (2017). https://doi.org/10.1016/j.fsi.2017.07.004

[68] Assefa, A., & Abunna, F. Maintenance of fish health in aquaculture: Review of epidemiological approaches for prevention and control of infectious disease of fish. *Veterinary Medicine International, 2018* (2018). https://doi.org/10.1155/2018/5432497

[69] Dadar, M., Shahali, Y., & Mojgani, N. (2022). Role of immunobiotic lactic acid bacteria as vaccine adjuvants. In *Probiotics in the Prevention and Management of Human Diseases*. INC. https://doi.org/10.1016/b978-0-12-823733-5.00012-x

[70] González-Stegmaier, R., Peña, A., Villarroel-Espíndola, F., Aguila, P., Oliver, C., MacLeod-Carey, D., … Figueroa, J. Full recombinant flagellin B from *Vibrio anguillarum* (rFLA) and its recombinant D1 domain (rND1) promote a pro-inflammatory state and improve vaccination against *P. salmonis* in Atlantic salmon (*S. salar*). *Developmental and Comparative Immunology, 117*(September 2020) (2021). https://doi.org/10.1016/j.dci.2020.103988

[71] Cui, X. et al. Early IgG response to foot and mouth disease vaccine formulated with a vegetable oil adjuvant. *Vaccines. 7*, 1–19 (2019).

[72] Vinay, T. N. et al. Inactivated vaccine against viral hemorrhagic septicemia (VHS) emulsified with squalene and aluminum hydroxide adjuvant provides long term protection in olive flounder (*Paralichthys olivaceus*). *Vaccine. 31*, 4603–4610 (2013).

[73] Matsumoto, M. et al. Adjuvant effect in aquaculture fish of cell-wall glycolipids isolated from acid-fast bacteria. *Developmental and Comparative Immunology. 85*, 142–149 (2018).

[74] Aminudin, A. et al. Effect of incorporating different concentrations of palm oil as adjuvant in fish vaccine. *International Journal of Biosciences. 12*, 35–41 (2018).

[75] Ravid-Peretz, S., Colorni, A., Sharon, G. & Ucko, M. Vaccination of European sea bass *Dicentrarchus labrax* with avirulent *Mycobacterium marinum* (iipA::kan mutant). *Fish & Shellfish Immunology. 90*, 317–327 (2019).

[76] Lin, C. F., Jiang, H. K., Chen, N. C., Wang, T. Y. & Chen, T. Y. Novel subunit vaccine with linear array epitope protect giant grouper against nervous necrosis virus infection. *Fish & Shellfish Immunology. 74*, 551–558 (2018).

[77] Lagos, L., Tandberg, J. I., Becker, M. I. & Winther-Larsen, H. C. Immunomodulatory properties of *Concholepas concholepas* hemocyanin against francisellosis in a zebrafish model. *Fish & Shellfish Immunology. 67*, 571–574 (2017).

[78] Kim, J. W., Shim, S. H., Lee, J. H., Jeong, J. M. & Park, C. I. Recombinant thioredoxin 1 protein: The immune-adjuvant effect of *Streptococcus iniae* and its safety in rock bream, *Oplegnathus fasciatus*. *Fish & Shellfish Immunology. 39*, 152–157 (2014).

Section IV

Technological Advancements and Efficacy of Fish Vaccination

9 Biotechnological Approaches to Vaccines

Thavasimuthu Citarasu, Ganapathi Uma, Ramamoorthy Sathish Kumar, Sugumar Vimal, and Mariavincent Michael Babu
Manonmaniam Sundaranar University

CONTENTS

9.1 INTRODUCTION

Fisheries and aquaculture are two areas in which the 2030 Sustainable Development Agenda aims to minimize resource depletion and ensure food security and nutrition for everyone without increasing the gap between developed and developing countries. Aquaculture has been a major focus for countries, although catch fishing productivity has been stable since the 1980s. Although catch fisheries grew at the fastest rate in the last decade, aquaculture outperformed it, generating an estimated

DOI: 10.1201/9781003388548-13

80 million tons of aquatic animals worth US$ 232 billion in the last decade alone.[1] The aquaculture industry has been plagued by an increased risk of infectious disease outbreaks as a result of this sector-wide intensification in production during the last several decades. Viral infections are thought to be responsible for up to 10% of the annual loss in worldwide aquaculture output, equivalent to more than $10 billion USD annually.[2] As a consequence, the worldwide emphasis should be on disease control in order to continue the high growth trajectory of aquaculture and accomplish the 2030 agenda's objectives.

Despite numerous initiatives to develop novel treatments, fish infections remain a significant financial problem in commercial aquaculture around the world. Despite the possibility of using antibiotics or chemotherapy to treat disease, there are some clear drawbacks, such as medication resistance and safety issues.[3] Considering the potential threats on one side and environmental issues on the other side, attention should be focused on developing alternative practices for solving problems. Among the different alternative practices, development of immunostimulation and vaccines with advanced protocols are effective to boost the immune system against bacterial, viral, and parasitic diseases. Recombinant vaccine production with different expression methods, production of edible antibodies, and its advanced delivery protocols may help to improve the efficiency and boost the immune system at a significant level. Recombinant vaccines have a number of advantages, including construction, heat stability, low production costs, simple design and long-term storage capacity, no risk of reversion, and stronger and more durable protection.

Vaccination aids aquaculture's environmental, social, and economic sustainability by effectively avoiding a broad range of bacterial and viral illnesses. The initial reports of fish vaccination for illness prevention in the 1940s[4] have led to the development of several vaccines, many of which have greatly reduced the impact of bacterial and viral diseases in fish.[5] Antibiotics are being replaced by vaccination for the immunization of millions of fish each year in different regions of the globe. Antibiotic usage in Norwegian salmon aquaculture, for example, has fallen dramatically since vaccinations were introduced,[6] and vaccination has emerged as the most practicable and long-term means of treating infectious fish diseases.[7] Antigens are often found in fish vaccines. This component activates the fish's innate and adaptive immune systems in response to an infection. Over 10,000 scientific articles on fish vaccines have been published in the past 20 years, demonstrating the progress of fish vaccine research in the 20th century. There have been a number of studies[8,9] that have looked at the potential and problems of creating vaccines in aquaculture as well as the history, developments, kinds, and modalities of administration of fish vaccines. History of fish vaccination and the early pioneers in the field were examined by Gudding and Van Muiswinkel.[10] In addition to delivery methods, several researchers have examined the use of adjuvants and immunostimulants in fish vaccines. Many of these studies have examined the protective effects of classic adjuvants and the potential of next generation adjuvants in addition to alternate delivery routes for vaccination (other than injection).[11] Also investigated were potential advances in inactivated, live-attenuated, and DNA vaccines, in addition to already used vaccination methods in large-scale fish farming operations.[12] A complete review of the current status of fish vaccination research is necessary because of new technologies and a lack of research articles on improvements in the area. The majority of the vaccines approved by the U.S. Department of Agriculture for use in aquaculture species are created using standard manufacturing methods that begin with growing the target pathogens.[2] These vaccines have shown to be effective in protecting fish against a wide range of dangerous infections.

Fish vaccines are built on a firm basis of microbiology and immunology. To keep pace with ever-improving knowledge of protective antigens in molecular biology and human genetics, new vaccination generations for both animals and people are being generated at breakneck speed.[13] Modern vaccine methods target specific pathogen components, such as subunit or recombinant DNA vaccines that incorporate diverse antigens manufactured using various expression systems.[14] RNA particle vaccines, for example, have been created throughout the world and seem to be more effective than standard vaccinations in terms of boosting immunity.[15]

The most important characteristics of aquaculture vaccines are their ability to (1) elicit humoral and/or cell-mediated immunological responses in the host; (2) induce long-lasting protective immune memory and trigger the host's memory B-cells and let the host detect and combat the antigen immediately on subsequent exposure, providing protection to the fish for at least a production cycle; and (3) induce long-lasting protective immune memory and preventative immunity.[16]

9.2 VACCINE GENERATION

9.2.1 First Generation of Vaccine

First-generation vaccinations include bacteria, viruses, and other microbes that cause illness. Microorganisms in attenuated or "weakened" vaccines are still capable of growing and expressing their immunogenic features, but they are no longer capable of causing sickness or inducing an immune response. Because of their self-replicating qualities, they do not need a booster dose of vaccination and provide the hosts with long-lasting protective immunity. Attenuation techniques include generic attenuation, chemical/heat attenuation, and continued transit of the pathogen in several heterologous systems (heterologous animals, tissue culture, embryonated eggs, and more). To render the germs in whole cell inactivated vaccines inactive, scientists have used heat, formaldehyde, and ultraviolet light. Procedures such as these destroy all types of microorganisms. Viruses that have been rendered immune or antigenic to the host nonetheless boost the host's defenses in reaction to the foreign structure they recognize while being inactive. Only humoral immune responses, not cell-mediated ones, are activated in the body.

9.2.2 Second Generation of Vaccine

The development of second-generation vaccines, also known as subunit vaccines, which contain defined protein antigens or recombinant protein components and are capable of inducing humoral and cell-mediated immune responses, particularly T helper antibody responses, was needed to minimize these risks. Because antigenic components cannot multiply in a host, subunit vaccines do not provide a risk of pathogenicity for either the host or nontarget animals.[17] Targeting specific microbial pathogens and allowing the presence of foreign chemicals, they may be freeze-dried to enable nonrefrigerated transportation or storage.[18] A subunit vaccine may be made in a number of ways. Specific immunogenic proteins can be created utilizing a variety of recombinant expression vectors or identified and purified directly from the target pathogen. Production of immunogenic proteins for bacterial and viral fish vaccines has been carried out using a variety of prokaryotic and eukaryotic cellular systems.[19] It is common for toxoid vaccines to be derived from exotoxin, a chemical secreted by bacteria, which is inactivated or reduced while maintaining its immunogenic properties. If an anatoxin is encountered by the immune system, antibodies are generated that latch onto and neutralize the toxin, known as an anatoxin, and they only stimulate humoral immune responses. They are a very safe kind of immunization and are less likely to create any negative effects. They are capable of eliciting an immune response and establishing immunological response and memory. An immunological adjuvant is needed for efficient antibody production after many rounds of subunit vaccines.[16]

9.2.3 Third Generation of Vaccine

DNA vaccines are third-generation vaccines that use a plasmid, a circular piece of bacterial DNA that has been genetically manipulated to produce one or two particular proteins (antigens) from a microbe. Immune cells are injected with vaccine DNA, which is "read" by the host cells' "inner machinery," resulting in the production of damaging proteins. The immune system is alerted when host cells digest these proteins and expose them on their surface. Immune responses, both humoral and cell-mediated, are induced by this. They have a robust and long-lasting resistance. The vector may encode a vaccine that can protect against many illnesses at the same time.[20] DNA vaccines may be made cheaply and easily utilizing the same manufacturing procedures.[21] An RNA vaccine is made consisting of a strand of mRNA that codes for a pathogenic microbe's virulence factor or protective antigen, similar to a DNA vaccine. Upon receiving the mRNA strand, host cells processed the genetic information to produce the antigenic protein. In cells, this antigen may be seen and identified by the host immune system, which leads to a variety of immunological reactions, including humoral and cell-mediated responses. Once the protein is synthesized, RNA vaccines, in contrast to DNA vaccines, dissolve and are not incorporated into the host genome.[16]

9.3 CONVENTIONAL VACCINES AND THEIR DRAWBACKS

Conventional vaccines are first-generation vaccines and include live, attenuated, and killed form of pathogenic microorganisms. Several studies have proved the positive effects of conventional vaccines for aquaculture operations against bacterial and viral pathogens. Unfortunately, they have several drawbacks, such as a weak or shorter immunity, reversal of virulence, ineffectiveness, heat liability, etc., despite the fact that they have produced immunity to the cultivable fin and shellfish species against microbial infection. The risk of new strains emerging from recombination of various strains in live-attenuated vaccines exists. In addition, they run the danger of switching back to virulent strains, which pose a major hazard to unintended animals and the aquatic environment. In order to retain an acceptable degree of protective immunity for a longer period of time and enhance their efficiency, inactivated vaccines require multiple booster doses, proper adjuvant, etc.[16] Because fish lack cellular immunity, adjuvants or several booster injections may be required to increase the immunogenicity of inactivated vaccines. In the presence of antigen-presenting cells (APCs), the immune system's elimination of activated T-cells and the onset of humoral immunity begin. Immune-enhancing adjuvant-induced toxicities, reduced immunogenicity due to protein denaturation, and systemic responses to a specific adjuvant are some of the negative consequences of using inactivated vaccines.[22] Viral infections and intracellular bacterial diseases are not usually prevented by inactivated vaccines.[23] A formalin-inactivated *Edwardsiella ictaluri* vaccine used in aquaculture was shown to have a constrained capacity.[24] Numerous studies have found that inactivated vaccinations don't give people enough immunity.[25] Physical or chemical treatments that result in denaturation of the proteins or damage to the nucleic acids are used to inactivate the vaccine virus in order to create killed vaccinations. It's possible to further purify the inactivated antigen and combine it with an adjuvant.[26] Although inactivated vaccines have better safety profiles, the destruction of pathogen reproduction prevents them from offering long-term protection that is reliable.[27] Although toxoid vaccinations mixed with traditional immune adjuvant are effective, high-level biosafety is sadly required.[28] Because of the drawbacks of traditional vaccines, it is advisable to develop improved vaccines based on genetics, immunology, chemistry, biotechnology, and molecular biology for robust and long-lasting immunity.

9.4 STATUS OF NON EXPRESSED VACCINES

9.4.1 Subunit Vaccines

The short, specialized proteins of a pathogen that are noninfectious because they cannot reproduce in the host are included in subunit vaccines. They are the immunogenic protective antigens that will promote immunity.[29] By directly extracting the native protein from pathogen cultures, subunit vaccines can be made. By including more than one protein in a subunit vaccine, it is possible to establish immunity against more than one strain or serotype of a bacterial or viral disease.[30] They have high level of safety and low production cost for making subunit vaccines and also get license easily.[31]

These preparations may not elicit a substantial immunological response as compared to dead or live whole cell preparations, for example. As a result, just a few components are when required while a large number of antigens are used to stimulate the immune system.[18] A few examples of subunit proteins are S/A layer proteins, extracellular products, outer-membrane proteins (OMPs), synthetic peptides, and lipopolysaccharides. In a study using a blue gourami (*Trichogaster richopterus*) and Freund's complete adjuvant (FCA) without FCA, the OMP of *A. hydrophila*, purified to 43 kDa, provided protection and evidence of an immune response.[32] Estuary cod (*Epinephelus coioides*) were vaccinated by Pang et al.,[33] with 60%–70% RPS using *V. harveyi* OmpN and FCA. The goldfish *Carassius auratus* was vaccinated with *A. hydrophila* culture extracellular products (ECPs) and OMPs, which improved hematological and immunological indices.[34] Synthetic peptide is an ingredient in subunit vaccinations because it is easy to make and easy to get a license for. They work better when combined with an immunoadjuvant given intravenously. Emmenegger et al.[35] used a variety of synthetic peptides to simulate the infectious hematopoietic necrosis virus (IHNV) glycoprotein's three neutralizing domains. It is possible for cell-penetrating peptides to cross the macrophage plasma membrane and deliver cargo proteins to the cytoplasm through this method of transport. Immunoadjuvants and a detailed knowledge of the epitope structure of pathogen-protective antigens are two limitations of this vaccination approach. In Saeed[36] and Saeed and Plumb[37], an lipopolysaccharide (LPS) extract of *E. ictaluri* was administered intraperitoneally to channel catfish, resulting in a survival rate of over 80% and a noticeable humoral immune response. When challenged, the *E. tarda* LPS-vaccinated eels and red sea bream displayed resistance to infection, a strong humoral immune response, and T-lymphocyte phagocytosis.[38] Chemical conjugation with a carrier protein may turn vaccine antigen LPS into an immunogen.[39] A number of eukaryotic animals, including fish, have been shown to have powerful innate immune systems when exposed to LPS.[40] Due to their lack of immunogenicity, nonexpressed subunit vaccines need several shots and the use of adjuvants in order to generate robust immune responses.

9.4.2 Biofilm Vaccines

Cells embedded in a matrix create biofilms (BFs), structures that help bacteria adapt to various micro niches and endure harsh environmental circumstances. They make pathogenic bacteria less susceptible to antibiotics and enable them to evade the immune system. For the purpose of creating vaccines that are effective in protecting against host organisms, BF antigens are desirable targets.[41] The BF vaccine's glycocalyx is thought to shield the antigen from being destroyed in the gut and enable its transportation in intact form to the immune-sensitive regions.[42] Bacterial cells within the BF and those located in its matrix, which includes polysaccharides, extracellular proteins, and DNA, are key sources of BF antigens.[43,44] The extracellular matrix, also known as extracellular polymeric substances, or EPS, which is made up of proteins, lipids, exopolysaccharides, and extracellular DNA (eDNA), surrounds the BF bacterium.[45] Biofilm antigens obtained from several bacterial pathogens were highly influenced to boost the immune system in fish species. Oral BF vaccines have been confirmed to have protective effects for many fish, such as *Lates calcarifer*,[46] *Labeo rohita*,[47] *Oreochromis* sp.,[48] *Cyprinus carpio*,[49] and *Mugil cephalus*.[50] *A. hydrophila* BF vaccine was given to three species of carp, including *Catla catla*, *Labeo rohita*, and *Cyprinus carpio*, reported by Azad et al.[51] Among the treated groups, catla produced the highest antibody and protective response, followed by rohu and common carp after 15 and 20 days post vaccination, respectively. *A. hydrophila* (AHV1) treated with goldfish *C. auratus* had improved survival, hematological, and immunological parameters against *A. hydrophila* challenge.[34] *Lactococcus garvieae* BF treated mullet, *Mugil cephalus* had improved survival and immune responses against *L. garvieae* challenge.[52] Biosurfactant obtained from Su et al.[53] vaccinated the BF of *Photobacterium damselae* to Giant Grouper, *E. lanceolatus*. The TLR 3, IL-1, and IL-8 genes were upregulated in the vaccinated fish 28 days after immunization, which increased survival and immunological parameter.

9.4.3 Egg Yolk Edible Vaccine IgY

The use of egg yolk antibodies in foods to inactivate certain pathogens, particularly enteric microorganisms, is seen as a possible alternative to antibiotics.[54,55] In recent years, IgY is the major antibody in chickens (*Gallus gallus domesticus*), and it is transferred to egg yolk to protect chicks against pathogens during the incubation period until laying.[56,57] Administration of IgY that highly influenced several aquatic fin and shellfish pathogens, including bacteria and virus, was documented. Oral administration of IgY could reduce the impact of *E. tarda* infection in Japanese eels.[58] Feeding Gibel carp (*C. auratus gibelio*) with a particular IgY immunoglobulin increased their survival and resistance to *A. hydrophila*.[59] When the antibody production was enough high and reached the body of the shrimp, the resistance to *V. harveyi* has increased.[60] The survival of *Metapenaeus ensis* shrimp exposed to WSSV infection was reported to be reduced by using specific IgY from hens inoculated with inactivated WSSV and DNA vaccine to neutralize WSSV.[61] Kumaran et al.[62] reported that anti-WSSV IgY-coated diet boosts the immune system by increasing total hemocyte count, oxyhemocyanin, phenol oxidase, lysozyme activity, and intracellular superoxide anion production in *Penaeus monodon*, against WSSV challenge. Kumaran et al.[63] developed and delivered the anti-*Vibrio harveyi* IgY in *Fenneropenaeus indicus* and had high immunogenicity. In order to protect the IgY from gastric enzymes, Punyokun et al.[64] delivered the anti-*Vibrio harveyi* IgY encapsulated with chitosan-alginate microcapsules and got successful results. The survival rate of *M. rosenbergii* infected with *E. cloacae* was dramatically increased by the oral injection of particular IgY.[65] IgYs provide advantages over mammalian antibody production in terms of cost, convenience, and invasiveness, making them a possible replacement for passive immunization applications. The IgY production and purification steps were given in Figure 9.1.

9.4.4 Genetically Altered Vaccine

By deleting, disrupting, or introducing the metabolic pathway or virulence gene that causes attenuation in pathogens, genetically modified technology has assisted in the production of live-attenuated vaccinations.[66] This modification aids in turning a pathogen into one that is nonpathogenic and produces a host's protective immune response.[67] This technology is employed for the genetic alteration of lipopolysaccharide, marker-less deletion system, and transposon mutagenesis for developing attenuated live vaccines against *Streptococcus iniae*,[68] *Vibrio anguillarum*,[67] *Flavobacterium columnare*,[66] *Edwardsiella ictaluri*,[69] *A. hydrophila*,[70] and *E. tarda*.[71] Even though this approach is effective, due to some of the safety concerns, the genetic altered vaccine is mostly not accepted for aquaculture operations.[72]

9.5 VECTOR EXPRESSED VACCINES

9.5.1 DNA Vaccine

A major advantage of DNA vaccines over standard immunization approaches is that they may stimulate both cellular and humoral immune responses that protect fish against illness. This method of activating an immune response using genetically edited DNA is possible.[73] Fish DNA vaccines have been studied and produced for a number of serious aquatic diseases since their initial publication in 1996. A circular plasmid dubbed a DNA vaccine contains an antigenic protein-producing gene from a disease of interest. In bacterial cells, the pathogen gene's promoter and terminator regions as well as the DNA vaccine's features enable it to be amplified to large numbers and utilized as a vaccine. Despite the discovery of other promoters, the cytomegalovirus (CMV) promoter is still used in almost all DNA vaccines to generate high levels of constitutive expression. The plasmid invades a small number of host cells after DNA vaccine injection into a live host, causing the pathogen gene to be expressed and a little quantity of antigenic protein to be generated. So as to activate a wide range of protective immune responses, including both innate and adaptive immunity, the antigenic protein folds and processes as it would in the event of an actual pathogen infection.[74] For various fish pathogens, the extent to which DNA vaccines are developed is dependent on several variables, including (1) how important a pathogen is economically, (2) the lack of a successful vaccine produced by another strategy, (3) the pathogen's ability to be cultured in the laboratory, (4) the pathogen's protective antigen(s), and (5) whether the pathogen is available.

Classic immunizations for bacterial fish illnesses, some of which may be administered by immersion, are quite effective. Thus, DNA vaccine research has focused on bacteria that are resistant to bacterins, which has led to the development of DNA vaccines. As a result, fewer DNA vaccines have been developed for more complicated illnesses like bacteria and cellular parasites since selecting effective antigens is more challenging. Infection with *Renibacterium salmoninarum*, which causes bacterial kidney disease (BKD) in salmonids, is an example of a bacterial-host association with complicated disease etiologies.[75] Anti-*Vibrio anguillarum* DNA vaccines that encoded a crucial outer-membrane protein were found to be ineffective.[76] Studies on *A. veronii* OMP48 found it to be a promising potential antigen, which led to the vaccines' efficacy.[77]

DNA vaccines for infectious hemopoietic necrosis and viral hemorrhagic septicemia were approved by the FDA for use in 1996 and 1998, respectively.[78,79] DNA vaccine plasmids have been shown to have a 65% survival rate in shrimp 15 days after inoculation, according to Rajeshkumar and coworkers.[80] WSV envelope proteins were shown to be protective for 25–50 days in *P. monodon* after vaccination with DNA vaccines that included VP28 and/or VP281.[81] For 7 or 14 days following immunization, *L. vannamei* were reported to be protected against WSSV infection by Li et al.[82] who delivered an intramuscular VP28-encoding DNA vaccine to the animals. Survival against *Mr*NV challenge was increased, viral load was reduced, and the immune system was strengthened to protect *M. rosenbergii* from *Mr*NV infection by MrNV-CP-RNA-2-pVAX1.[83] Current research on DNA vaccines for fish parasitic pathogens is ongoing. Most recent research has focused on identifying prospective antigen genes that might be employed as DNA vaccines due to the increased level of complexity in these organisms. Protozoan ciliate *Ichthyophthirius multifiliis* (Ich) has been widely studied, and researchers have found antigens (i-antigens) that protect the host against the parasite.[84] Protease genes from a hemoflagellate parasite, Cryptobiasalmositica, have recently been used in an effort to create an anti-virus for the salmon parasite, Cryptobiasoma.[85] To their credit, rainbow trout that had been administered a DNA vaccination that expressed the metalloprotease virulence component recovered more quickly than control fish, even if the virus had not been completely destroyed during the challenge. For the first time, the efficacy of a fish parasite DNA immunization was shown in this investigation.

9.5.2 RNA Vaccine

Antigen-specific immune responses, especially cell-mediated responses, are generally stronger in RNA vaccines than in ordinary plasmid DNA vaccines. There is a new kind of vaccine that is less expensive, safer, and faster to produce than standard vaccinations: the mRNA vaccine.[86] mRNA is a cellular protein-production blueprint generated from genetic information. Degradation occurs quickly when the protein is no longer required. DNA is unaffected by mRNA immunization because it does not enter the nucleus.[87,88] There are two types of RNA vaccines, namely nonreplicating mRNA and *in vivo* self-replicating mRNA. Dendritic cells that may transmit antigens to other immune cells in order to help activate an immune response are known as in vitro dendritic cell nonreplicating mRNA vaccines. A patient's immune system is stimulated by the administration of RNA vaccine-infected cells that are subsequently administered back to the patient.[89] When using RNA vaccines, they must be noninfectious and can be cleared by the body's natural mechanisms. RNA is a significant stimulant for the immune system, which is enhanced by RNA.[86] Because of their increased effectiveness, constant immunological response, and stability, they are well tolerated by healthy persons.[89] Replicon vaccines based on the salmonid alphavirus 3 (SAV3) genome have been developed by Kalsen et al.[90] that protect against infectious salmon anemia. It was shown that ISAV hemagglutinin-esterase was particularly effective against *Salmo salar*, which is a virus.[91] Viral infections in farmed fish may be prevented using mRNA vaccines, which have proved to be an effective and reliable method.[92]

9.5.3 Recombinant Subunit Vaccine

Recombinant subunit vaccines are small, targeted proteins of a microbial pathogen that lack the capacity to multiply in the host, making them noninfectious. In addition, they could need several immunoadjuvant booster shots to guarantee long-lasting protective immunity.[18] To more fully understand and generate the identified antigen, it is frequently essential to clone the gene encoding the protein. Heterologous protein expression has frequently been carried out in *E. coli*, but this method has some drawbacks in terms of the yield, folding, and posttranslational modifications of the recombinant protein.[93] These issues can be resolved by using alternative expression systems like yeast, mammalian, and insect cell lines. For high-quality subunit vaccine production, the mammalian and insect cell line expression systems offer a number of benefits, including proteolytic cleavage, glycosylation, secretion, folding, phosphorylation, acylation, and amidation.[94] It is feasible to induce immunity against more than one strain or serotype of a bacterial or viral disease by including multiple proteins in a subunit vaccination.[95] The pathogen and host are connected via OMPs. Thirumalaikumar et al.[96] expressed recombinant subunit OMP and hemolysin proteins through pET-30b vector by the bacterial expression system and delivered to *Labeo rohita*. The treated fishes had higher survival and boosted the immune system. *A. hydrophila* S-layer recombinant protein vaccinated carps had the highest protection and survival.[97] *V. harveyi* recombinant OMP (VhhP2) subunit vaccine treated Japanese flounder had elicited strong immunoprotection.[98] Viral hemorrhagic septicemia of trout (VHSV) protein fragments expressed in yeast *S. cerevisia* and treated to rainbow trout, *Oncorhynchus mykiss*, had higher immunological response.[31] Wei and Xu[99] used a baculovirus vector to express the WSSV rVP28 protein in silkworms (*Bombyx mori*) and examined the effectiveness of the vaccine in *Procambarus clarkii*. The results showed that the crayfish that had been given the rVP28 had considerably higher survival rates when exposed to the virus orally at 35 and 75 days after vaccination (dpv), respectively, of 54.16 and 59.26%. Caipang et al.[100] expressed the WSSV VP28 in *Brevibacillus brevis* and delivered to *Penaeus japonicus* and challenged with WSSV. The shrimp fed the purified protein at the dose of 50 mg had 72.5% survival. *Penaeus chinensis* treated with recombinant WSSV VP19 and VP466 subunit vaccines produced in Sf21 insect cells utilizing a baculovirus expression system fared better after WSSV challenge.[101] Another recombinant WSSV envelope protein, rVP41A (VP292), was demonstrated to offer considerable protection for 30 days by Vaseeharan et al.[102] Namikoshi et al.[103] proved the quasi-immune response of *P. japonicas* against WSSV by delivering recombinant WSSV rVP26 and rVP28 with the immunostimulant β, 3-glucan, and *Vibrio penaeicida*, and the vaccinated groups had more than 95% survival. Two structural genes of WSSV VP19 and VP466 (VP28) of WSSV were cloned and expressed in Sf21 insect cells using baculovirus expression system and further its efficacy was studied by both intramuscular and oral routes and the vaccinated shrimp had more than 50% survival.[101] According to Maftuch et al.,[104] tiger shrimp were given two doses of the OMPs from *V. alginolyticus*, which had immunomodulatory effects and increased protection against *V. harveyi*. Adult *M. rosenbergii* that had been injected with recombinant *M. rosenbergii* nodavirus capsid protein (r-MCP) responded well to the *Mr*NV challenge.[105] The baculovirus-expressed *M. rosenbergii* nodavirus (*Mr*NV) capsid protein produces protective immunity and increases survival against *Mr*NV challenge in *M.rosenbergii*.[106] VLPs, subunit vaccine components, are generated when viral capsid proteins self-assemble into particles that match the virus' natural structure.[107] Because VLPs lack genetic material, reversion mutations or pathogenic infection cannot occur, unlike with genuine viral particles.[108] By recognizing and generating strong cellular and humoral responses to repeated subunits, VLPs, which cannot replicate in the receiver, may boost both the innate and adaptive immune responses of the recipient.[109] To produce heterologous proteins, such as VLPs, baculovirus expression provides a cost-effective and efficient technique.[110] *E. lanceolatus* showed a better resistance to infection after being exposed to varied concentrations of GNNV virus-like particles (VLPs).[111] Researchers Chien[112] and Cho[113] have produced oral VLP vaccinations to protect grouper against NNV. The IPNV capsid protein VP2, produced in yeast, self-assembles into subviral particles (SVPs), and the injection of the SVPs into rainbow trout causes an immunological response, as reported by Dhar et al.[114] In order to protect shrimp from the virus that causes white spot syndrome, Jariyapong et al.[115] also created *Mr*NV VLPs in *E. coli* and employed them as a vehicle to transmit double-stranded RNA (WSSV).

9.5.4 Reverse Vaccine

Methods based on bioinformatics predict the immunogenic sequences of pathogens to be used in vaccine production. An infectious organism's whole antigenic repertoire may be exploited to find and test prospective vaccine candidates together with several defensive targets.[116] Using this method, more vaccine candidates, including previously unknown and/or unusual proteins, may be discovered. In addition, rational design's capacity to make candidate antigens stronger might improve defenses against diseases using changeable antigens.[117] Modern bioinformatics methods are used to predict suitable antigens for reverse vaccination. Extracellular and periplasmic areas, cell walls, cytoplasm, and outer-membrane parts of proteins should all be taken into account when making a judgment about using them as vaccine candidates. To examine transmembrane helices, adsorption potentials, and subcellular localization, we utilized PSORTb V3.0.2 server.[118] The settings were manually modified. PSI BLAST and the NCBI Conserved Domains Database (CDD) were used to identify putative domains for all of the most frequently encountered antigens.[119] For the prediction of few antigens such as B-cell epitopes, BediPred linear epitope prediction tools from the Immune Epitope Database (IEDB) analysis resource are used. For a few antigens in the research, linear epitope prediction methods such as BepiPred will be used to detect B-cell epitopes.[120] We found that CTLPred's artificial neural network and stabilized matrix approach implementation could effectively predict peptide's ability to attach to MHC molecules in cytotoxic T-cells.[121] VirusPloc is a useful tool for predicting the subcellular location of viral proteins in infected cells and in the host, as well as in cells infected by pharmacological targets.[122] For membrane proteins, we can reliably predict the transmembrane helices using the TMHMM 2.0 server, a hidden Markov model-based approach. TMHMM 2.0 server.[123] It is possible that the disulfide bond and molecular docking investigations may throw light on how modeling proteins and receptor molecules interact. The ClusPro v2.0 server is a widely used tool for protein-protein docking. In order to get the most accurate protein structure and its activities, Raptorx is the best server.[124,125] Many bacterial, viral, and eukaryotic diseases are now being studied using this technology, and in every case, it has been successful in producing novel antigens for the creation of new vaccines.[126] A promising method for finding recombinant vaccines for infectious, parasitic, and even metabolic illnesses is reverse vaccinology. There is a clear need to create stronger, safer, and better defined vaccines that can combine many antigens to create vaccinations that can protect against various pathogen strains. Recombinant vaccines satisfy this need, making them particularly appealing for use as animal vaccinations. Vaccine mixtures are a helpful immunization alternative for animals.[127] The development of fish vaccines against a variety of important bacterial diseases, such as *Photobacterium damselae* subsp. *piscicida*, has increased the importance of reverse vaccination,[128] *Streptococcus agalactiae*,[129] *E. tarda*, and *F. columnare*.[130] Several vaccines for fish RNA viruses, such as the Novirhabdoviruses IHNV and VHSV,[131,132] the Alphavirus SDV,[133] the Aquabirnavirus IPNV[134] and also for several Betanodavirus species such as for the striped jack nervous necrosis virus (SJNNV) have been developed.[135] Based on reverse vaccinology, Chukwu-Osazuwa et al.[119] determined the common and distinctive antigens for *Piscirickettsia salmonis*, *Aeromonas salmonicida*, *Yersinia ruckeri*, *Vibrio anguillarum*, and *Moritella viscosa*. They came to the conclusion that 80 of these discovered antigens resembled exposed OMPs, while 74 of them matched secreted proteins (OMPs). Reverse vaccinology has been utilized to find novel antigens for fish diseases including *V. anguillarum* and *Photobacterium damselae* subsp. piscicida.[128,136] Using the primary capsid protein VP2 and RNA-dependent RNA polymerase (RdRp) genes as well as immunodominant T-cell epitopes and immunoinformatics, Islam et al.[121] developed a vaccine against *Lates calcarifer* birnavirus (MABV). The vaccine was nonallergenic, had a high level of immunogenicity, and had good solubility. Based on conserved proteins in *Streptococcus agalactiae*, Kawasaki[137] developed candidate serotype-independent preventive vaccines using a pan-genome reverse vaccinology technique (Group B Streptococcus; GBS). *Vibrio parahaemolyticus* multivalent vaccination was created by Wang et al.[138] using a bioinformatics technique and the OMP protein for signaling peptides, transmembrane (TM)-helix, and subcellular location.

9.6 EXPRESSION METHODS FOR VACCINE DEVELOPMENT

9.6.1 Bacterial Expression System

Many parameters, such as expression levels, posttranslational changes, scale-up, and manufacturing costs, are taken into consideration for the development of better antigen expression for vaccines.[139] Among the different expression systems, bacterial expression is one of the easiest and high levels of expression for producing recombinant proteins and antibodies. *E. coli* and *B. subtilis* are the two bacteria that are most frequently utilized as expression hosts for recombinant proteins. When the conditions are appropriate, they reproduce incredibly quickly at a low cost and with a high expression level, making it an effective and affordable method for mass production. Since their entire genomes have been sequenced, information about their genetic histories and metabolic pathways is available. Bacteria are much easier to transfect than eukaryotes due to their basic architecture. The expression process is made simpler and faster by these advantageous qualities. Recombinant protein is easily expressed in bacteria by inserting DNA coding for the desired protein into a plasmid expression vector, which is then transformed into a bacterial cell. To create the desired protein, the transformants are multiplied, induced with the proper inducers, and purified. *E. coli* is one of the dominant strains for the most widely used expression platform and high levels of heterologous protein expression among the various bacterial hosts. There are numerous methods for improving protein expression in bacteria, including strain selection, transcriptional and translational control, and the addition of signal sequences that direct the protein to the periplasm or culture media.[140] Since the bacterial host may quickly digest foreign proteins, the host can be modified to slow down proteolysis.[141,142] Other bacteria, such *Bacillus brevis*,[143] *Vibrio cholerae*,[144] and *Salmonella typhimurium*,[145] are sometimes helpful for the expression of antigens. Bacterial expression systems have several benefits, including basic physiology that makes it simple to grow and modify, quick generation times due to how quickly bacteria grow and multiply, high product yields of up to 10% of mass, low cost, and many more. Its doubling time in glucose-salts media, under ideal environmental conditions, is approximately 20 minutes.[146] A liquid culture of *E. coli* is thought to have a theoretical density limit of 200 g dry cell weight/L, or around 1×10^{13} viable bacteria/mL.[147] The bacterial expression mechanism has a few drawbacks as well, including the lack of posttranslational modification enzymes, the generated proteins' frequent improper folding, which renders them biologically inactive, the manufactured proteins' frequent toxicity to bacteria, and others. Bacteria cannot process cellular transport and posttranslational changes because they lack a nucleus, a Golgi apparatus, and an endoplasmic reticulum. As a result, membrane-bound proteins like glycophorins, certain enzymes, and so forth cannot be produced using bacterial expression systems.

9.6.2 Yeast Expression System

Yeast is the most economical eukaryotic expression system for large-scale production of both endogenous and heterologous proteins.[148,149] High cell densities, ease of growth and manipulation, proteolysis, well-adapted for secreting proteins into the culture supernatant, which facilitates purification and posttranslational modifications such as glycosylations, are just a few advantages that yeast expression has over bacterial expression system.[139] In addition, due to their nonpathogenic nature, comprehensive genome sequence, well-established genetics, and intrinsic natural adjuvant, yeast species are the best model system for creating vaccines. For expressing the recombinant proteins, yeast cells such as *Saccharomyces cerevisiae*, *Pichia pastoris*, *Hansenula polymorpha*, *Yarrowia lipolytica*, *Arxula adeninivorans*, *Kluyveromyces lactis*, and *Schizosaccharomyces pombe* are excellent.[150] One of the well-characterized types of signal transduction at the biochemical, genetic, and molecular biology levels is the galactose induction system in yeasts, which serves as a model transcriptional induction system in eukaryotes.[151] *P. pastoris* has a more effective expression system than Saccharomyces because it has a strong oxidase promoter called AOX1, which assisted to post-translate proteins.[139] With the help of Pichia, greater concentrations of vaccine antigens have been made from a range of bacterial and viral proteins.[152] In *S. cerevisiae*, heterologous glycoprotein synthesis typically leads in a hyper-mannosylated glycan structure, which may result in decreased activity and increased immunogenicity. For a higher yield or better recombinant product quality, this yeast species' various selective promoter elements and mutations were examined.[153] It is common practice to overexpress heterologous proteins using well-characterized inducible or constitutive promoters with strong transcriptional activity seen in diverse yeast strains.[154] The well-known and potent promoters GAL1, GAL10, JUB1, SNR52, MET17, TDH3, TPI1, ENO1, and PDC1 have been employed for high-level expression of foreign genes in *S. cerevisiae*.[155,156] In addition, promising for enhancing the expression of recombinant proteins in *S. cerevisiae* are dual promoter systems.[157] Yeast expression system has several advantages compared to bacterial and baculovirus mode of expression for veterinary and fish vaccines. Some of the advantages are: there is no need of purification, largely scalable, easily deliverable for oral vaccine, low cost, etc.[158] In addition, yeast expression system is naturally robust to severe conditions and have rigid cell walls, which indicates that recombinant antigens produced in yeast may have a higher probability of coming into touch with the immune cells of the gastrointestinal (GI) tract in intact form.[159] The extra benefits of yeast cell walls acting as immune stimulators make them stronger candidates for vaccines. Because yeast cells contain polysaccharides such as mannan, beta-1, and 3-D-glucan (BGs), it was thought that yeast cells were immunogenic. Zhu et al.[160] expressed hemolysin of *V. harveyi* SF-1 through yeast cells and immunized with turbot and the immunized fishes had significant protection. Jha et al.[161] expressed WSSV protein in eukaryotic expression systems such as yeast *Pichia pastoris* and delivered to *Procambarus clarkia*. Viral capsid protein (VP2) of infectious pancreatic necrosis (IPN) expressed through *S. cerevisiae* and immunized with rainbow trout. The immunized fishes had improved immunity and reduction of viral load after IPN challenge.[162] The IPNV capsid protein VP2, produced in yeast, self-assembles into SVPs, and the injection of SVPs into rainbow trout causes an immunological response, as shown by Dhar et al.[114] Mao et al. [163] successfully immunized sea bass *Lateolabrax japonicas* with *V. harveyi* OmpK, which expressed in *P. pastoris* GS115. Recombinant main capsid protein (r-MCPMCP) of RBIV was generated in yeast after the virus was challenged in rock bream, *Oplegnathus fasciatus*, and this helped to reduce viral load and develop gut mucosal protection.[164] Ananphongmanee et al.[165] successfully displayed WSSV pVP28 through the yeast expression in *S. cerevisiae* and *P. pastoris*. Nervous necrosis virus (NNV) VLPs expressed through *S. cerevisiae* and delivered to *E. septemfasciatus* had higher survival rate and improved immunity against NNV challenge.[166] Cyprinid herpesvirus 2 (CyHV-2) recombinant truncated proteins, TCID50 and tORF25, expressed through *P. pastoris* and vaccinated with *C. gibelio*. The vaccinated fishes had improved survival and upregulated the immune genes.[167]

9.6.3 Baculovirus Expression

Recombinant baculoviruses with circular double-stranded DNA genomes have been widely used to produce recombinant proteins in insect cells and larvae. The *Autographa californica* multiple nucleopolyhedrovirus (AcMNPV) is the most often used baculovirus species in biotechnological applications.[168] A baculovirus genome's enormous genetic storage capacity allows for the inclusion of many genes.[169] Inserting foreign genes downstream of the polyhedron promoter will result in high-level expression upon infection of permissive cells. The polyhedron protein is replaced by the foreign gene of interest and integrated into the viral genome by homologous recombination.[139] The *Spodoptera frugiperda* (Sf-9 and Sf-21) cell lines are capable of producing the recombinant virus easily, and after infection, the replication cycle is accompanied by the production of the foreign protein driven by either the polyhedron or p10 promoters.[170] This protein is expressed at very high levels so that it can package assembled viral particles into occlusion bodies that protect them from proteolytic degradation by the decomposing host.[139] Both insect cells and larvae can effectively synthesize heterologous proteins.[171] Compared to other expression systems, such as the bacterial expression system, they have a number of advantages. Utilizing the function of the polyhedron protein in the viral life cycle, high-level production of a foreign protein, often in a soluble form, in baculoviruses is possible.[139] Glycosylation, fatty acid acylation, phosphorylation, and proteolytic processing are among the posttranslational alterations that insect cells infected with baculovirus are capable of doing. Since baculovirus vectors are simple to make, there is no risk of endotoxin contamination, unlike with bacterial expression. Two WSSV structural genes, VP19 and VP466 (VP28), were cloned and expressed in Sf21 insect cells using a baculovirus expression system. The efficacy of the vaccine was further investigated using both intramuscular and oral delivery methods, and the vaccinated shrimp had a survival rate of greater than 50%.[101] *Brevibacillus brevis* was modified to express the WSSV VP28, which was then given to *P. japonicus* and challenged with the virus by Caipang et al.[100] Around 72.5% of the shrimp fed pure protein at a dose of 50 mg survived. In shrimp treated by Syed and Kwang[172] using oral and immersion vaccinations, which expressed VP28 on the surface of a baculovirus, protection against WSSV challenge was seen 15 days after the final booster. White spot syndrome virus (WSSV) protein VP28 was delivered in a recombinant form inside the spores of the gram-positive bacteria *B. subtilis* using a unique technique by Fu et al.[173] The baculovirus-expressed *M. rosenbergii* nodavirus (*Mr*NV) capsid protein produces protective immunity and increases survival against *Mr*NV challenge in *M. rosenbergii*.[106] Recombinant baculovirus BmNPV-VP35-VP4 was created using a Bac-to-Bac baculovirus expression method to express the VP35 and VP4 proteins from grass carp reovirus (GCRV) type II, and the expressed protein was then inoculated with *Ctenopharyngodon idella*. The immune-stimulated fish exhibited a greater percentage of survival, better immunity, and an upregulation of immune genes.[174]

9.6.4 Mammalian Cell Line Expression

The various vector systems, including plasmid-based expression vectors, adenovirus vectors, vaccinia vectors, retroviral vectors, and baculovirus as vectors, that transfer the gene into mammalian cells were investigated.[175] Viral-mediated transduction is an effective method for achieving protein expression in mammalian cells.[176] Recombinant baculoviruses are used in this technology to easily transduce mammalian cells and produce milligram amounts of protein.[177] The benefits include a wide range of posttranslationally modified proteins with a higher degree of flexibility than other expression systems, correct protein folding, and genuine glycosylation.[178] Mammalian cells carry out posttranslational alterations such as glycosylation, carbohydrate trimming, and proteolytic processing of the propeptide.[179] Other alterations may involve carboxylation of glutamic acid residues,[180,181] hydroxylation of aspartic acid and asparagine residues,[182] sulfation of tyrosine residues,[183] phosphorylation of proteins via cell receptor–protein interaction,[184] fatty acid acylation,[185] and correct assembly of multimeric proteins. Chen et al.[186] expressed the structural proteins (glycoprotein) of spring viremia of carp virus (SVCV) through mammalian cell lines. Red sea bream iridovirus's (RSIV) transmembrane domain (pCMV-569) and capsid protein (pCMV-MCP) were expressed using the pCI-neo mammalian expression vector by Marlowe et al.[187] The proteins immunized *Pagrus major* had improved cellular immune response and upregulated the MHC class I transcripts. Wolf et al.[91] used pCI mammalian expression vector to express hemagglutinin-esterase of infectious salmon anemia virus (ISAV). The Atlantic salmon (*Salmo salar*) used to immunize the protein provided good protection from an upcoming ISAV test. Kayansamruaj et al.[188] developed α-enolase-based DNA vaccination against *Streptococcus iniae* using the mammalian expression vector pCI-neo. The immune-relevant genes (IL-1, TNF-, COX-2, IL-12, and IL-13R-1) were upregulated in *Oreochromis niloticus* at 7 days after immunization, and it had the best survival rate.

9.6.5 Micro-Algal Expression

A novel, affordable, and highly productive vaccine approach for many infectious diseases, including fish disease, is the algae-based vaccine. Algae have been found to collect quickly, fold a variety of vaccine antigens correctly, and produce recombinant algal fusion proteins that can increase the antigenicity of vaccinations that are administered orally. It has been suggested that a chloroplast expression system is a promising method of producing oral vaccinations.[189] For the purpose of preventing and controlling infectious diseases, *Chlamydomonas reinhardtii*, *Dunaliella salina*, cyanobacteria, and other microalgae have been treated to express antigen genes in chloroplasts.[190] Algal expression involves physically introducing foreign DNA into the nuclear or chloroplast genome through transformation using electroporation, glass beads, or silicon carbide whiskers, microparticle bombardment, or agrobacterium-mediated DNA transfer.[191] Then, the foreign gene is integrated into the chloroplast genome by homologous recombination.[192] Algal expression has a number of benefits, including a reduction in the requirement for costlier purification, expensive fermentation, cold storage, and shipping, as well as simplicity, safety, and sterile administration.[193] In addition, they have health-promoting advantages due to their high protein, fat, and nutritional contents, as well as the potential immunogenicity of their pigments for the hosts.[194] Through the use of *C. reinhardtii*'s transgenic technology, Siripornadulsil et al.[195] developed the p57 antigen of *Renibacterium salmoninarum*, the culprit behind BKD. Different forms of oral WSSV VP28 delivery systems have been created employing the cyanobacterium *Anabaena* sp., which had a 68% survival rate in crayfish against the WSSV challenge, and *Dunaliella salina*, which had a 59% survival rate.[196,197] A hairpin DNA cassette encoding dsRNA targeted at the yellow head virus (YHV) was successfully integrated in the *C. reinhardtii* nucleus by Somchai et al.[198] The transgenic algae fed *L. vannamei* had partial protection against YHV challenge. Transgenic *Anabaena sp.* PCC 7120, which expressed WSSV VP28 fed *L. vannamei* PL, had higher resistance against WSSV than the control.[199]

9.7 ROLE OF ADJUVANT IN VACCINE DEVELOPMENT

Because these viruses are so difficult to create vaccines for, they are a significant source of the infectious illnesses that presently afflict the world's fisheries. Consequently, DNA vaccines are most likely to provide long-term protection against infectious diseases. Studies on fish have revealed that a DNA plasmid that creates an immunogenic antigen is much more successful than those conducted on other animals, such as mammals.[200] In contrast to other animals, such as humans, fish have demonstrated exceptional reactions to intramuscular injection of a DNA plasmid generating an immunogenic antigen. MHC classes I and II process the antigen, resulting in both humoral and cellular immune responses that are effective in either method.[201] The addition of adjuvants, such as alternative immunization techniques that allow for mass vaccination, may still improve several properties of DNA vaccines despite their benefits over conventional immunizations. It is also necessary to use rational vaccine design principles while developing fish vaccines, using the optimal antigen in combination with a custom adjuvant system to build vaccines that may elicit a stronger immune response against a specific virus while having minimal side effects.

Adjuvants have been used extensively to increase the potency and durability of some immune responses to specific antigens, as well as to enhance vaccine efficacy. They can also help to reduce the dosage numbers (boosters) or the amount of antigen needed per dosage.[202] Adjuvant, which means "to help," is therefore typically thought of as a compound that boosts a vaccine's effectiveness or capacity to prevent infection and death (efficacy). But today's researchers agree that adjuvants could play a bigger role in determining the type of adaptive response that will be made in response to a particular infection. The class of structurally varied compounds known as adjuvants can change the innate immunogenicity of an antigen.[200] It is possible to classify them based on their chemical composition or physical attributes, but new classification schemes have focused on the immunological reactions they trigger because even very closely related compounds appear to have different immunomodulating potentials, even though the precise mechanism of action is unknown for many of them. It is possible to classify them based on their chemical composition or physical attributes, but new classification schemes have focused on the immunological reactions they trigger because even very closely related compounds appear to have different immunomodulating potentials, even though the precise mechanism of action is unknown for many of them. Adjuvant categorization based on signal 1 and signal 2 facilitation has already been extensively recognized by the scientific community. T- and B-cells in the adaptive immune system must be activated by both an antigen presentation (signal 1) and extra secondary signals, according to the two-signal paradigm (signal 2).

The inclusion of an antigen into an emulsion frequently results in a delayed release, which enhances the immunogenicity. An emulsion is a dispersion of a liquid, known as the dispersed phase, in a second liquid, known as the continuous phase, where the first liquid is not miscible. These vaccine formulation phases use oil and water (antigenic medium). The emulsions are stabilized by the addition of surfactants. Commercial vaccines of different viral and bacterial vaccines have been developed by Pharmaq with names like Alpha Ject and Alpha Marine Vaccine with the use of mineral oil as adjuvant for intraperitoneal route of injections.[200] MSD animal health has produced vaccines against *Aeromonas salmonicida* and salmonid alphavirus by using adjuvant Montanide ISA711, Drakeol 6VR, and Montanide ISA763A. Recently, *L. rohita*, *Cirrhinus mrigala*, and *C. idella* were administered varied dosages of a Montanide oil-based inactivated vaccine for *A. hydrophila* to test its effectiveness.[203]

For experimental application, FCA has shown to be the most popular and successful adjuvant. An oil-surfactant mixture and mycobacteria burnt to death are the components of FCA. Pre-injection water dissolution creates a stable W/O emulsion for antigen injection. Using FCA and antigens, the MyD88 pathway is principally responsible for inducing robust Th1 and Th17 responses. As a result of a number of serious adverse effects, such as injection-site granuloma, FCA has only been used in animal studies. There has been a mixed response to the use of FCA in fish when it has been applied. The use of FCA did not substantially improve the protection provided by formalin-killed *S. iniae* cells against an experimental challenge, for example.[204] For the blue gourami (*Trichogaster trichopterus*), an *A. hydrophila* OMP with a molecular mass of 43 kDa was shown to have no effect on the immune system.[205] Toxicology of Freund's incomplete adjuvant (FIA), which lacks mycobacterial components in the W/O emulsion, has generally replaced FCA. Despite the fact that its toxicity has been greatly reduced, this adjuvant is still extraordinarily efficient at immunizing.

9.8 SAFETY DELIVERY METHODS USING NANOMATERIALS

Small particles known as "nondimensional materials" are widely used in applications such as controlled medication release, target-based drug delivery, and to improve biocompatibility and biodegradability.[206,207] There is currently a great deal of interest in research explorations on drug delivery to successfully deploy drug molecules and therapeutic agents to their intended locations for the treatment of various ailments.[208] Because of their adaptable physiochemical and structural characteristics, the use of nanoparticles (NPs) as drug carriers can have significant benefits over conventional therapy.[209] Proteins, hydrophobic and hydrophilic medications, vaccinations, biological macromolecules, and other substances are all delivered via NPs. NPs can be made to circulate throughout the body for a long time or to be administered specifically to the artery walls, lymphatic system, spleen, lungs, brain, and liver.[210]

NPs have been utilized to deliver peptides, proteins, genes, and vaccines to treat inflammation, cardiovascular illness, infectious disease, cerebral disease, several types of cancer, ophthalmic disease, and diabetes.[211] Here, several current investigations of significant biocompatible NPs were reviewed in order to learn more about the precise and safe distribution of different medicinal substances. The nanosystems used in vaccine delivery may improve adjuvant or antigen uptake by APCs, resulting in greater immune responses. Peptides, cell lysates, DNA from plasmids or viruses, proteins, and lipopeptides have all been produced as antigens using PLGA NPs. More effective immune responses are provided by vaccinations with sustained antigen release.[211,212] Inorganic NPs called magnetic nanoparticles (MNP) are adaptable stimuli-sensitive and responsive to an applied magnetic field. MNPs are at the forefront of being used for various biomedical purposes because they have appealing properties that have potential use in biomedicine for treating diseases and disorders. They have the singular property of changing their magnetic behavior in the presence of an external magnetic field where they behave as small magnets but lose their active behavior when the magnetic field is removed.[213] Natural polymers can be created through physical and chemical processes, or they can be altered to create novel materials using developing nanotechnology that are extremely biocompatible for drug delivery. Other widely utilized natural polymer components for drug delivery include chitosan, starch, alginate, cellulose, hyaluronic acid, and chondroitin sulphate.[214]

Numerous biomaterial investigations have raised a lot of interest in chitosan. These intriguing attributes of this polymer include biodegradability, excellent biocompatibility, the ability to be functionalized, antibacterial capabilities, and natural processability.[215] For instance, mesoporous silica NPs treated with chitosan were found to be effective at delivering the medication methotrexate for the treatment of breast cancer.[216] Collagen is a useful delivery system for a variety of substances, including DNA, medications, proteins, and growth hormones.[217] Strong bioactivity, effective in vivo absorption, hemostatic characteristics, and low immunogenicity are all present.[218]

Many studies have employed CS/TPP NPs instead of chitosan to contain peptides, proteins, pDNA, and siRNA.[219–225] During the aqueous phase of the chitosan nanogel-making technique, a chemical cross-linker known as tripolyphosphate (TPP) was injected.[220] Chitosan-TPP NPs have been demonstrated to be more efficient vector siRNA delivery vehicles than chitosan-based siRNA complexes.[224] In order to evaluate these NPs for their potential as gene delivery nanocarriers, they were embedded with plasmid DNA and short oligonucleotides using an ionic gelation approach.[223,226] Due to its ability to penetrate deeper into tissues, chitosan-TPP is better able to deliver proteins and medications to the body. Researchers have discovered a nonviral way of gene delivery by employing chitosan tripolyphosphate (CS/TPP) NPs for oral gene injection in fish, shrimp, and other aquatic species. For these collagen NPs, when retinol was supplied through dermal methods, it was shown to be more stable and migrate through the skin more quickly compared to when it was taken orally.[217,227,228]

9.9 CONCLUSION AND FUTURE DIRECTION

Among the different generations of vaccine development, the third generation has several advantages like strong and long-lasting immunity. The production of effective recombinant vaccines made possible by the use of cutting-edge biotechnological technologies and expression systems will end nearly all disease outbreaks. The recombinant subunit vaccines produced through yeast, baculovirus expression system have several advantages including proper posttranslational modification and protein folding that gives strong and effective immunity compared to the bacterial mode of expression. RNA vaccines also have so many advantages compared to DNA vaccines; they have self-amplifying replicons and that would translate directly in the cytoplasm. The edible vaccine like IgY has been widely used for immunotherapy against infectious diseases, especially in aquaculture operations. They are a viable alternative for applications involving passive vaccination due to their numerous benefits, including lower cost, greater convenience, and less invasiveness than antibody generation in mammals. One of the most efficient ways to deliver vaccines for improved host animal immunity against pathogenic interruption is through efficient vaccine delivery systems. The current delivery system has many drawbacks and the advance delivery systems solve the drawbacks of the previous delivery protocols. Encapsulation of the vaccines with biodegradable NPs like alginate and chitosan are more effective for controlling release and preventing gastric enzymes' degradation. The fish/shrimp larvae benefit greatly from the delivery of vaccinations via transgenic microalgae. It has been demonstrated to aggregate quickly, fold several vaccine antigens correctly, and produce recombinant algal fusion proteins, all of which can increase antigenicity for efficient orally administered vaccinations. This strategy can provide robust defense, strengthen immunity, and provide economic, environmental, and safety benefits, which are especially appealing to farmers and boost their economic standing.

DISCLOSURE STATEMENT

The authors declare that they have no conflicts of interest.

REFERENCES

[1] FAO. The State of World Fisheries and Aquaculture 2018 - Meeting the sustainable development goals. Rome. Licence: CC BYNC-SA 3.0 IGO. https://www.fao.org/documents/card/en/c/I9540EN/

[2] Adams. Progress, challenges and opportunities in fish vaccine development. 90:210–214. ISSN: 1050-4648. https://doi.org/10.1016/j.fsi.2019.04.066

[3] S. Sneeringer, M. Bowman, and M. Clancy. The US and EU Animal Pharmaceutical Industries in the Age of Antibiotic Resistance; USDA Economic Research Service Report Number 264; USDA: Washington, DC, USA, May 2019. https://www.ers.usda.gov/webdocs/publications/93179/err-264.pdf?v=1498

[4] S.F. Snieszko, and S.B. Friddle. Prophylaxis of furunculosis in brook trout (*Salvelinus fontinalis*) by oral immunization and sulfamerazine. Prog. 11:161–168. ISSN: 1522-2055. https://doi.org/10.1577/1548-8640(1949)11[161:POFIBT]2.0.CO;2

[5] R. Gudding, and T. Goodrich. The history of fish vaccination. In *Fish vaccination*, R. Gudding, A. Lillehaug, and O. Evensen, Eds., John. Wiley & Sons, Inc.: New. York, NY, USA, pp. 1–11. ISBN: 9780470674550. https://doi.org/10.1002/9781118806913.ch1

[6] H.D. Rodger. Fish disease causing economic impact in global aquaculture. In: *Fish vaccines*, 1st ed.; A. Adams, Ed.; Springer: Basel, Switzerland, pp. 1–34. ISBN: 978-3-0348-0980-1. https://link.springer.com/chapter/10.1007/978-3-0348-0980-1_1

[7] M.C. Horzinek, V.E.C.J. Schijns, M. Denis, P. Desmettre, and L.A. Babiuk. General description of vaccines. In *Veterinary vaccinology*, P.P. Pastoret, J. Blancou, P. Vannier, and C. Verschueren, Eds.; Elsevier Press: Amsterdam, the Netherlands, pp. 132–152. ISBN 9780444819680 0444819681. https://www.worldcat.org/title/veterinary-vaccinology/oclc/36672023

[8] P.S. Sudheesh, and K.D. Cain. Prospects and challenges of developing and commercializing immersion vaccines for aquaculture. 1:1–20. ISSN: 2572-7168. https://doi.org/10.18103/ibr.v1i1.1313

[9] K.D. Thompson, and R.J. Roberts. 2013. Fish vaccines, In A. Adams, Ed.; Ist Edition, Springer: Basel, Switzerland

[10] Gudding, R, and W.B.Van Muiswinkel. A history of fish vaccination: Science-based disease prevention in aquaculture. 35:1683–1688. ISSN: 1050-4648. https://doi.org/10.1016/j.fsi.2013.09.031

[11] K.P. Plant, and S.E. LaPatra. Advances in fish vaccine delivery. 35: 1256–1262. ISSN: 1879-0089. https://doi.org/10.1016/j.dci.2011.03.007.

[12] B.E. Brudeseth, R. Wiulsrod, B.N. Fredriksen, K. Lindmo, K.E. Lokling, M. Bordevik, N. Steine, A. Klevan, and K. Gravningen. Status and future perspectives of vaccines for industrialised fin-fish farming. 35:1759–1768. ISSN: 1050-4648. https://doi.org/10.1016/j.fsi.2013.05.029

[13] Brun, J. Barcena, E. Blanco, B. Borrego, D. Dory, J.M. Escribano, G. Le Gall-Recule, J. Ortego, and L.K. Dixon. Current strategies for subunit and genetic viral veterinary vaccine development. 157:1–12. ISSN: 0168-1702. https://doi.org/10.1016/j.virusres.2011.02.006

[14] V. Cimica, and J.M. Galarza. Adjuvant formulations for virus-like particle (VLP) based vaccines. 183:99–108. ISSN: 0271-9142. https://doi.org/10.1016/j.clim.2017.08.004

[15] K.M. Frietze, D.S. Peabody, and B. Chackerian. Engineering virus-like particles as vaccine platforms. 18:44–49. ISSN: 1879-6257. https://doi.org/10.1016/j.coviro.2016.03.001

[16] M.K. Bedekar, and S. Kole. Fundamentals of fish faccination, Methods Mol Biol. 2411:147–173. ISSN: 1064-3745. https://doi.org/10.1007/978-1-0716-1888-2_9
[17] M. Hansson, P.A.K. Nygren, S. Ståhl. Design and production of recombinant subunit vaccines. 32: 95–107. ISSN: 0885-4513. https://doi.org/10.1042/ba20000034
[18] L. Holten-Andersen, T.M. Doherty, K.S. Korsholm, P. Andersen. Combination of the cationic surfactant dimethyl dioctadecyl ammonium bromide and synthetic mycobacterial cord factor as an efficient adjuvant for tuberculosis subunit vaccines. 72: 1617. ISSN: 0019-9567. https://doi.org/10.1128/iai.72.3.1608-1617.2004
[19] J. Ma, T.J. Bruce, E.M. Jones, K.D. Cain. A review of fish vaccine development strategies: Conventional methods and modern biotechnological approaches. 7: 569. ISSN: 2076-2607. https://doi.org/10.3390/microorganisms7110569
[20] S. LaPatra, S. Kao, E.B. Erhardt and I. Salinas. Evaluation of dual nasal delivery of infectious hematopoietic necrosis virus and enteric red mouth vaccines in rainbow trout (*Oncorhynchus mykiss*). 33(6):771–776. ISSN: 0264-410X ULR https://doi.org/10.1016/j.vaccine.2014.12.055
[21] L.B. Hølvold, A.I. Myhr, and R.A. Dalmo. Strategies and hurdles using DNA vaccines to fish. 45(1): 21. ISSN: 0928-4249. https://doi.org/10.1186/1297-9716-45-21
[22] A.D. Pasquale, S. Preiss, F.T. Silva, and N. Garcon. Vaccine adjuvants: From 1920 to 2015 and beyond. 3: 320–343. ISSN: 0264-410X. https://doi.org/10.3390%2Fvaccines3020320
[23] R.A. Seder, and A.V. Hill. Vaccines against intracellular infections requiring cellular immunity. 406:793. ISSN: 1476-4687. https://doi.org/10.1038/35021239
[24] K.E. Nusbaum, and E.E. Morrison. Entry of 35 S-labeled *Edwardsiella ictaluri* into channel catfsh (*Ictalutus punctatus*). 8: 146–149. ISSN: 1548-8667. https://doi.org/10.1577/1548-8667(1996)008%3C0146:CEOSLE%3E2.3.CO;2
[25] Ø. Evensen, and J.A.C. Leong. DNA vaccines against viral diseases of farmed fish. 35: 1751–1758. ISSN: 1050-4648. https://doi.org/10.1016/j.fsi.2013.10.021
[26] P. van Gelder, and B. Makoschey. (2012). Production of viral vac-cines for veterinary use. 125: 103–109 ISSN: 0005-9366. http://dx.doi.org/10.2376/0005-9366-125-103
[27] H. W. Cho, C. R. Howard, H.W. Lee. Review of an inactivated vaccine against hantaviruses. 45:328–333. ISSN: 0300-5526. http://dx.doi.org/10.1159/000067925
[28] H. Arimitsu, J.C. Lee, Y. Sakaguchi, Y. Hayakawa, M. Hayashi, and M. Nakaura. Vaccination with recombinant whole heavy chain fragments of *Clostridium botulinum* type C and D neurotoxins. 11(3): 496–502. ISSN: 1556-6811. http://dx.doi.org/10.1128/CDLI.11.3.496-502.2004
[29] J.C. Leong, and J.L. Fryer. Viral vaccines for aquaculture. 3: 225–240. ISSN: 0959-8030. https://doi.org/10.1016/0959-8030(93)90036-B
[30] O. Dellagostin, A.A. Grassmann, D. D. Hartwig, S. R. Félix, É. F. da Silva, and J.A. A. McBride. Recombinant vaccines against Leptospirosis. 7(11): 1215–1224. ISSN: 2164-5515. ULR https://doi.org/10.4161/hv.7.11.17944
[31] Estepa, A, M. Thiry, and J.M. Coll. Recombinant protein fragments from haemorrhagic septicaemia rhabdovirus stimulate trout leukocyte anamnestic responses *in vitro.* 75(6): 1329–1338. ISSN: 0022-1317. https://doi.org/10.1099/0022-1317-75-6-1329
[32] H.M. Fang, K.C. Ling, R. Ge, and Y.M. Sin. Enhancement of protective immunity in blue gourami, *Trichogaster trichopterus* (Pallas), against *Aeromonas hydrophila* and *Vibrio anguillarum* by *A. hydrophila* major adhesin. 23: 137–145. ISSN: 1078-5884. https://doi.org/10.1046/j.1365-2761.2000.00229.x
[33] H.Y. Pang, Y. Li, Z.H. Wu, J.C. Jian, Y.S. Lu, S.H. Cai. Immunoproteomic analysis and identification of novel immunogenic proteins from *Vibrio harveyi.* 109(5): 1800–1809. ISSN: 0021-8847. https://doi.org/10.1111/j.1365-2672.2010.04808.x
[34] V. Thanga Viji, K. Deepa, S. Velmurugan, M. Birdilla Selva Donio, J. Adlin Jenifer, M. Michael Babu, and T. Citarasu. Vaccination strategies to protect goldfish *Carassius auratus* against *Aeromonas hydrophila* infection. 104: 45–57. ISSN: 1616-1580. https://doi.org/10.3354/dao02581
[35] E. Emmenegger, M. Landolt, S. LaPatra, J. Winton. Immunogenicity of synthetic peptides representing antigenic determinants on the infectious hematopoietic necrosis virus glycoprotein. 28:175–184. ISSN: 0928-4249. https://pubmed.ncbi.nlm.nih.gov/8581009/
[36] M.O. Saeed. Chemical characterization of the lipopolysacchande of *Edwardsiella ictaluri* and the immune response of channel catfish to this fracbon and to whole cell antigen with histopathological comparisons. PhD dissertation, Auburn University, Alabama
[37] M. Saeed, J. Plumb. Serological detection of *Edwardsiella ictaluri* Hawke lipopolysaccharide antibody in serum of channel catfish, *Ictalurus punctatus* Rafinesque. J Fish Dis 10:205–209. ISSN: 1078-5884. https://doi.org/10.1111/j.1365-2761.1987.tb01062.x

[38] F. Salati, Y. Ikeda, and R. Kusuda. Effect of *Edwnrdsiella rarda* lipopolysaccharide immunization on phagocytosis in eel. 53: 201–204. ISSN: 0021-5392. https://www.jstage.jst.go.jp/article/suisan1932/53/2/53_2_201/_pdf

[39] H. Zhu, C. S. Rollier, and A. J. Pollard. Recent advances in lipopolysaccharide-based glycoconjugate vaccines, 20(12): 1515–1538. ISSN: 1476-0584. https://DOI.ORG/10.1080/14760584.2021.1984889

[40] D.B. Iliev, G.W. Goetz, S. Mackenzie, J.V. Planas, and F.W. Goetz. Pathogen-associated gene expression profiles in rainbow trout macrophages. 20(12):1515–1538. ISSN: 1744-117X. https://doi.org/10.1016/j.cbd.2006.10.003

[41] A. Loera-Muro, A. Guerrero-Barrera, D.N.Y. Tremblay, S. Hathroubi, C. Angulo. Bacterial biofilm-derived antigens: a new strategy for vaccine development against infectious diseases. 20(4):385–396. ISSN: 1476-0584. https://doi.org/10.1080/14760584.2021.1892492

[42] I.S. Azad, K. M Shankar, C.V. Mohan, and B. Kalita. Biofilm vaccine of *Aeromonas hydrophila* – Standardization of dose and duration for oral vaccination of carps. 9: 519–528. https://doi.org/10.1006/fsim.1998.0206

[43] H. Koo, R. Allan, R. P. Howlin, P. Stoodley, and L. Hall-Stoodley. Targeting microbial biofilms: current and prospective therapeutic strategies. 15: 740–755. ISSN: 1740-1526. https://doi.org/10.1038/nrmicro.2017.99

[44] A. Loera-Muro, and C. Angulo. New trends in innovative vaccine development against *Actinobacillus pleuropneumoniae*. 217: 66–75. ISSN: 0378-1135. https://doi.org/10.1016/j.vetmic.2018.02.028

[45] H.C. Flemming, J. Wingender. The biofilm matrix. **8:** 623–633. ISSN: 1740-1526. https://doi.org/10.1038/nrmicro2415

[46] D. Nayak, A. Asha, K. Shankar, and C. Mohan. Evaluation of biofilm of *Aeromonas hydrophila* for oral vaccination of *Clarias batrachus*: A carnivore model. 16: 613–619. ISSN: 1050-4648. https://doi.org/10.1016/j.fsi.2003.09.012

[47] B. T. Kaur, A. Naveen Kumar, S. Tyagi, A. Holeyappa, and N. K. Singh. Identification of novel vaccine candidates in the whole-cell *Aeromonas hydrophila* biofilm vaccine through reverse vaccinology approach. 114:132–141. ISSN: 1050-4648. https://doi.org/10.1016/j.fsi.2021.04.019

[48] M. Kahieshesfandiari, M.Y. Sabri, M.Y. Ina-Salwany, M.D. Hassan, O. Noraini, A.A. Ajadi, and A.I. Isiaku, Streptococcosis in *Oreochromis* sp.: Is feed-based biofilm vaccine of *Streptococcus agalactiae* effective? 27: 817–832. ISSN: 1573-143X. https://doi.org/10.1007/s10499-019-00372-8

[49] M.K. Ram, B.N. Kumar, S.R. Poojary, P. Abhiman, P. Patil, K. Ramesh, and K. Shankar, Evaluation of biofilm of *Vibrio anguillarum* for oral vaccination of Asian seabass, *Lates calcarifer* (Bloch, 1790). 94: 746–751. ISSN: 1050-4648. https://doi.org/10.1016/j.fsi.2019.09.053

[50] S. Kitiyodom, C. Trullàs, C. Rodkhum, K. D. Thompson, T. Katagiri, S. Temisak, K. Namdee, T. Yata, and N. Pirarat. Modulation of the mucosal immune response of red tilapia (*Oreochromis sp.*) against columnaris disease using a biomimetic-mucoadhesive nanovaccine. 112: 81–91. ISSN: 1050-4648. https://doi.org/10.1016/j.fsi.2021.02.017

[51] I. Azad, K. Shankar, C. Mohan, and B. Kalita. Uptake and processing of biofilm and free-cell vaccines of *Aeromonas hydrophila* in Indian major carps and common carp following oral vaccination antigen localization by a monoclonal antibody. 43:103–108. ISSN: 0177-5103. https://doi.org/ 10.3354/dao043103.

[52] F.J. Su, and M.M. Chen. Protective efficacy of novel oral biofilm vaccines against *Lactococcus garvieae* infection in mullet, *Mugil cephalus*. 9:844. ISSN: 2076-393X. https://doi.org/ 10.3390/vaccines9080844

[53] F.J. Su, M.M. Chen. Protective efficacy of novel oral biofilm vaccines against *Photobacterium damselae* subsp. damselae infection in giant grouper, *Epinephelus lanceolatus*. 10:207. ISSN: 2076-393X. https://doi.org/10.3390/ vaccines10020207.

[54] M.J. Zorriehzahra, R. Tiwari, S. Sachan, K. Karthik, Y.S. Malik, M. Dadar, M. Sarwar, M. Sayab and K. Dhama. Avian egg yolk antibodies (IgY) and their potential therapeutic applications for countering infectious diseases of fish and aquatic animals. 12:760–768. eISSN: 1812-5700 pISSN: 1811-7775. https://doi.org/10.3923/ijp.2016.760.768.

[55] Y. Xu, X. Li, L. Jin, Y. Zhen, Y. Lu, S. Li, S. J. You, and L. Wang. Application of chicken egg yolk immunoglobulins in the control of terrestrial and aquatic animal diseases: A review. 29:860–868. ISSN: 0734-9750. https://doi.org/10.1016/j.biotechadv.2011.07.003.

[56] G.W. Warr, K.E. Magor, and D.A. Higgins. IgY: Clues to the origins of modern antibodies. 16:392–398. ISSN: 0167-5699. https://doi.org/10.1016/0167-5699(95)80008-5.

[57] L.S. Munhoz, G.D.A. Vargas, G. Fischer, M. Lima, P.A. Esteves, and S.O. Hubner. 2014. Avian IgY antibodies: characteristics and applications in immunodiagnostic. 44:153–160. ISSN: 0103-8478. https://doi.org/10.1590/S0103-84782014000100025.

[58] M.A. Gutierrez, T. Miyazaki, H. Hatta, and M. Kim. Protective properties of egg yolk IgY containing anti-*Edwardsiella tarda* antibody against paracolo disease in the Japanese eel, *Anguilla japonica* Temminck & Schlegel. 16:113–122. ISSN: 1365-2761. https://doi.org/10.1111/j.1365-2761.1993.tb00854.x

[59] X.L. Li, J.B. Shuai, and W.H. Fang. Protection of *Carassius auratus Gibelio* against infection by *Aeromonas hydrophila* using specific immunoglobulins from hen egg yolk. 7:922–928. ISSN: 1673-1581. https://doi.org/10.1631/jzus.2006.B0922.

[60] P. Srisapoome, K. Punyokun, R. Hongprayoon, and N. Areechon. Passive immunization of anti-*Vibrio harveyi* egg yolk immunoglobulin against luminous disease in black tiger shrimp (*Penaeus monodon*). Proceedings of the 44th Kasetsart University Annual Conference (pp. 365374). Bangkok: Kasetsart University.

[61] Y.N. Lu, J.J. Liu, L.J. Jin, X.Y. Li, Y.H. Zhen, and H.Y. Xue. Passive protection of shrimp against white spot syndrome virus (WSSV) using specific antibody from egg yolk of chickens immunized with inactivated virus or a WSSV-DNA vaccine. 25:604–610. ISSN: 1050-4648. https://doi.org/10.1016/j.fsi.2008.08.010

[62] T. Kumaran, M. Michaelbabu, T. Selvaraj, S. Albindhas, and T. Citarasu. Production of anti WSSV IgY edible antibody using herbal immunoadjuvant *Asparagus racemosus* and its immunological influence against WSSV infection in *Penaeus monodon*. 2:1–5. ISSN: 2070-1667. https://doi.org/10.3923/joafsnu.2010.1.5.

[63] T. Kumaran, E. Thirumalaikumar, C. Lelin, P. Palanikumar, M. Michaelbabu, and T. Citarasu. Physicochemical properties of anti-*vibrio harveyi* egg yolk antibody (IgY) and its immunological influence in Indian white shrimp *Fenneropenaeus indicus*. 74:349–362. ISSN: 1050-4648. https://doi.org/10.1016/j.fsi.2017.12.062.

[64] K. Punyokun, R. Hongprayoon, P. Srisapoome, and T. Sirinarumitre. Influence of chitosan-alginate microcapsules containing anti-*Vibrio harveyi* IgY in the gastrointestinal tract simulation. 9:110. ISSN: 1913-1844. http://dx.doi.org/10.5539/mas.v9n12p110.

[65] X. Gao, H. Zhang, Q. Jiang, N. Chen, X. Li, X. Liu, H. Yang, W. Wanhong, and X. Zhang. *Enterobacter cloacae* associated with mass mortality in zoea of giant freshwater prawns *Macrobrachium rosenbergii* and control with specific chicken egg yolk immunoglobulins (IgY). 501:331–337. ISSN: 0044-8486. https://doi.org/10.1016/j.aquaculture.2018.11.050

[66] Craig A. Shoemaker, Phillip H. Klesius, Joyce J. Evans, and Covadonga R. Arias. Use of modified live vaccines in Aquaculture. 40(5):573–585. ISSN: 1749-7345. https://doi.org/10.1111/j.1749-7345.2009.00279.x.

[67] X. Liu, H. Wu, Q. Liu, Q. Wang, J. Xiao, and X. Chang. Profiling immune response in zebrafish intestine, skin, spleen and kidney bath-vaccinated with a live attenuated *Vibrio anguillarum* vaccine. 18:235–242. ISSN: 1050-4648. https://doi.org/10.1016/j.fsi.2015.04.028.

[68] J.T. Buchanan, J.A. Stannard, X. Lauth, V.E. Ostland, H.C. Powell, M.E. Westerman, and V. Nizet. *Streptococcus iniae* phosphoglucomutase is a virulence factor and target for vaccine development. Infection and Immunity 73(10):6935–6944. ISSN: 1098-5522. https://doi.org/10.1128/IAI.73.10.6935-6944.2005

[69] M.L. Lawrence, and M.M. Banes. Tissue persistence and vaccine efficacy of an O polysaccharide mutant strain of *Edwardsiella ictaluri*. 17:228–232. ISSN: 1548-8667. https://doi.org/ 10.1128/IAI.73.10.6935-6944.2005.

[70] J.C Leong, E. Anderson, L.M. Bootland, P.W. Chiou, M. Johnson, C. Kim, D. Mourich, and G. Trobridge. Fish vaccine antigens produced or delivered by recombinant DNA technologies. 90:267–277. ISSN: 0301-5149.

[71] A. Igarashi, and T. Iida. A vaccination trial using live cells of *Edwardsiella tarda* in tilapia. 37:145–148. ISSN: 0388-788X. https://doi.org/ 10.3147/jsfp.37.145

[72] M. Marsden, L. Vaughan, R. Fitzpatrick, T. Foster, and C. Secombes. Potency testing of a live, genetically attenuated vaccine for salmonids. Vaccine 16: 1087–1094. ISSN: 2076-393X. https://doi.org/10.1016/S0264-410X(98)80103-5.

[73] V.B. Pereiro, M. Zurita-Turk, T.D.L. Saraiva, C.B.D. Castro, B.M. Souza, P.M. Agresti, F.A. Lima, V.N. Pfeiffer, M.S.P. Azevedo, C.S. Rocha, D.S. Pontes, V. Azevedo, and A. Miyoshi. DNA vaccines approach: from concepts to applications. 4:50–71. ISSN: 2160-5815. http://doi:10.4236/wjv.2014.42008.

[74] M. Gomez-Chiarri, and L.A. Chiaverini. Evaluation of eukaryotic promoters for the construction of DNA vaccines for aquaculture. 15:121–124. ISSN: 0735-0651. https:doi.org/10.1016/s1050-3862(99)00014-5.

[75] M. R. Delghandi, M. El-Matbouli, and S. Menanteau-Ledouble. *Renibacterium salmoninarum*-The causative agent of bacterial kidney disease in salmonid fish. 9:845. ISSN: 20760817. https://doi.org/10.3390/pathogens9100845.

[76] J. Xing, H. Xu, X. Tang, X. Sheng, and W. Zhan. A DNA vaccine encoding the VAA gene of *Vibrio anguillarum* induces a protective immune response in flounder. 10:499. ISSN: 1664-3224. https://doi.org/10.3389/fimmu.2019.00499

[77] R.C. Vazquez-Juarez, H.A. Barrera-Saldana, N.Y. Hernández-Saavedra, M. Gomez-Chiarri, and F. Ascencio. Molecular cloning, sequencing and characterization of omp48, the gene encoding for an antigenic outer membrane protein from *Aeromonas veronii*. 94(5):908–918. ISSN: 1365-2672. https://doi.org/10.1046/j.1365-2672.2003.01928.x.

[78] A. Ballesteros, Natalia, Alonso Marta, Saint-Jean, Sylvia Rodríguez, Perez-Prieto, and I. Sara. An oral DNA vaccine against infectious haematopoietic necrosis virus (IHNV) encapsulated in alginate microspheres induces dose-dependent immune responses and significant protection in rainbow trout *(Oncorrhynchus mykiss*). 45(2):877–888. ISSN: 1050-4648. https://doi.org/10.1016/j.fsi.2015.05.045

[79] P.E. Mc Lauchlan, B. Collet, E. Ingerslev, C.J. Secombes, N. Lorenzen, and A.E. Ellis. DNA vaccination against viral haemorrhagic septicaemia (VHS) in rainbow trout: Size, dose, route of injection and duration of protection—early protection correlates with Mx expression. 15(1):0–50. ISSN: 1050-4648. https://doi.org/10.1016/s1050-4648(02)00137-7.

[80] S. Rajeshkumar, C. Venkatesan, M. Sarathi, V. Sarathbabu, J. Thomas, K.A. Basha, and A.S. Hameed. Oral delivery of DNA construct using chitosan nanoparticles to protect the shrimp from white spot syndrome virus (WSSV). 26:429–437. ISSN: 1050-4648. https://doi.org/10.1016/j.fsi.2009.01.003.

[81] N. Rout, S. Kumar, S. Jaganmohan, amd V. Murugan. DNA vaccines encoding viral envelope proteins confer protective immunity against WSSV in black tiger shrimp. 25:2778–2786. ISSN: 2076-393X. https://doi.org/10.1016/j.vaccine.2006.12.056.

[82] X. Li, Q. Liu, L. Hou, and J. Huang. Effect of VP28 DNA vaccine on white spot syndrome virus in *Litopenaeus vannamei*. 18:1035–1044. ISSN: 1573-143X, 0967-6120. http://doi:10.1007/s10499-010-9321-z.

[83] T. Citarasu, C. Lelin, E. Thirumalaikumar, M.M. Babu, and V.N. Vakharia. *Macrobrachium rosenbergii* nodavirus (*Mr*NV)-CP-RNA-2 DNA vaccine confers protective immunity in giant freshwater prawn *Macrobrachium rosenbergii* against *Mr*NV infection. 86:319–326. ISSN: 1050-4648. https://doi.org/10.1016/j.fsi.2018.11.049.

[84] Xuting Wang, and Harry W. Dickerson. Surface ımmobilization antigen of the parasitic ciliate *Ichthyophthirius multifiliis* elicits protective ımmunity in channel catfish (*Ictalurus punctatus*). 176–181. ISSN: 1071-412X. https://doi.org:/10.1128/CDLI.9.1.

[85] P.T.K. Woo. Immunological and therapeutic strategies against salmonid cryptobiosis. 1–9. ISSN: 2314-6133. https://doi.org/10.1155/2010/341783

[86] N. Pardi, M. Hogan and F. Porter. mRNA vaccines — A new era in vaccinology. 17:261–279. ISSN: 1474-1776. https://doi.org/10.1038/nrd.2017.243.

[87] R. Verbeke, I. Lentacker, S.C. De Smedt, and H. Dewitte. The dawn of mRNA vaccines: The COVID-19 case. 333:511–520. ISSN: 0168-3659. https://doi.org/10.1016/j.jconrel.2021.03.043

[88] S. Jain, A. Venkataraman, M.E. Wechsler, and N.A. Peppas. Messenger RNA-based vaccines: Past, present, and future directions in the context of the COVID-19 pandemic. 9(179):114000. ISSN: 0169-409X. https://doi.org/10.1016/j.addr.2021.114000.

[89] R. Weiss, S. Scheiblhofer, and J. Thalhamer. Generation and evaluation of prophylactic mRNA vaccines against allergy. 1499:123–139. ISSN: 1064-3745. https://doi.org/10.1007/978-1-4939-6481-9_7.

[90] M. Karlsen, S. Villoing, E. Rimstad, and A. Nylund. Characterization of untranslated regions of the salmonid alphavirus 3 (SAV3) genome and construction of a SAV3 based replicon. 6:173. ISSN: 1743-422X. https://doi.org/10.1186/1743-422X-6-173

[91] A. Wolf, K. Hodneland, P. Frost, S. Braaen, E. Rimstad. A hemagglutinin-esterase-expressing salmonid alphavirus replicon protects Atlantic salmon (*Salmo salar*) against infectious salmon anemia (ISA). 31(4):661–669. ISSN: 2076-393X. https://doi.org/10.1016/j.vaccine.2012.11.045

[92] G. V. S. Deva, R. Krishnan, K.U. Shyam, and Megha K Bedekar. mRNA vaccination – A way forward for aquaculture vaccine development. July – August 2017 Aquaculture Times May-June 2018

[93] A.I. Heinson, C.H. Woelk, and M.L. Newell. The promise of reverse vaccinology. 7(2): 85–89. ISSN: 1876-3405. https://doi.org/10.1093/inthealth/ihv002.

[94] J.M. Vlak, and R.J. Keus. Baculovirus expression vector system for production of viral vaccines. 14:91–128. ISSN: 0734-9750. https://pubmed.ncbi.nlm.nih.gov/2222839/

[95] O.A. Dellagostin, A.A. Grassmann, D.D. Hartwig, S.R. Felix, E.F. da Silva, and A.J.A. McBride. Recombinant vaccines against leptospirosis. 7(11):1215–1224. ISSN: 2076-393X. http://dx.doi.org/10.4161/hv.7.11.17944.

[96] E. Thirumalaikumar, R. Sathishkumar, S. Vimal, M. Michael Babu, G. Uma, A. M. Lusiastuti, T. Citarasu, Efficacy of recombinant subunit OMP and hly vaccines against *Aeromonas hydrophila* in Rohu (*Labeo rohita*). ISSN: 0140-7775. https://doi.org/10.1111/jfd.13682.
[97] S. Poobalane, K. D. Thompson, L. Ardo, N. Verjan, H.Han, G. Jeney, I. Hirono, T. Aoki, and A. Adams. Production and efficacy of an *Aeromonas hydrophila* recombinant S-layer protein vaccine for fish 28(20):3540–3547. ISSN: 2076-393X. https://doi.org/10.1016/j.vaccine.2010.03.011
[98] K. Sun, W. Zhang, J. Hou, and L. Sun, 2009. Immunoprotective analysis of VhhP2, A *Vibrio harveyi* vaccine candidate. 27:2733–2740. ISSN: 2076-393X. https://doi.org/10.1016/j.vaccine.2009.03.012.
[99] K.Q. Wei, and Z.R. Xu, 2005. Effect of white spot syndrome virus envelope protein Vp28 expressed in silkworm (*Bombyx mori*) pupae on disease resistance in *Procambarus clarkii*. 38(3):190–198. https://pubmed.ncbi.nlm.nih.gov/16044911/
[100] C.M.A. Caipang, N. Verjan, E.L. Ooi, H. Kondo, I. Hirono, T. Aoki, H. Kiyono, and Y. Yuki. Enhanced survival of shrimp, *Penaeus* (*Marsupenaeus*) *japonicus* from white spot syndrome disease after oral administration of recombinant VP28 expressed in *Brevibacillus brevis*. 25:315–320. ISSN: 1095-9947. https://doi.org/10.1016/j.fsi.2008.04.012.
[101] Y.M. Ha, G. Soo-Jung, N. Thi-Hoai, C.H. Ra, K.H. Kim, Y.K. Nam, and S.K. Kim. Vaccination of shrimp *(Penaeus chinensis*) against white spot syndrome virus (WSSV). 18:964–967. ISSN: 1738-8872. https://pubmed.ncbi.nlm.nih.gov/18633299/
[102] B. Vaseeharan, T. Prem Anand, T. Murugan, and J.C. Chen. Shrimp vaccination trials with the VP292 protein of white spot syndrome virus. 43:137–142. ISSN: 0266-8254, 1472-765X. http://doi:10.1111/j.1472-765x.2006.01941.x.
[103] A. Namikoshi, J.L. Wu, T. Yamashita, T. Nishizawa, T. Nishioka, M. Arimoto, and K. Muroga. Vaccination trials with *Penaeus japonicus* to induce resistance to white spot syndrome virus. 229:25–35. ISSN: 0044-8486. https://doi.org/10.1016/S0044-8486(03)00363-6
[104] E. Maftuch Prasetio, A. Sudianto, M. Rozik, R. Nurdiyani, E. Sanusi, H. Nursyam, F. Fariedah, Marsoedi, and Murachman. Improvement of innate immune responses and defense activity in tiger shrimp (*Penaeus monodon* Fab.) by intramuscular administration of the outer membrane protein *Vibrio alginolyticus*. 3(2):432. ISSN: 2193-1801. https://doi.org/ 10.1186/2193-1801-2-432.
[105] M.A. Farook, N. Sundar Raj, N. Madan, S. Vimal, S. Abdul Majeed, G. Taju, T. Rajkumar, S. Santhoshkumar, S. Sivakumar, and A.S. Sahul Hameed. Immunomodulatory effect of recombinant *Macrobrachium rosenbergii* nodavirus capsid protein (r-MCP) against white tail disease of giant freshwater prawn, *Macrobrachium rosenbergii* (de Man, 1879). 433:395–403. ISSN: 0044-8486. https://doi.org/10.1016/j.aquaculture.2014.07.004.
[106] T. Citarasu, C. Lelin, M.M. Babu, S.B. Anand, A.A. Nathan, and V.N. Vakharia. Oral vaccination of *Macrobrachium rosenbergii* with baculovirus-expressed *M. rosenbergii* nodavirus (*Mr*NV) capsid protein induces protective immunity against *Mr*NV challenge. 86:1123–1129. ISSN: 1095-9947. https://doi.org/10.1016/j.fsi.2018.12.010
[107] J.A. Rosenthal, L. Chen, J.L. Baker, D. Putnam, and M.P. DeLisa. Pathogen-like particles: Biomimetic vaccine carriers engineered at the nanoscale. 28:51–58. ISSN: 0958-1669. https://doi.org/10.1016/j.copbio.2013.11.005
[108] R. Noad, and P. Roy. Virus-like particles as immunogens. 11:438–444. ISSN: 0966-842X. https://doi.org.10.1016/s0966-842x(03)00208-7.
[109] S.A. Keller, M. Bauer, V. Manolova, S. Muntwiler, P. Saudan, and M.F. Bachmann. Cutting edge: Limited specialization of dendritic cell subsets for. MHC class. II-associated presentation of viral particles. 184:26–29. ISSN: 0022-1767. https://doi.org/10.4049/jimmunol.0901540.
[110] R.B. Shivappa, P.E. McAllister, G.H. Edwards, N. Santi, O. Evensen, and V.N. Vakharia. Development of a subunit vaccine for infectious pancreatic necrosis virus using a baculovirus insect/larvae system. 121:165–174. ISSN: 0012-1606. https://europepmc.org/article/med/15962479
[111] W. Liu, C. Hsu, C. Chang, H. Chen, and C. Lin. Immune response against grouper nervous necrosis virus by vaccination of virus-like particles, 24:6282–6287. ISSN: 2076-393X. https://doi, org/10.1016/j.vaccine.2006.05.073
[112] M.H. Chien, S.Y. Wu, and C.H. Lin. Oral immunization with cell-free self-assembly virus-like particles against orange-spotted grouper nervous necrosis virus in grouper larvae, *Epinephelus coioides*. 197:69–75. ISSN: 0165-2427. https://doi.org/10.1016/j.vetimm.2018.01.012
[113] S.Y. Cho, H.J. Kim, N.T. Lan, H.J. Han, D.C. Lee, J.Y. Hwang, M.G. Kwon, B.K. Kang, S.Y. Han, and H. Moon. Oral vaccination through voluntary consumption of the convict grouper *Epinephelus septemfasciatus* with yeast producing the capsid protein of red-spotted grouper nervous necrosis virus. 204:159–164. ISSN: 0378-1135. https://doi.org/10.1016/j.vetmic.2017.04.022.

[114] A.K. Dhar, R.M. Bowers, C.G. Rowe, and F.T. Allnutt. Expression of a foreign epitope on infectious pancreatic necrosis virus VP2 capsid protein subviral particle (SVP) and immunogenicity in rainbow trout. 85:525–531. ISSN: 0166-3542. https://doi.org/ 10.1016/j.antiviral.2009.12.009.

[115] P. Jariyapong, C. Chotwiwatthanakun, S. Direkbusarakom, I. Hirono, S. Wuthisuthimethavee, and W. Weerachatyanukul. Delivery of double stranded RNA by *Macrobrachium rosenbergii* nodavirus-like particles to protect shrimp from white spot syndrome virus, Aquaculture 435:86–91. ISSN: 0044-8486. https:doi.org/ 10.1016/j.aquaculture.2014.09.034

[116] S. Bambini and R. Rappuoli, The use of genomics in microbial vaccine development. 14(5–6):252–260. ISSN: 1359-6446. https://doi.org/10.1016/j.drudis.2008.12.007

[117] K.L. Seib, X. Zhao, and R. Rappuoli. Developing vaccines in the era of genomics: A decade of reverse vaccinology. 18(5):109–16. ISSN: 1198-743X. https://dx.doi.org/10.1111/j.1469-0691.2012.03939.x.

[118] Rappuoli, R. Reverse vaccinology. 3:445–450. ISSN: 1369-5274. https://doi.org/10.1016/s1369-5274(00)00119-3.

[119] J. Chukwu-Osazuwa, T. Cao, I. Vasquez, H. Gnanagobal, A. Hossain, V.I. Machimbirike, and J. Santander. Comparative reverse vaccinology of *Piscirickettsia salmonis*, *Aeromonas salmonicida*, *Yersinia ruckeri*, *Vibrio anguillarum* and *Moritella viscose*. 10: 473. ISSN: 2076-393X. https://doi.org/10.3390/ vaccines10030473.

[120] J.E.P. Larsen, O. Lund, and M. Nielsen. Improved method for predicting linear B-cell epitopes. 2:2. ISSN: 1745–7580. https://doi.org/10.1186/1745-7580-2-2.

[121] S.K. Injamamul Islam, Moslema Jahan Mou, and Saloa Sanjida. Application of reverse vaccinology for designing of an mRNA vaccine against re-emerging marine birnavirus affecting fish species. 30:100948. ISSN: 2352–9148. https://doi.org/10.1016/j.imu.2022.100948.

[122] S. Moller, M.D. Croning, and R. Apweiler. Evaluation of methods for the prediction of membrane spanning regions. 17(7):646–53. ISSN: 1367-4803. https://doi.org/ 10.1093/bioinformatics/17.7.646.

[123] D.B. Craig, and A.A. Dombkowski. Disulfide by Design 2.0: a web-based tool for disulfide engineering in proteins. 14:346. ISSN: 1471-2105. https://doi.org/10.1186/1471-2105-14-346.

[124] J. Xu, M. Mc Partlon, and J. Li. Improved protein structure prediction by deep learning irrespective of co-evolution information. 3:601–609. ISSN: 2522-5839. https://doi.org/10.1038/s42256-021-00348-5.

[125] D. Kozakov, D. R. Hall, B. Xia, K. A. Porter, D. Padhorny, C. Yueh, D. Beglov, and S. Vajda. The ClusPro web server for protein-protein docking. 12(2):255–78. ISSN: 1754-2189. https:doi.org/10.1038/nprot.2016.169.

[126] F. Bagnoli, B. Baudner, R.P.N. Mishra, E. Bartolini, L. Fiaschi, P. Mariotti, V. Nardi-Dei, P. Boucher, and R. Rappuoli. Designing the next generation of vaccines for global public health. 15(9):545–566. ISSN: 1557-8100. http://doi.org/10.1089/omi.2010.0127.

[127] J. Sergio, O.A. Dellagostin. The development of veterinary vaccines: A review of traditional methods and modern biotechnology approaches. 1(1):S2452072117300667. ISSN: 2452-0721. https://doi.org/10.1016/j.biori.2017.10.001.

[128] F. Andreoni, G. Amagliani, and M. Magnani. Selection of vaccine candidates for fish pasteurellosis using reverse vaccinology and an *in vitro* screening approach vaccine design. 1404:181–192. ISSN: 1064-3745. https://doi.org/10.1007/978-1-4939-3389-1_12

[129] W. Li, H.Q. Wang, R.Z. He, Y.W. Li, Y.L. Su, and A.X. Li. Major surfome and secretome profile of *Streptococcus agalactiae* from Nile tilapia *(Oreochromis niloticus)*: insight into vaccine development. 55:737–746. ISSN: 1095-9947. https://doi.org/10.1016/j.fsi.2016.06.006.

[130] R. Mahendran, S. Jeyabaskar, G. Sitharaman, R.D. Michael, and A.V. Paul. Computer-aided vaccine designing approach against fish pathogens *Edwardsiella tarda* and *Flavobacterium columnare* using bioinformatics softwares. ISSN: 1177-8881. 10:1703–1714. https://doi.org/10.2147/DDDT.S95691.

[131] S. Biacchesi, A. Lamoureux, E. Merour, J. Bernard, and M. Bremont. Limited interference at the early stage of infection between two recombinant novirhabdoviruses: Viral hemorrhagic septicemia virus and infectious hematopoietic necrosis virus. 84(19):10038–10050. ISSN: 0022-538X. https://doi.org:10.1128/JVI.00343-10.

[132] A. Ammayappan, S.E. Lapatra, and V.N. Vakharia. A vaccinia-virus-free reverse genetics system for infectious hematopoietic necrosis virus. 167(2):132–139. ISSN: 0166-0934. https://doi.org/10.1016/j.jviromet.2010.03.023.

[133] C. Moriette, M. Leberre, A. Lamoureux, T.L. Lai, and M. Bremont. Recovery of a recombinant salmonid alphavirus fully attenuated and protective for rainbow trout. 80(8):4088–4098. ISSN: 0022-538X. https://doi.org/10.1128/JVI.80.8.4088-4098.2006.

[134] K. Yao, and V.N. Vakharia. Generation of infectious pancreatic necrosis virus from cloned cDNA. 72(11):8913–8920. ISSN: 0022-538X. https://doi.org/10.1128/JVI.72.11.8913-8920.

[135] T. Iwamoto, K. Mise, K. Mori, M. Arimoto, T. Nakai, and T. Okuno. Establishment of an infectious RNA transcription system for Striped jack nervous necrosis virus, the type species of the betanodaviruses. 82(Pt11):2653–2662. ISSN: 1465-2099. https:doi.org/10.1099/0022-1317-82-11-2653.

[136] P. Baliga, M. Shekar, and M.N. Venugopal. Potential outer membrane protein candidates for vaccine development against the pathogen *Vibrio anguillarum*: A reverse vaccinology based identification. 75:368–377. ISSN: 0343-8651. https://doi.org/10.1007/s00284-017-1390-z.

[137] M. Kawasaki. A pan genome reverse vaccinology approach to prevent *Streptococcus agalactiae* infection in farmed tilapia, A thesis submitted for the degree of Doctor of Philosophy at The University of Queensland, Australia

[138] W. Wang, J. Liu, S. Guo, L. Liu, Q. Yuan, L. Guo, and S. Pan. Identification of *Vibrio parahaemolyticus* and *Vibrio* spp. specific outer membrane proteins by reverse vaccinology and surface proteome. 11:625315. ISSN: 1664-302X. https://doi.org/10.3389/fmicb.2020.625315.

[139] M.T. Dertzbaugh. Genetically engineered vaccines: An overview. 39(2):100–113. ISSN: 0147-619X. https://doi.org/10.1006/plas.1997.1329.

[140] S.C. Makrides. Strategies for achieving high-level expression of genes in *Escherichia coli*. 60(3):512–538. https://doi.org/10.1128/mr.60.3.512-538.1996

[141] M. Murby, M. Uhlén, and S. Ståhl. Upstream strategies to minimize proteolytic degradation upon recombinant production in *Escherichia coli*. 7:129–136. ISSN: 1096-0279. https://doi.org/10.1006/prep.1996.0018.

[142] L.A. Babiuk. Broadening the approaches to developing more effective vaccines. 17:1587–1595. ISSN: 2076-393X. https://doi.org/10.1016/s0264-410x(98)00419-8

[143] Y. Ichikawa, H. Yamagata, K. Tochikubo, and S. Udaka. Very efficient extracellular production of cholera toxin B subunit using *Bacillus brevis*. 111:219–224. ISSN: 0378-1097. https://doi.org/10.1111/j.1574-6968.1993.tb06389.x.

[144] J.F. Viret, S.J. Cryz, S.J., Jr. and D. Favre. Expression of *Shigella sonnei* lipopolysaccharide in *Vibrio cholerae*. 19:949–963. ISSN: 0950-382X. https://doi.org/10.1046/j.1365-2958.1996.435967.x.

[145] A. Martin-Gallardo, E. Fleischer, S.A. Doyle, R. Arumugham, P.L. Collins, S.W. Hildreth, and P. R. Paradiso. Expression of the G glycoprotein gene of human respiratory syncytial virus in *Salmonella typhimurium*. 74:453–458. ISSN: 1465-2099. https://doi.org/10.1099/0022-1317-74-3-453.

[146] G. Sezonov, D. Joseleau-Petit, and R. D'Ari, *Escherichia coli* physiology in Luria-Bertani broth. 189:8746–8749. ISSN: 0021-9193. https:doi.org/10.1128/JB.01 368-07.

[147] J. Shiloach, and R. Fass. Growing *E. coli* to high cell density-A historical perspective on method development. 23(5):345–357. ISSN: 1873–1899. https://doi.org/10.1016/j.biotechadv.2005.04.004.

[148] L. Phan, L.W. Schoenfeld, L. Valasek, K.H. Nielsen, and A.G. Hinnebusch. A subcomplex of three eIF3 subunits binds eIF1 and eIF5 and stimulates ribosome binding of mRNA and tRNA(i)Met. 20:2954–2965. ISSN: 1460-2075. https://doi.org/10.1093/emboj/20.11.2954.

[149] V.K. Ton, and R. Rao. Functional expression of heterologous proteins in yeast: Insights into Ca2þ signaling and Ca2þ-transporting ATPases. 287:C580–C589. ISSN: 0363-6143 https://doi.org/10.1152/ajpcell.00135.2004

[150] R. Baghban, S. Farajnia, M. Rajabibazl, Y. Ghasemi, A. A. Mafi, R. Hoseinpoor, L. Rahbarnia and M. Aria. Yeast expression systems: Overview and recent advances. 61(5):365–384. ISSN: 1559-0305. https://doi.org/10.1007/s12033-019-00164-8.

[151] D. Lohr, P.Venkov, and J. Zlatanova. Transcriptional regulation in the yeast GAL gene family: a complex genetic network. 9(9):777–87. ISSN: 0892-6638. https://doi.org/10.1096/fasebj.9.9.7601342

[152] X. Zhu, S.Wu, and G. J. Letchworth 3rd. Yeast-secreted bovine herpesvirus type 1 glycoprotein D has authentic conformational structure and immunogenicity. 15: 679–88. ISSN: 0264-410X. https://doi.org/10.1016/s0264-410x(96)00234-4

[153] G. Gellissen, and C.P. Hollenberg. Application of yeasts in gene expression studies: a comparison of *Saccharomyces cerevisiae*, *Hansenula polymorpha* and *Kluyveromyces lactis* – A review. 190:87–97. ISSN: 0378-1119 https://doi.org/10.1016/s0378-1119(97)00020-6

[154] S. Partow, V. Siewers, S. Bjørn, J. Nielsen, and J. Maury. Characterization of different promoters for designing a new expression vector in *Saccharomyces cerevisiae*. 27: 955–964. ISSN: 1097-0061. https://doi.org/10.1002/yea.1806

[155] F. Machens, S. Balazadeh, B. Mueller-Roeber, K. Messerschmidt. Synthetic promoters and transcription factors for heterologous protein expression in *Saccharomyces cerevisiae*. 19(5): 63. ISSN: 2296-4185. https://doi.org/10.3389/fbioe.2017.00063

[156] A. Rantasalo, E. Czeizler, R. Virtanen, J. Rousu, H. Lähdesmäki, and M. Penttilä. Synthetic Transcription Amplifier System for Orthogonal Control of Gene Expression in *Saccharomyces cerevisiae*. 11(2): e0148320. ISSN: 1932-6203 https://doi.org/10.1371/journal.pone.0148320

[157] S. Öztürk, B.G. Ergün, and P. Çalık. Double promoter expression systems for recombinant protein production by industrial microorganisms. 101: 7459–7475. ISSN: 0175-7598. https://doi.org/10.1007/s00253-017-8487-y
[158] M W Nasser, V Pooja, M Z Abdin and S K Jain, 2003. Evaluation of Yeast as an Expression System, Indian Journal of Biotechnology 2: 477–493. ISSN: 09725849.
[159] H.J. Kim, J.Y. Lee, H.A. Kang, Y. Lee, E.J. Park, and H.J. Kim. Oral immunization with whole yeast producing viral capsid antigen provokes a stronger humoral immune response than purified viral capsid antigen. 58(3): 285–291. ISSN: 1472-765X. https://doi.org/10.1111/lam.12188
[160] K. Zhu, Z. Chi, J. Li, F. Zhang, M. Li, H. N. Yasoda, L. and Wu. The surface display of haemolysin from *Vibrio harveyi* on yeast cells and their potential applications as live vaccine in marine fish. 24(35–36): 6046–6052. ISSN: 0264-410X. https://doi.org/10.1016/j.vaccine.2006.05.043
[161] R. K. Jha; Z. R. Xu, S. J. Bai, J. Y. Sun, W. F. Li, J. Shen. Protection of *Procambarus clarkii* against white spot syndrome virus using recombinant oral vaccine expressed in *Pichia pastoris*. 22(4): 295–307. ISSN: 1050-4648. https://doi.org/10.1016/j.fsi.2006.04.006
[162] F.C. T. Allnutt, R. M. Bowers, C. G. Rowe, V. N. Vakharia, S. E. LaPatra, A. K. Dhar. Antigenicity of infectious pancreatic necrosis virus VP2 subviral particles expressed in yeast. 25(26): 4880–4888. ISSN: 0264-410X. https://doi.org/10.1016/j.vaccine.2007.04.068
[163] Z. Mao, C. He, Y. Qiu, and J. Chen. Expression of *Vibrio harveyi* ompK in the yeast *Pichia pastoris:* The first step in developing an oral vaccine against vibriosis? 318(3–4): 268–272.ISSN 0044-8486. http://dx.doi.org/10.1016/j.aquaculture.2011.05.047
[164] J. Y. Seo, H.J. Chung, T.J. Kim. Codon-optimized expression of fish iridovirus capsid protein in yeast and its application as an oral vaccine candidate. 36(9): 763–768. ISSN: 1365-2761. https://doi.org/10.1111/jfd.12037
[165] V. Ananphongmanee, J. Srisala, K. Sritunyalucksana, C. Boonchird. Yeast surface display of two proteins previously shown to be protective against White Spot Syndrome Virus (WSSV) in Shrimp. 17:10(6):e0128764. ISSN: 1932-6203. https://doi.org/10.1371/journal.pone.0128764
[166] G.R.Wi, J.Y. Hwang, M.G. Kwon, H.J. Kim, H.A. Kang, and H.J. Kim. Protective immunity against nervous necrosis virus in convict grouper *Epinephelus septemfasciatus* following vaccination with virus-like particles produced in yeast *Saccharomyces cerevisiae*. 177(1–2):214–218. ISSN: 0378-1135. https://doi.org/10.1016/j.vetmic.2015.02.021
[167] Y. Zhou, N. Jiang, J. Ma, Y. Fan, L. Zhang, J. Xu, and L. Zeng. Protective immunity in gibel carp, *Carassius gibelio* of the truncated proteins of cyprinid herpesvirus 2 expressed in *Pichia pastoris*. 47(2): 1024–1031. ISSN: 10504648. https://doi.org/10.1016/j.fsi.2015.11.012
[168] A. Shrestha, K. Bao, Y.Chen, W. Chen, P. Wang, Z. Fei, and G. W. Blissard, Global Analysis of Baculovirus *Autographa californica* Multiple Nucleopolyhedrovirus Gene Expression in the Midgut of the Lepidopteran Host *Trichoplusia ni*. 92(23): e01277-18. https://doi.org/10.1128/JVI.01277-18
[169] M.M. van Oers. Vaccines for viral and parasitic diseases produced with baculovirus vectors. 68: 193–253. https://doi.org/10.1016%2FS0065-3527(06)68006-8
[170] J.L. Vaughn, R.H. Goodwin, G.J. Tompkins, P.McCawley. The establishment of two cell lines from the insect *Spodoptera frugiperda* (Lepidoptera; Noctuidae). 13:213–7. ISSN: 1965-1984. https://doi.org/10.1007/bf02615077
[171] S. Geisse, H. Gram, B. Kleuser, and H.P. Kocher, Eukaryotic expression systems: a comparison. *Protein*. 8: 271–282. ISSN: 1046-5928. https://doi.org/10.1006/prep.1996.0101
[172] S. Syed Musthaq, and J. Kwang. Oral vaccination of baculovirus-expressed VP28 displays enhanced protection against White Spot Syndrome Virus in *Penaeus monodon*. 6(11): e26428. ISSN: 1932-6203. https://doi.org/10.1371/journal.pone.0026428
[173] F. Ling-Lin; W. Yanbo, W. Zheng-Cun, and L. Wei-Fen. *In vivo* assessment for oral delivery of *Bacillus subtilis* harboring a viral protein (VP28) against white spot syndrome virus in *Litopenaeus vannamei*. 322–323: 33–38. ISSN: 0044-8486. https://doi.org/10.1016%2Fj.aquaculture.2011.09.036
[174] C. Mu, Q. Zhong, Y. Meng, Y. Zhou, N. Jiang, W. Liu, Y. Li, M. Xue, L. Zeng, V.N. Vakharia, and Y. Fan; (2021). Oral Vaccination of Grass Carp *(Ctenopharyngodon idella)* with Baculovirus-Expressed Grass Carp Reovirus (GCRV) Proteins Induces Protective Immunity against GCRV Infection. 9(1): 41. ISSN: 0264-410X. https://doi.org/10.3390/vaccines9010041
[175] K. H. Khan. Gene expression in mammalian cells and its applications. 2013, 3(2): 257–263. ISSN: 2228-5881. http://dx.doi.org/10.5681/apb.2013.042
[176] F.M Boyce, N.L Bucher. Baculovirus-mediated gene transfer into mammalian cells. 93(6): 2348–2352. https://doi.org/10.1073/pnas.93.6.2348
[177] Dukkipati, H.H Park, D. Waghray, S. Fischer, K.C. Garcia. BacMam system for high-level expression of recombinant soluble and membrane glycoproteins for structural studies. 62(2):160–70. ISSN: 1046-5928. https://doi.org/10.1016/j.pep.2008.08.004

[178] L. Chu, D.K. Robinson. Industrial choices for protein production by large-scale cell culture. 12(2):180–187. ISSN: 0958-1669. https://doi.org/10.1016/S0958-1669(00)00197-X

[179] C.B. Hirschberg, M.D Snider. Topography of glycosylation in the rough endoplasmic reticulum and Golgi apparatus. 56:63–89. ISSN: 0066-4154. https://doi.org/10.1146/annurev.bi.56.070187.000431

[180] J.W. Suttie. Vitamin K–dependent carboxylase. 54:459–477. ISSN: 0066-4154. https://doi.org/10.1146/annurev.bi.54.070185.002331

[181] R.J Kaufman, L.C Wasley, B.C. Furie, B. Furie, C. Schoemaker. Expression, purification and characterization of recombinant γcarboxylated factor IX synthesized in Chinese hamster ovary cells. 261:9622–9628. ISSN: 0021-9258. https://doi.org/10.1016/S0021-9258(18)67559-3

[182] J. Stenflo, E. Holme, S. Lindstedt, N. Chandramuli, L. Huang, J. Tam, R. Merrifield. Hydroxylation of aspartic acid in domains homologous to the epidermal growth factor precursor is catalyzed by a 2-oxoglutarate-dependent dioxygenase. 86(2):444–447. ISSN: 0369-3236. https://doi.org/10.1073/pnas.86.2.444

[183] P.A Baeuerle, W.B Huttner. Tyrosine sulfation is a trans-Golgi-specific protein modification. 105(6):2655–2664. ISSN: 1540-8140. https://doi.org/10.1083/jcb.105.6.2655

[184] G. Martin, JR. Myers, Xian Jian Sun, F. Morris, White. The IRS-1 signaling system. 19(7): 289–293. ISSN: 1362-4326. doi:10.1016/0968-0004(94)90007-8

[185] R.A.J McIlhinney. The fats of life: The importance and function of protein acylation. 15:387–391. ISSN: 1362-4326. https://doi.org/10.1016/0968-0004(90)90237-6

[186] W.C Chen, E.L Wang, F. Luo, K.G Song, G.X Wang. Identification and pathogenicity analysis of a novel highly virulent strain of spring viraemia of carp virus and production of its structural proteins by the mammalian expression system. 547: 737487. ISSN: 1873-5622. https://doi.org/10.1016/j.aquaculture.2021.737487

[187] C.M.A Caipang, T. Takano, I. Hirono, T. Aoki. Genetic vaccines protect red seabream, Pagrus major, upon challenge with red seabream iridovirus (RSIV). 21(2): 130–138. ISSN: 1050-4648. https://doi.org/10.1016/j.fsi.2005.10.012

[188] P. Kayansamruaj, H.T. Dong, N. Pirarat, D. Nilubol, C. Rodkhum. Efficacy of α-enolase-based DNA vaccine against pathogenic *Streptococcus iniae* in Nile tilapia (*Oreochromis niloticus*). 468: 102–106. ISSN: 1873-5622. https://doi.org/10.1016/j.aquaculture.2016.10.001

[189] Davoodi-Semiromi, M. Schreiber, S. Nalapalli, D. Verma, N.D Singh, R.K. Banks, H. Daniell. Chloroplast-derived vaccine antigens confer dual immunity against cholera and malaria by oral or injectable delivery. 8(2): 1467–7644. ISSN: 223-242. https://doi.org/10.1111/j.1467-7652.2009.00479.x

[190] F. Ma, E. Zhang, Q. Li, Q. Xu, J. Ou, H. Yin, G. Zhang. A Plant-produced recombinant fusion protein-based Newcastle disease subunit vaccine and rapid differential diagnosis platform. 8(1): 122. ISSN: 2076-393X. https://doi.org/10.3390/vaccines8010122

[191] L. Doron, N.A Segal, M. Shapira, Transgene expression in microalgae—from tools to applications. 7: 505. ISSN: 1664-462X. https://doi.org/10.3389/fpls.2016.00505

[192] Spicer, S. Purton. Genetic engineering of microalgae current status and future Prospects. 139–163. ISBN: 9781315373706

[193] J.L Clarke, L. Paruch, M.O Dobrica, I. Caras, C. Tucureanu, A. Onu, et al. Lettuce-produced hepatitis C virus E1E2 heterodimer triggers immune responses in mice and antibody production after oral vaccination. 15: 1611–1621. ISSN: 1467-7644. https://doi.org/10.1111/pbi.12743

[194] P. Charoonnart, S. Purton, V. Saksmerprome. Applications of microalgal biotechnology for disease control in aquaculture. 7(2):24. ISSN: 2079-7737. doi: 10.3390/biology7020024

[195] S. Siripornadulsil, K. Dabrowski, R. Sayre. Microalgal vaccines. 616:122–128. ISSN: 2512-2908. https://doi.org/10.1007/978-0-387-75532-8_11

[196] S. Feng, W. Feng, L. Zhao, H. Gu, Q. Li, K. Shi, S. Guo, N. Zhang. Preparation of transgenic *Dunaliella salina* for immunization against white spot syndrome virus in crayfish. 159: 519–525. ISSN: 0304-8608. https://doi.org/10.1007/s00705-013-1856-7

[197] Kiataramgul, S. Maneenin, S. Purton, N. Areechon, I. Hirono, T.W Brocklehurst, S. Unajak. An oral delivery system for controlling white spot syndrome virus infection in shrimp using transgenic microalgae. 521: 735022. ISSN: 1873-5622. https://doi.org/10.1016/j.aquaculture.2020.735022

[198] P. Somchai, S. Jitrakorn, S. Thitamadee, M. Meetam, V. Saksmerprome. Use of microalgae *Chlamydomonas reinhardtii* for production of double-stranded RNA against shrimp virus. 3:178–183. ISSN: 2352-5134. https://doi.org/10.1016/j.aqrep.2016.03.003

[199] X.H Jia, C.L Zhang, D.J Shi, M.M Zhuang, X. Wang, R. Jia, Z.Y Zhang, J. Huang, Y.H Sun, W.Y Qian. Oral administration of Anabaena-expressed VP28 for both drug and food against white spot syndrome virus in shrimp. 28:1001–1009. ISSN: 1573-5176. https://doi.org/10.1007/s10811-015-0607-4

[200] C. Tafalla, J. Bøgwald, R.A Dalmo. Adjuvants and immunostimulants in fish vaccines: current knowledge and future perspectives. 35(6):1740–1750. ISSN: 1095-9947. https://doi.org/10.1016/j.fsi.2013.02.029

[201] F. Azmi, A.A.H Ahmad Fuaad, M. Skwarczynski, I. Toth. Recent progress in adjuvant discovery for peptide-based subunit vaccines. 10(3): 778–796. ISSN: 2164-554X. https://doi.org/10.4161/hv.27332

[202] E. Wangkaghart, S. Deville, B. Wang, P. Srisapoome, T. Wang, C.J Secombes. Immune response and protective efficacy of two new adjuvants, Montanide™ ISA 763B VG and Montanide™ GEL02, administered with a *Streptococcus agalactiae* ghost vaccine in Nile tilapia (*Oreochromis niloticus*). 116:19–29. ISSN: 1095-9947. https://doi.org/10.1016/j.fsi.2021.06.014

[203] F. Sughra, M. Hafeez-ur-Rehman, F. Abbas, I. Altaf, S. Aslam, A. Ali, M. Khalid, G. Mustafa, S.M. Azam. Evaluation of oil-based inactivated vaccine against *Aeromonas hydrophila* administered to *Labeo rohita*, *Cirrhinus mrigala* and *Ctenopharyngodon idella* at different concentrations: Immune response, immersion challenge, growth performance and histopathology. 21:100885. ISSN: 2352-5134. https://doi.org/10.1016/j.aqrep.2021.100885

[204] M. Soltani, M. Alishahi, S. Mirzargar, G. Nikbakht. Vaccination of rainbow trout against *Streptococcus iniae* infection: comparison of different routes of administration and different vaccines. 7(1):129–140. ISSN: 2322-5696. http://hdl.handle.net/1834/11186

[205] H.M Fang, R. Ge, Y.M Sin. Cloning, characterisation and expression of *Aeromonas hydrophila* major adhesin. 16(5):645–658. ISSN: 1095-9947. https://doi.org/10.1016/J.FSI.2003.10.003

[206] J.K Patra, G. Das, L.F Fraceto, E.V.R Campos, M.D.P Rodriguez-Torres, L.S Acosta-Torres, L.A Diaz-Torres, R. Grillo, M.K Swamy, S. Sharma, S. Habtemariam, H.S Shin. Nano based drug delivery systems: recent developments and future prospects. 16(1): 1–33. ISSN: 1477-3155. https://doi.org/10.1186/S12951-018-0392-8

[207] Thakuria, B. Kataria, D. Gupta. Nanoparticle-based methodologies for targeted drug delivery—an insight. 23(4): 1–30. ISSN: 1388-0764. https://doi.org/10.1007/s11051-021-05190-9

[208] T. Sahu, Y.K. Ratre, S. Chauhan, L.V.K.S Bhaskar, M.P Nair, H.K Verma. Nanotechnology based drug delivery system: Current strategies and emerging therapeutic potential for medical science. 63:102487. ISSN: 1773-2247. https://doi.org/10.1016/j.jddst.2021.102487

[209] N.A Mohamed, I. Marei, S. Crovella, H. Abou-Saleh. Recent Developments in Nanomaterials-Based Drug Delivery and Upgrading Treatment of Cardiovascular Diseases. 23(3): 1404. ISSN: 1422-0067. https://doi.org/10.3390/ijms23031404

[210] M. Anwar, F. Muhammad, B. Akhtar. Biodegradable nanoparticles as drug delivery devices. 64: 102638. ISSN: 1773-2247. https://doi.org/10.1016/J.JDDST.2021.102638

[211] M. Anwar, F. Muhammad, B. Akhtar. Biodegradable nanoparticles as drug delivery devices. 64: 102638. ISSN: 1773-2247. https://doi.org/10.1016/j.jddst.2021.102638

[212] T. Knuschke, V. Sokolova, O. Rotan, M. Wadwa, M. Tenbusch, W. Hansen, P. Staeheli, M. Epple, J. Buer, A.M. Westendorf. Immunization with biodegradable nanoparticles efficiently induces cellular immunity and protects against influenza virus infection. 190(12): 6221–6229. ISSN: 1550-6606. https://doi.org/10.4049/JIMMUNOL.1202654

[213] V. Chandrakala, V. Aruna, G. Angajala. Review on metal nanoparticles as nanocarriers: current challenges and perspectives in drug delivery systems. 1–23. ISSN: 2522-574X. https://doi.org/10.1007/s42247-021-00335-x.

[214] J. Han, D. Zhao, D. Li, X. Wang, Z. Jin, K. Zhao. Polymer-based nanomaterials and applications for vaccines and drugs. 10:1. ISSN: 0171-709X. https://doi.org/10.3390/polym10010031

[215] S. Cao, Y. Deng, L. Zhang, M. Aleahmad. Chitosan nanoparticles, as biological macromolecule-based drug delivery systems to improve the healing potential of artificial neural guidance channels: A review. 201:569–579. ISSN: 1879-0003. https://doi.org/10.1016/j.ijbiomac.2022.01.017

[216] J. Jhaveri, Z. Raichura, T. Khan, M. Momin, A. Omri, R. Stancanelli, S. Tommasini, C. Anna Ventura, V. Crupi, D. Majolino. Chitosan Nanoparticles-Insight into Properties, Functionalization and Applications in Drug Delivery and Theranostics. 26(2):272. ISSN: 1420-3049. https://doi.org/10.3390/MOLECULES26020272

[217] Arun, P. Malrautu, A. Laha, H. Luo, S. Ramakrishna. Collagen nanoparticles in drug delivery systems and tissue engineering. 11:23. ISSN: 2523-3971. https://doi.org/10.3390/app112311369

[218] K.H Khan. DNA vaccines: roles against diseases. 3(1):26–35. ISSN: 2248-2997. https://doi.org/10.11599%2Fgerms.2013.1034

[219] P. Calvo, C. Remunan Lopez, J.L Vila Jato, M.J Alonso. Chitosan and chitosan/ethylene oxide-propylene oxide block copolymer nanoparticles as novel carriers for proteins and vaccines. 14:1431–1436. ISSN: 0724-8741. https://doi.org/10.1023/A:1012128907225

[220] P. Calvo, C. Remuñan Lopez, J.L Vila Jato, M.J Alonso. Novel hydrophilic chitosan-polyethylene oxide nanoparticles as protein carriers. *Journal of Applied Polymer Science*, 63:125–132. ISSN: 1097-4628. https://doi.org/10.1002/(SICI)1097-4628(19970103)63:1%3C125::AID-APP13%3E3.0.CO;2-4

[221] M. Cuna, M. Alonso Sande, C. Remunan-Lopez, M.J Alonso. Development of chitosan/glucomannan nanoparticles as carrier for oral protein administration. 136: 2887–2895.

[222] Vila, A**. Sanchez,** K. Janes, I. Behrens, T. Kissel, J.L.V Jato, M.J Alonso. Low molecular weight chitosan nanoparticles as new carriers for nasal vaccine delivery in mice. 57: 123–131. ISSN: 1873-3441. https://doi.org/10.1016/j.ejpb.2003.09.006.

[223] Q. Gan, T. Wang, P. McCarron. Modulation of surface charge, particle size and morphological properties of chitosan–TPP nanoparticles intended for gene delivery. 44: 65–73. ISSN: 1873-4367. https://doi.org/10.1016/j.colsurfb.2005.06.001

[224] H. Katas, H.O Alpar. Development and characterisation of chitosan nanoparticles for siRNA delivery. 115:216–225. ISSN: 1873-4995. https://doi.org/10.1016/j.jconrel.2006.07.021

[225] S.L Wang, H.H Yao, L.L Guo, L. Dong, S.G Li, Y.P Gu, Z.H Qin. Selection of optimal sites for TGFB1 gene silencing by chitosan-TPP nanoparticle-mediated delivery of shRNA. **190:** 8–14. ISSN: 1873-4456. https://doi.org/10.1016/j.cancergencyto.2008.10.013

[226] N. Csaba, M. KopingHoggard, M.J. Alonso. Ionicallycrosslinked chitosan/tripolyphosphate nanoparticles for oligonucleotide and plasmid DNA delivery. 382: 05–14. ISSN: 1873-3476. https://doi.org/10.1016/j.ijpharm.2009.07.028

[227] S. Vimal, S.A Majeed, K.S.N NambiMadan, N. Farook, M.A Venkatesan, A.S.S. Hameed. Delivery of DNA vaccine using chitosan-tripolyphosphate (CS/TPP) nanoparticles in Asian sea bass, *Lates calcarifer* (Bloch, 1790) for protection against nodavirus infection. 420: 240–246. ISSN: 1873-5622. https://doi.org/10.1016/j.aquaculture.2013.11.017

[228] S. Vimal, G. Taju, K.S.N Nambi, S.A Majeed, V.S Babu, M. Ravi, A.S.S Hameed. Synthesis and characterization of CS/TPP nanoparticles for oral delivery of gene in fish. 358: 14–22. ISSN: 1873-5622. https://doi.org/10.1016/j.aquaculture.2012.06.012.

10 Safety and Efficacy of Vaccines in Aquaculture

Eakapol Wangkahart
Mahasarakham University

Po-Tsang Lee
National Taiwan Ocean University

Chou-Min Chong
Universiti Putra Malaysia

Fernando Yamamoto
Mississippi State University

CONTENTS

10.1 INTRODUCTION

The major objective of vaccination in aquaculture is to prevent disease and increase the specific immune response in fish against a particular pathogen. The main benefit of using vaccines in aquaculture is that they are a preventative measure, in contrast to trying to cure a disease after infection. Fish vaccination is currently an important component of fish health management, and vaccination has been widely accepted as the most viable strategy for controlling infectious diseases threatening the aquaculture industry globally (1,2). Vaccines can also have a significantly positive impact on antibiotic therapy in fish farming, resulting in a rapid reduction in antibiotic usage (3). Fish vaccines, on the other hand, must be safe, be inexpensive to produce, and promote long-lasting protection. Furthermore, for large aquaculture operations, the vaccinations must be stable and simple to be delivered (4). Vaccines are currently available for more than 17 distinct fish species, protecting them from more than 22 different bacterial and viral diseases (5).

DOI: 10.1201/9781003388548-14

10.2 VACCINE ADMINISTRATION IN AQUACULTURE

Vaccines are administered to fish through different routes, including immersion, injection, and oral administration (4,6). The most common method for commercial fish vaccination is the injection, which offers long-term protection and enables the use of adjuvants (7,8). Injection vaccinations can be administered by intramuscular injection (i.m.) or intraperitoneal (i.p.) injection, with the latter being the most used method, allowing to avoid fillet damage. Injection vaccination, on the other hand, has the disadvantage of being time-consuming and requiring skilled personnel. Immersion vaccination is an effective method of mass immunization for small fish, and can be used to protect them against a variety of diseases (9). This process is less labor-intensive as well as less stressful for the fish (10). Oral vaccinations have also been developed that are intended to be used in conjunction with the diet. This results in a very simple administration method, improved safety, and a significant reduction in the stress imposed on the fish. Their effectiveness, however, is restricted, most likely due to vaccine breakdown in the fish digestive system (11). Tables 10.1–10.3 provide a review of the advantages and disadvantages of each immunization strategy. The mode of delivery is, thus, determined by a variety of parameters, including the size of the fish, the duration of protection required, the type of disease and likelihood of encounter, and the cost (12).

TABLE 10.1
Advantages and Disadvantages of Injection Vaccination

Advantages	Disadvantages
• The most popular vaccine administration method in fish • Extremely effective at eliciting both cellular and humoral immune response • Incompatible with small fish	• Requires advanced machinery or a large number of highly skilled individuals • Excessive handling of stress and risk of fungal infection following vaccination • Local reaction

TABLE 10.2
Advantages and Disadvantages of Immersion Vaccination

Advantages	Disadvantages
• It can be used to mass-vaccinate fish of various sizes • Fish are less stressed • Reduced labor costs • The vaccination team is at a lower risk	• The large amount of vaccine required and lower level of vaccine required • Protection and immunity duration are reduced

TABLE 10.3
Advantages and Disadvantages of Oral Vaccination

Advantages	Disadvantages
• Vaccine in combination with fish feed • The simplest method for bulk immunization of fish of all sizes • Reduces labor costs and eliminates stress	• Large quantities of antigen required • Requires all fish to be feeding • It is difficult to ensure that the proper dosage will be delivered

Several vaccinations for bacterial and viral diseases have been found to have good protective effectiveness (13,14). The relative percent survival rate is commonly used to determine vaccine effectiveness (RPS) (15), calculated as follows:

$$RPS = 1 - \frac{(\text{vaccinated fish mortality } \%)}{(\text{control fish mortality } \%)} \times 100.$$

10.3 TYPES OF FISH VACCINES AND THEIR EFFICACIES

Vaccine research has become a critical tool for preventing diseases in farmed fish (4). Vaccines come in a variety of forms, including inactivated vaccines, live attenuated vaccines, DNA vaccines, and subunit vaccines (16).

10.3.1 Inactivated Whole-Cell Vaccines

For many years, inactivated whole-cell vaccines have been widely utilized, and they have been shown to be effective against bacterial infections in fish (5). The most extensively used and commercially manufactured vaccines for controlling diseases in aquaculture are formalin-inactivated vaccines. The vaccine against enteric red mouth (ERM) disease, which consists of killed whole cells, was initially licensed and commercialized in the late 1970s (17). The safety of formalin treatment, which is a dependable and inexpensive technique for killing pathogens, is an advantage of inactivated bacterial vaccinations (4). However, if injected, this form of vaccine may require an adjuvant to elicit an appropriate long-term immune response while, if administered through the mucosal route, it may require booster immunization (13–18). Despite promoting a specific immune response against a pathogen, these vaccinations do not mount a response specific to a target, but many epitopes. Thus, even though the immune system of the fish can build memory cells against the disease, it may not be as effective as a live attenuated vaccine or a DNA vaccine. These fish vaccinations are currently available in North America, Chile, Japan, Australia, and Europe (5,19). Examples of successful inactivated whole-cell vaccines and their efficacies in aquaculture are listed in Table 10.4.

10.3.2 Live Attenuated Vaccines

Several live attenuated vaccines against bacterial diseases have recently been produced (20–22). They have been shown to be more effective than inactivated whole-cell vaccines at eliciting a protective immune response, with more persistent memory responses, greater survival rates, and more robust cellular and humoral immune responses (23). Unlike inactivated vaccines, a live attenuated vaccine can still undergo some replication *in vivo*, triggering the immune responses in a manner similar to that of a natural infection, but not properly expressing the virulence factors (4). Although attenuated live vaccines can induce an effective immune response, they may pose a danger of reverting to a virulent phenotype (4); however, it has been proven that live attenuated vaccines are an effective strategy against bacterial infectious diseases. Table 10.4 lists a number of successful examples of their use in aquaculture.

10.3.3 Deoxyribonucleic Acid (DNA) Vaccines

DNA vaccines have recently been produced and are a viable alternative to whole-cell immunizations (24,25). DNA vaccines are composed of a strong promoter expression plasmid, the gene of interest, and a polyadenylation/transcriptional termination sequence. This vaccine is given to the host as purified plasmid DNA, and the expression of the encoded protein(s) is induced in the host cells, resulting in strong and long-lasting immunity (1,26). DNA vaccines have been shown, in a number of trials, to elicit robust innate and adaptive immunity (27–29). They have been developed for several fish species, to promote immunity against bacterial and viral diseases (5). DNA vaccines may be more effective in the production of immune responses against viral pathogens, as the antigen(s) are formed intracellularly and likely presented by the MHC I presentation pathway, which is comparable to the reaction during viral infections (4). Recent research has shown that the antigen type, vaccine dose, and duration of expression have an impact on the generated immune responses and vaccine efficacy (1). Table 10.4 details a number of successful instances of DNA vaccine applications in aquaculture.

10.3.4 Subunit Vaccines

The development of subunit vaccines for aquaculture has been made possible by advancements in molecular biology and genetic engineering. They are primarily recombinant proteins used as purified antigens, and rely on the identification of vaccine candidates that can elicit specific protective immunity against the disease in question (30). Subunit vaccines are safer than conventional vaccines in experimental animals (i.e., inactivated whole-cell or live attenuated vaccines) (31). However, despite their effectiveness in preventing diseases, subunit vaccines have several limitations, including their high cost of production and the fact that, as single pure proteins, they typically lack self-adjuvant ability, produce short-lived immune responses, and have poor immunogenicity (32,33). They have, however, been demonstrated to generate long-lasting immune responses against diseases when given regularly over a period of time (34,35). Table 10.4 provides a number of successful instances of subunit vaccine usage in aquaculture.

TABLE 10.4
Examples of Fish Vaccines

Pathogens	Fish Species	Administration Method	Efficacy as Relative Percent Survival (RPS)	Remark	References
		Inactivated vaccines			
Salmonid alphavirus	Inactivated vaccine based on a strain of SAV subtype 3-ALV405	i.p. injection	98.5%	Protection against SAV-induced mortality	(36)
				Induce specific antibodies	(37)
Vibrio anguillarum	Turbot, *Scophthalmus maximus*	i.p. injection	90%	Induce specific antibodies	(37)
A novel hydrogen peroxide-inactivated vaccine	Nile tilapia, *Oreochromis niloticus*	i.p. injection	59.3%		
		Live attenuate vaccines			
Aeromonas hydrophila	Common carp, *Cyprinus carpio*	i.p. injection	37–83%	Induction of serum-specific IgM and upregulation of immune-related genes after vaccination	(38)
	Zebrafish, *Danio rerio*		70–100%		(39)
Edwardsiella spp.	Turbot, *Scophthalmus maximus*	immersion	83–89%	The novel vaccine manifested satisfactory immune protection on turbot	(40)

(Continued)

TABLE 10.4 (*Continued*)
Examples of Fish Vaccines

Pathogens	Fish Species	Administration Method	Efficacy as Relative Percent Survival (RPS)	Remark	References
E. ictaluri *E. piscicida*	Channel catfish, *I. punctatus*, Hybrid catfish, *I. punctatus* × *I. furcatus*	Oral	85-88%	Cross-protection of live attenuated *E. ictaluri* vaccine against *E. piscicida*	(41)
		DNA vaccines			
Vibrio anguillarum	Flounder, *Paralichthys olivaceus*	i.m. injection	50%	Induction of the T lymphocyte response, followed by B lymphocyte induction of specific antibodies	(42)
Nocardia spp.	Hybrid snakehead	i.m. injection	79.33%	DNA vaccine show protection against *N. seriolae* infection	(43)
		Subunit vaccines			
Edwardsiella ictaluri	Nile tilapia, *Oreochromis niloticus*	i.p. injection	42%	Antigenicity of the chimeric protein was exhibited by strong reactivity with serum	(30)
Infectious spleen and kidney necrosis virus	Mandarin fish, *Siniperca chuatsi*	immersion	86.7%	Induction of the immune protective effect of naked subunit vaccine	(34)
Aeromonas hydrophila	Rainbow trout, *Oncorhynchus mykiss*	immersion and i.p. injection	21.56–78.88%	*A. hydrophila* glycoproteins with $Al(OH)_3$ and ginseng could be used as a safe and effective vaccine for fish	(44)
S. agalactiae	Nile tilapia, *Oreochromis niloticus*	i.p. injection	80–93%	Ghost vaccine can enhance cellular and humoral immune responses and show high immunoprotection	(45)
A. salmonicida	Rainbow trout, *O. mykiss*	i.p. injection	46–86%	High survival post-vaccination and a few novel candidate proteins, and the survival can be linked to the antibody-producing protective qualities.	(33)
V. harveyi	Golden pompano, *Trachinotus ovatus*	i.p. injection	52.39%	Fish had both innate and adaptive immune responses	(32)

10.4 ADJUVANTS FOR FISH VACCINES AND THE PROTECTIVE EFFICACIES

Vaccines are frequently unable to provide protection on their own, requiring the inclusion of adjuvants to improve the vaccine efficacy and for pathogen detection (9,46). Adjuvants can help to reduce the doses required (e.g., boosters) or the amount of antigen required at each dosage (47). Since 1990, oil-adjuvant vaccines have been applied for aquaculture (48), which has been deemed a success for the control of many (mainly bacterial) fish diseases.

Adjuvants are classified based on their chemical or physical components, as well as the immunological reactions they can elicit (9). The activation of specific T and B lymphocytes, which form the adaptive immune system, is a key requirement of vaccine adjuvants (49). However, it is now widely accepted that an efficient immune response against infections begins with the recognition of pathogen-associated microbial patterns (PAMPs) by pathogen recognition receptors (PRRs), such as Toll-like receptors (TLRs), NOD-like receptors (NLRs), and RIG-I-like receptors (RLRs), which are predominantly found on cells of the innate immune system (50). As a result, molecules that can interact with these receptors can be a promising adjuvant candidate. Hence, the main functions of vaccine adjuvants are to help in initiating the innate immune response by boosting the immunogenicity of antigens that enhance the protective adaptive immune response, thus helping to reduce the demand for booster immunizations (9,46). Antigen presentation to T-cells is facilitated by antigen-presenting cell maturation and, when activated, T-cells provide an antigen-specific boost to B-cell proliferation and antibody production. Furthermore, some vaccine adjuvants tested in fish enable Th cells to differentiate into T-cell lineages, such as Th1, Th2, Treg, and Th17 cells (46). It is believed that inducing $CD4^+$ and $CD8^+$ T-cell differentiation is crucial for long-term protection. Understanding these responses in fish may also aid in the development of new adjuvants that stimulate specific immunological pathways to achieve the desired immune response and improve protective immunity.

At present, several fish vaccines are frequently combined with adjuvants (37,51), including Montanide ISA763A, CpG/Poly I:C (52), chitosan oligosaccharide (53), squalene-aluminum hydroxide (54), and liposome (55). Mineral oil Montanide™ ISA 761 VG and nonmineral oil Montanide™ ISA 763A VG are known to promote both cell-mediated and humoral immunity, and are widely used in commercial fish (18,56,57). Flagellin has also been reported to induce a strong antibody response, and may be a potent immunostimulant and vaccine adjuvant (58). The white oil-adjuvant Marcol 52 has been widely used in farmed animals, including chickens, swine, and calves, as well as in fish vaccines (37,59–61). Two new adjuvants, Montanide™ ISA 763B VG and Montanide™ GEL02, could improve the immune response by increasing the efficacy of vaccination and providing long-term protection (45).

10.5 SIDE EFFECTS OF FISH VACCINATION

The most often used adjuvants in fish vaccines especially oil-based adjuvants have been demonstrated to produce long-term immune responses to a variety of antigens (62). However, in injected fish, they can potentially cause serious negative effects. Intraperitoneally injected vaccines with oil-based adjuvants can cause adhesions and granulomatous lesions at the injection site and in abdominal organs. According to recent studies, such immunizations in fish can induce systemic autoimmunity, but not all oil-based adjuvants cause injection site lesions (63). Adjuvants based on aluminum have been reported to be less harmful than oil-based adjuvants (64). Previous studies reported that a combination of squalene and aluminum hydroxide had no influence on growth and caused no adverse effects in Japanese flounder (65). Microsphere-based vaccines have been found to be safe for fish. Several studies reported that immunization with vaccines based on microspheres did not show negative influences in turbot (66) and Atlantic salmon (67); however, in Indian major carp, only a local reaction occurring at the point of injection was found (68). The concentration of the adjuvant is also significant in the production of lesions. At specific concentrations of adjuvants, such as aluminum hydroxide and quillaja saponin, they are considered safe for fish (65). In mammals, saponin-based adjuvants produce a positive immunological response (69), although it is unknown as to whether they make the fish immune system more responsive to antigens.

Apart from the production of specific antibodies in the fish serum, little is known about the mechanism of the immune response that vaccinations elicit in fish. Untreated fish have a mixture of different cell types in their peritoneal cavity, which differ in amount relying on the fish species, where lymphocytes, macrophages, and granulocytes are the most prevalent types (70,71). Neutrophils or comparable cells constitute the majority of the granulocyte population. A limited fraction of other cells, such as eosinophilic granules, have been demonstrated in several fish species (72). The number of lymphocytes increases when vaccinations or other immune stimulations are administered into the peritoneal cavity of the fish (73,74). Previous reports showed a dramatic increase in the number of neutrophils and macrophages in the peritoneal cavity of European sea bass within 2–3 hours after stimulation, accounting for more than 90% of the cell population after 6 hours (70). The migratory dynamics of cells, on the other hand, differ with types of fish and with the mode of immunostimulation. Induction of a macrophage-dominated response was reported in Atlantic salmon injected with glucan, while a neutrophil-dominated response was found in salmon injected with glycogen and Freund's incomplete adjuvant (FIA) (75); nevertheless, the mitigation of neutrophils is normally faster than macrophages. After i.p. injection of yeast cells into gilthead seabream, acidophil granulocytes' (a neutrophil analog) levels increased after 4 hours, while monocyte-macrophages increased from 24 to 72 hours (76). Sea bass injected intraperitoneally with various vaccines showed a significant elevation in neutrophils after 24 hours, as well as an increase in macrophages on days 2–4 and even up to day 7 post-vaccination (72). It has been reported that lymphocytes play a role in the inflammatory response and may be the most numerous leucocytes in fish stimulated with *A. salmonicida* (77). The effects of adjuvants administered intraperitoneally in fish have been studied extensively; however, little is known about in what way i.p. vaccines influence the establishment of adjuvant-induced internal lesions. In addition, various reports investigated how intraperitoneally injected vaccinations affect the fish inflammatory response, which evaluated changes in the quantity or activity of peritoneal leucocytes. Vaccines, on the other hand, tend to stick to internal organs, which is likely where the majority of the inflammatory reaction occurs.

10.6 THE POTENCY AND SAFETY TESTING OF FISH VACCINES

During vaccine development, *in vivo* experiments are frequently used. In the absence of a comprehensive understanding of the vaccine candidate or knowledge of the vaccine's important attributes, *in vivo* assays provide a biological response to the candidate vaccine and the ability to measure the potency of the vaccine (45,78). However, for early screening and testing of novel vaccination candidates/preparations, as well as batch potency and safety studies, *in vitro* cell culture techniques have been developed to potentially replace *in vivo* pathogen exposure. Over the last decade, several efforts have been made to develop alternative potency testing methods for fish vaccines, largely because these experiments generally involve subjecting fish to lethal challenges, inoculating the animals with the target pathogen, and a control group. In order to reduce the number of experimental animals used in these types of studies, a number of successful efforts have been reported (79). The immunological mechanisms that contribute to vaccination-mediated protection may be revealed by the mode of action after stimulation by the vaccine candidates. Therefore, *in vitro* cell culture systems can be a good platform for selecting prospective vaccine candidates, assessing the induced immune responses with isolated cells, and providing a low-cost alternative to animal testing in aquaculture vaccine development.

Nevertheless, after screening potential vaccines, a vaccine safety test is still conducted to ensure that the product of interest will cause no harm to the animals or to the end consumers. To determine whether the immunizations have any *in vivo* side effects, fish mortality must be monitored. During the first week, fish must be monitored daily for acute side effects, such as behavior changes, lesions around the injection site, and sudden death caused by toxicity. Ultimately, the fish may be euthanized and necropsied following vaccination, to evaluate long-term negative effects such as internal lesions or adhesions.

10.7 THE STABILITY OF VACCINE PREPARATION

The first requirement in developing a robust vaccine stability system is to ensure that the potency technique indicates stability. This is mainly performed by introducing samples to the mildest conditions, in order to cause structural changes in the antigens. This may require testing a variety of antigen concentrations, buffers, pH, and temperature, and all these variables through time. Testing against an accelerated (e.g., 25°C) or stressed (e.g., 37°C) environment for a period of 2–4 weeks is a simple protocol for a vaccine developed to be kept at 4°C. Due to these relatively mild conditions, if the efficacy of the vaccine is reduced when compared to untreated controls, it may be appropriate to build more information as to whether the vaccine is stable when samples are accidentally exposed to room temperature during transportation (80).

For assays with considerable variability, it is difficult to thoroughly evaluate the results of potency assays' results with time also being a source of variance during stability studies. Therefore, to boost confidence in the mean or "real" value, a replication approach should always be implemented. Of course, the cost of replication (in terms of materials and animals) can be rather high, which must be taken into account before commencing stability tests. In order to ensure that the vaccine pharmaceutical product's shelf-life is properly established for clinical or commercial usage, a potency test needs to be included in stability studies at the last time-points. Verification of the stability of the chemical ingredient is critical early on in preclinical research, to ensure that no changes occur during storage prior to drug product manufacture.

10.8 CONCLUSIONS AND FUTURE PERSPECTIVES

Vaccination is a practical method for protecting cultured fish against infectious diseases and has been shown to be effective in the control of infections in most trials. However, fish vaccination is still a developing sector in the aquaculture industry. Some studies in this field are underway, as many of the scientific findings have yet to be translated to practical conditions. In some cases, several factors influence the success of immunization in fish, including antigen dose, potency and safety, stability, water temperature, density, seasonal factors, salinity, photoperiod, and handling stress. Many vaccines are only partially effective, and researchers are actively looking for ways to increase their performance. Adjuvants are generally accepted as the most effective techniques for improving the efficacy of fish vaccines, providing prolonged protection and modulating the immunogenicity of the components. However, when testing new vaccine formulations, it is paramount to conduct studies evaluating side effects. In future investigations, it will be important to design and analyze the side effects of prospective adjuvants. Also, further research on the efficacy of adjuvanted vaccinations is needed to evaluate the RPS when applying higher challenge doses, to assess the potential of cross-protection across isolates from different serotypes, and assess the influence of various infection routes (e.g., immersion, cohabitation). It is necessary to determine the relationship between antigen concentration and protection, as well as to select appropriate adjuvants to mount the desired specific immune response and promote a sustainable aquaculture production.

ACKNOWLEDGMENT

This work (Grant No. RGNS 63-149) was supported by Office of the Permanent Secretary, Ministry of Higher Education, Science, Research and Innovation (OPS MHESI), Thailand Science Research and Innovation (TSRI) and Mahasarakham University.

DISCLOSURE STATEMENT

The authors declare that they have no conflicts of interest.

REFERENCES

[1] Evensen, Ø., & Leong, J. A. C. (2013). DNA vaccines against viral diseases of farmed fish. *Fish & Shellfish Immunology*, *35*(6), 1751–1758. https://doi.org/10.1016/j.fsi.2013.10.021

[2] Adams, A. (2019). Progress, challenges and opportunities in fish vaccine development. *Fish & Shellfish Immunology*, *90*, 210–214. https://doi.org/10.1016/j.fsi.2019.04.066

[3] Cuesta, A., Chaves-Pozo, E., de Las Heras, A. I., Saint-Jean, S. R., Perez-Prieto, S., & Tafalla, C. (2010). An active DNA vaccine against infectious pancreatic necrosis virus (IPNV) with a different mode of action than fish rhabdovirus DNA vaccines. *Vaccine*, *28*(19), 3291–3300. https://doi.org/10.1016/j.vaccine.2010.02.106

[4] Austin, B. (2012). Developments in vaccination against fish bacterial disease. In *Infectious disease in aquaculture* (pp. 218–243). Woodhead Publishing, Cambridge. https://doi.org/10.1533/9780857095732.2.218

[5] Brudeseth, B. E., Wiulsrød, R., Fredriksen, B. N., Lindmo, K., Løkling, K. E., Bordevik, M., & Gravningen, K. (2013). Status and future perspectives of vaccines for industrialised fin-fish farming. *Fish & Shellfish Immunology*, *35*(6), 1759–1768. https://doi.org/10.1016/j.fsi.2013.05.029

[6] Vinay, T. N., Bhat, S., Gon Choudhury, T., Paria, A., Jung, M. H., Shivani Kallappa, G., & Jung, S. J. (2018). Recent advances in application of nanoparticles in fish vaccine delivery. *Reviews in Fisheries Science & Aquaculture*, *26*(1), 29–41. https://doi.org/10.1080/23308249.2017.1334625

[7] Sommerset, I., Krossøy, B., Biering, E., & Frost, P. (2005). Vaccines for fish in aquaculture. *Expert Review of Vaccines*, *4*(1), 89–101. https://doi.org/10.1586/14760584.4.1.89

[8] Evensen, O. (2009). Development in fish vaccinology with focus on delivery methodologies, adjuvants and formulations. *Options Mediterraneennes*, *86*, 177–186. https://www.researchgate.net/profile/Oystein-Evensen/publication/242541465_Development_in_fish_vaccinology_with_focus_on_delivery_methodologies_adjuvants_and_formulations/links/02e7e52812a4d12aa8000000/Development-in-fish-vaccinology-with-focus-on-delivery-methodologies-adjuvants-and-formulations.pdf

[9] Bøgwald, J., & Dalmo, R. A. (2012). Developments in adjuvants for fish vaccines. In *Infectious disease in aquaculture* (pp. 244–274). Woodhead Publishing, Cambridge. https://doi.org/10.1533/9780857095732.2.244

[10] Villumsen, K. R., & Raida, M. K. (2013). Long-lasting protection induced by bath vaccination against *Aeromonas salmonicida* subsp. *salmonicida* in rainbow trout. *Fish & Shellfish Immunology*, *35*(5), 1649–1653. https://doi.org/10.1016/j.fsi.2013.09.013

[11] Vervarcke, S., Ollevier, F., Kinget, R., & Michoel, A. (2004). Oral vaccination of African catfish with *Vibrio anguillarum* O_2: Effect on antigen uptake and immune response by absorption enhancers in lag time coated pellets. *Fish & Shellfish Immunology*, *16*(3), 407–414. https://doi.org/10.1016/j.fsi.2003.07.002

[12] Deshmukh, S., Raida, M. K., Dalsgaard, I., Chettri, J. K., Kania, P. W., & Buchmann, K. (2012). Comparative protection of two different commercial vaccines against *Yersinia ruckeri* serotype O1 and biotype 2 in rainbow trout (*Oncorhynchus mykiss*). *Veterinary Immunology and Immunopathology*, *145*(1–2), 379–385. https://doi.org/10.1016/j.vetimm.2011.12.014

[13] Chettri, J. K., Deshmukh, S., Holten-Andersen, L., Jafaar, R. M., Dalsgaard, I., & Buchmann, K. (2013). Comparative evaluation of administration methods for a vaccine protecting rainbow trout against *Yersinia ruckeri* O1 biotype 2 infections. *Veterinary Immunology and Immunopathology*, *154*(1–2), 42–47. https://doi.org/10.1016/j.vetimm.2013.04.001

[14] Chettri, J. K., Jaafar, R. M., Skov, J., Kania, P. W., Dalsgaard, I., & Buchmann, K. (2015). Booster immersion vaccination using diluted *Yersinia ruckeri* bacterin confers protection against ERM in rainbow trout. *Aquaculture*, *440*, 1–5. https://doi.org/10.1016/j.aquaculture.2015.01.027

[15] Amend, D. F. (1981). Potency testing of fish vaccines. In *Fish biologics: serodiagnostics and vaccines*, 447–454.

[16] Gudding, R., & Van Muiswinkel, W. B. (2013). A history of fish vaccination: Science-based disease prevention in aquaculture. *Fish & Shellfish Immunology*, *35*(6), 1683–1688. https://doi.org/10.1016/j.fsi.2013.09.031

[17] Amend, D. F., Johnson, K. A., Croy, T. R., & McCarthy, D. H. (1983). Some factors affecting the potency of *Yersinia ruckeri* bacterins. *Journal of Fish Diseases*, *6*(4), 337–344. https://doi.org/10.1111/j.1365-2761.1983.tb00085.x

[18] Jaafar, R. M., Chettri, J. K., Dalsgaard, I., Al-Jubury, A., Kania, P. W., Skov, J., & Buchmann, K. (2015). Effects of adjuvant Montanide™ ISA 763 A VG in rainbow trout injection vaccinated against *Yersinia ruckeri*. *Fish & Shellfish Immunology*, *47*(2), 797–806. https://doi.org/10.1016/j.fsi.2015.10.023

[19] Wangkahart, E., Secombes, C. J., & Wang, T. (2019). Dissecting the immune pathways stimulated following injection vaccination of rainbow trout (*Oncorhynchus mykiss*) against enteric redmouth disease (ERM). *Fish & Shellfish Immunology*, *85*, 18–30. https://doi.org/10.1016/j.fsi.2017.07.056
[20] Gliniewicz, K., Plant, K. P., LaPatra, S. E., LaFrentz, B. R., Cain, K., Snekvik, K. R., & Call, D. R. (2012). Comparative proteomic analysis of virulent and rifampicin-attenuated *Flavobacterium psychrophilum*. *Journal of Fish Diseases*, *35*(7), 529–539. https://doi.org/10.1111/j.1365-2761.2012.01378.x
[21] Rømer Villumsen, K., Dalsgaard, I., Holten-Andersen, L., & Raida, M. K. (2012). Potential role of specific antibodies as important vaccine induced protective mechanism against Aeromonas salmonicida in rainbow trout. *PLoS One*, *7*(1): e46733. https://doi.org/10.1371/journal.pone.0046733
[22] Yu, L. P., Hu, Y. H., Sun, B. G., & Sun, L. (2012). C312M: An attenuated *Vibrio anguillarum* strain that induces immunoprotection as an oral and immersion vaccine. *Diseases of Aquatic Organisms*, *102*(1), 33–42. https://doi.org/10.3354/dao02544
[23] Boutier, M., Ronsmans, M., Ouyang, P., Fournier, G., Reschner, A., Rakus, K., & Vanderplasschen, A. (2015). Rational development of an attenuated recombinant cyprinid herpesvirus 3 vaccine using prokaryotic mutagenesis and *in vivo* bioluminescent imaging. *PLoS Pathogens*, *11*(2), e1004690. https://doi.org/10.1371/journal.ppat.1004690
[24] Jiao, X. D., Zhang, M., Hu, Y. H., & Sun, L. (2009). Construction and evaluation of DNA vaccines encoding *Edwardsiella tarda* antigens. *Vaccine*, *27*(38), 5195–5202. https://doi.org/10.1016/j.vaccine.2009.06.071
[25] Chaves-Pozo, E., Cuesta, A., & Tafalla, C. (2010). Antiviral DNA vaccination in rainbow trout (*Oncorhynchus mykiss*) affects the immune response in the ovary and partially blocks its capacity to support viral replication in vitro. *Fish & Shellfish Immunology*, *29*(4), 579–586. https://doi.org/10.1016/j.fsi.2010.06.006
[26] Liu, L., Gong, Y. X., Liu, G. L., Zhu, B., & Wang, G. X. (2016). Protective immunity of grass carp immunized with DNA vaccine against *Aeromonas hydrophila* by using carbon nanotubes as a carrier molecule. *Fish & Shellfish Immunology*, *55*, 516–522. https://doi.org/10.1016/j.fsi.2014.11.026
[27] Liu, X., Xu, J., Zhang, H., Liu, Q., Xiao, J., & Zhang, Y. (2016). Design and evaluation of an *Edwardsiella tarda* DNA vaccine co-encoding antigenic and adjuvant peptide. *Fish & Shellfish Immunology*, *59*, 189–195. https://doi.org/10.1016/j.fsi.2016.10.029
[28] Huang, Y., Cai, S., Pang, H., Jian, J., & Wu, Z. (2017). Immunogenicity and efficacy of DNA vaccine encoding antigenic AcfA via addition of the molecular adjuvant Myd88 against *Vibrio alginolyticus* in *Epinephelus coioides*. *Fish & Shellfish Immunology*, *66*, 71–77. https://doi.org/10.1016/j.fsi.2017.05.021
[29] Ma, Y. P., Ke, H., Liang, Z. L., Ma, J. Y., Hao, L., & Liu, Z. X. (2017). Protective efficacy of cationic-PLGA microspheres loaded with DNA vaccine encoding the sip gene of *Streptococcus agalactiae* in tilapia. *Fish & Shellfish Immunology*, *66*, 345–353. https://doi.org/10.1016/j.fsi.2017.05.003
[30] Machimbirike, V. I., Pornputtapong, N., Senapin, S., Wangkahart, E., Srisapoome, P., Khunrae, P., & Rattanarojpong, T. (2022). A multi-epitope chimeric protein elicited a strong antibody response and partial protection against *Edwardsiella ictaluri* in Nile tilapia. *Journal of Fish Diseases*, *45*(1), 1–18. https://doi.org/10.1111/jfd.13525
[31] Wang, E., Qin, Z., Yu, Z., Ai, X., Wang, K., Yang, Q., & Lai, W. (2018). Molecular characterization, phylogenetic, expression, and protective immunity analysis of OmpF, a promising candidate immunogen against *Yersinia ruckeri* infection in channel catfish. *Frontiers in Immunology*, *9*, 2003. https://doi.org/10.3389/fimmu.2018.02003
[32] Sun, Y., Liu, C. S., & Sun, L. (2011). Comparative study of the immune effect of an *Edwardsiella tarda* antigen in two forms: Subunit vaccine vs DNA vaccine. *Vaccine*, *29*(11), 2051–2057. https://doi.org/10.1016/j.vaccine.2011.01.013
[33] Marana, M. H., Jørgensen, L. V. G., Skov, J., Chettri, J. K., Holm Mattsson, A., Dalsgaard, I., & Buchmann, K. (2017). Subunit vaccine candidates against *Aeromonas salmonicida* in rainbow trout *Oncorhynchus mykiss*. *PLoS One*, *12*(2), e0171944. https://doi.org/10.1371/journal.pone.0171944
[34] Zhao, Z., Xiong, Y., Zhang, C., Jia, Y. J., Qiu, D. K., Wang, G. X., & Zhu, B. (2020). Optimization of the efficacy of a SWCNTs-based subunit vaccine against infectious spleen and kidney necrosis virus in mandarin fish. *Fish & Shellfish Immunology*, *106*, 190–196. https://doi.org/10.1016/j.fsi.2020.07.062
[35] Lueangyangyuen, A., Senapin, S., Dong, H. T., Unajak, S., Wangkahart, E., & Khunrae, P. (2022). Expression and purification of S5196-272 and S6200-317 proteins from Tilapia Lake Virus (TiLV) and their potential use as vaccines. *Protein Expression and Purification*, *190*, 106013. https://doi.org/10.1016/j.pep.2021.106013
[36] Karlsen, M., Tingbø, T., Solbakk, I. T., Evensen, Ø., Furevik, A., & Aas-Eng, A. (2012). Efficacy and safety of an inactivated vaccine against Salmonid alphavirus (family Togaviridae). *Vaccine*, *30*(38), 5688–5694. https://doi.org/10.1016/j.vaccine.2012.05.069

[37] Li, J., Tang, L., Li, S., Li, G., & Mo, Z. (2020). The efficacy and side-effects of oil-based adjuvants emulsified *Vibrio anguillarum* bivalent inactivated vaccine in turbot (*Scophthalmus maximus*) under production mode. *Aquaculture*, *524*, 735259. https://doi.org/10.1016/j.aquaculture.2020.735259

[38] Jiang, X., Zhang, C., Zhao, Y., Kong, X., Pei, C., Li, L., … & Li, X. (2016). Immune effects of the vaccine of live attenuated *Aeromonas hydrophila* screened by rifampicin on common carp (*Cyprinus carpio* L). *Vaccine*, *34*(27), 3087–3092. https://doi.org/10.1016/j.vaccine.2016.04.075

[39] Zhang, Z., Wu, H., Xiao, J., Wang, Q., Liu, Q., & Zhang, Y. (2012). Immune responses of zebrafish (Danio rerio) induced by bath-vaccination with a live attenuated *Vibrio anguillarum* vaccine candidate. *Fish & Shellfish Immunology*, *33*(1), 36–41. https://doi.org/10.1016/j.fsi.2012.03.031

[40] Yan, Y., Mu, W., Zhang, L., Guan, L., Liu, Q., & Zhang, Y. (2013). Asd-based balanced-lethal system in attenuated *Edwardsiella tarda* to express a heterologous antigen for a multivalent bacterial vaccine. *Fish & Shellfish Immunology*, *34*(5), 1188–1194. https://doi.org/10.1016/j.fsi.2013.01.027

[41] Griffin, M. J., Greenway, T. E., Byars, T. S., Ware, C., Aarattuthodiyil, S., Kumar, G., & Wise, D. J. (2020). Cross-protective potential of a live-attenuated *Edwardsiella ictaluri* vaccine against *Edwardsiella piscicida* in channel (*Ictalurus punctatus*) and channel × blue (*Ictalurus furcatus*) hybrid catfish. *Journal of the World Aquaculture Society*, *51*(3), 740–749.

[42] Xing, J., Xu, H., Tang, X., Sheng, X., & Zhan, W. (2019). A DNA vaccine encoding the VAA gene of *Vibrio anguillarum* induces a protective immune response in flounder. *Frontiers in Immunology*, *10*, 499. https://doi.org/10.3389/fimmu.2019.00499

[43] Chen, J., Tan, W., Wang, W., Hou, S., Chen, G., Xia, L., & Lu, Y. (2019). Identification of common antigens of three pathogenic *Nocardia* species and development of DNA vaccine against fish nocardiosis. *Fish & Shellfish Immunology*, *95*, 357–367. https://doi.org/10.1016/j.fsi.2019.09.038

[44] Ciftci, A., Onuk, E. E., Ciftci, G., Fındık, A., Söğüt, M. Ü., Didinen, B. I., & Altun, S. O. N. E. R. (2016). Development and validation of glycoprotein-based native-subunit vaccine for fish against *Aeromonas hydrophila*. *Journal of Fish Diseases*, *39*(8), 981–992. https://doi.org/10.1111/jfd.12499

[45] Wangkaghart, E., Deville, S., Wang, B., Srisapoome, P., Wang, T., & Secombes, C. J. (2021). Immune response and protective efficacy of two new adjuvants, Montanide™ ISA 763B VG and Montanide™ GEL02, administered with a *Streptococcus agalactiae* ghost vaccine in Nile tilapia (*Oreochromis niloticus*). *Fish & Shellfish Immunology*, *116*, 19–29. https://doi.org/10.1016/j.fsi.2021.06.014

[46] Tafalla, C., Bøgwald, J., & Dalmo, R. A. (2013). Adjuvants and immunostimulants in fish vaccines: Current knowledge and future perspectives. *Fish & Shellfish Immunology*, *35*(6), 1740–1750. https://doi.org/10.1016/j.fsi.2013.02.029

[47] Wang, C., Hu, Y. H., Chi, H., & Sun, L. (2013). The major fimbrial subunit protein of *Edwardsiella tarda*: Vaccine potential, adjuvant effect, and involvement in host infection. *Fish & Shellfish Immunology*, *35*(3), 858–865. https://doi.org/10.1016/j.fsi.2013.06.021

[48] Ribeiro, C., & Schijns, V. E. (2010). Immunology of vaccine adjuvants. In *Vaccine adjuvants* (pp. 1–14). Humana Press, Totowa, NJ. https://doi.org/10.1007/978-1-60761-585-9_1

[49] Schijns, V. E. (2001). Induction and direction of immune responses by vaccine adjuvants. *Critical Reviews in Immunology*, *21*(1–3), 75–85. https://doi.org/10.1615/CritRevImmunol.v21.i1-3.50

[50] Awate, S., Babiuk, L. A. B., & Mutwiri, G. (2013). Mechanisms of action of adjuvants. *Frontiers in Immunology*, *4*, 114. https://doi.org/10.3389/fimmu.2013.00114

[51] Labarca, C. C., Makhutu, M., Lumsdon, A. E., Thompson, K. D., Jung, R., Kloas, W., & Knopf, K. (2015). The adjuvant effect of low frequency ultrasound when applied with an inactivated *Aeromonas salmonicida* vaccine to rainbow trout (*Oncorhynchus mykiss*). *Vaccine*, *33*(11), 1369–1374. https://doi.org/10.1016/j.vaccine.2015.01.027

[52] Thim, H. L., Villoing, S., McLoughlin, M., Christie, K. E., Grove, S., Frost, P., & Jørgensen, J. B. (2014). Vaccine adjuvants in fish vaccines make a difference: Comparing three adjuvants (Montanide ISA763A oil, CpG/Poly I: C combo and VHSV glycoprotein) alone or in combination formulated with an inactivated whole salmonid alphavirus antigen. *Vaccines*, *2*(2), 228–251. https://doi.org/10.3390/vaccines2020228

[53] Wei, G., Cai, S., Wu, Y., Ma, S., & Huang, Y. (2020). Immune effect of *Vibrio harveyi* formalin-killed cells vaccine combined with chitosan oligosaccharide and astragalus polysaccharides in ♀ *Epinephelus fuscoguttatus* × ♂ *Epinephelus lanceolatus*. *Fish & Shellfish Immunology*, *98*, 186–192. https://doi.org/10.1016/j.fsi.2020.01.015

[54] Dar, S. A., Kole, S., Shin, S. M., Jeong, H. J., & Jung, S. J. (2021). Comparative study on antigen persistence and immunoprotective efficacy of intramuscular and intraperitoneal injections of squalene-aluminium hydroxide (Sq+Al) adjuvanted viral hemorrhagic septicaemia virus vaccine in olive flounder (*Paralichthys olivaceus*). *Vaccine*, *39*(47), 6866–6875. https://doi.org/10.1016/j.vaccine.2021.10.026

[55] Rømer Villumsen, K., Kania, P. W., Christensen, D., Koppang, E. O., & Bojesen, A. M. (2020). Injection vaccines formulated with nucleotide, liposomal or mineral oil adjuvants induce distinct differences in immunogenicity in rainbow trout. *Vaccines*, *8*(1), 103. https://doi.org/10.3390/vaccines8010103

[56] Xu, W., Jiao, C., Bao, P., Liu, Q., Wang, P., Zhang, R., ... & Zhang, Y. (2019). Efficacy of Montanide™ ISA 763 A VG as aquatic adjuvant administrated with an inactivated *Vibrio harveyi* vaccine in turbot (*Scophthalmus maximus* L.). *Fish & Shellfish Immunology*, *84*, 56–61. https://doi.org/10.1016/j.fsi.2018.09.024

[57] Veenstra, K. A., Wang, T., Russell, K. S., Tubbs, L., Arous, J. B., & Secombes, C. J. (2021). Montanide™ ISA 763A VG and ISA 761 VG induce different immune pathway responses in rainbow trout (*Oncorhynchus mykiss*) when used as adjuvant for an *Aeromonas salmonicida* bacterin. *Fish & Shellfish Immunology*, *114*, 171–183. https://doi.org/10.1016/j.fsi.2021.04.024

[58] Wangkahart, E., Secombes, C. J., & Wang, T. (2019). Studies on the use of flagellin as an immunostimulant and vaccine adjuvant in fish aquaculture. *Frontiers in Immunology*, *9*, 3054. https://doi.org/10.3389/fimmu.2018.03054

[59] Muneer, R., Akhtar, S., & Afzal, M. (1994). Evaluation of three oil-adjuvant vaccines against *Pasteurella multocida* in buffalo calves. *Revue Scientifique et Technique (International Office of Epizootics)*, *13*(3), 837–843. https://doi.org/10.20506/rst.13.3.806

[60] Pitcovski, J., Gutter, B., Gallili, G., Goldway, M., Perelman, B., Gross, G., ... & Michael, A. (2003). Development and large-scale use of recombinant VP2 vaccine for the prevention of infectious bursal disease of chickens. *Vaccine*, *21*(32), 4736–4743. https://doi.org/10.1016/S0264-410X(03)00525-5

[61] Dong, Y., Weng, S., He, J., & Dong, C. (2013). Field trial tests of FKC vaccines against RSIV genotype Megalocytivirus in cage-cultured mandarin fish (*Siniperca chuatsi*) in an inland reservoir. *Fish & Shellfish Immunology*, *35*(5), 1598–1603. https://doi.org/10.1016/j.fsi.2013.09.005

[62] Noia, M., Domínguez, B., Leiro, J., Blanco-Méndez, J., Luzardo-Álvarez, A., & Lamas, J. (2014). Inflammatory responses and side effects generated by several adjuvant-containing vaccines in turbot. *Fish & Shellfish Immunology*, *38*(1), 244–254. https://doi.org/10.1016/j.fsi.2014.03.020

[63] Mutoloki, S., Alexandersen, S., Gravningen, K., & Evensen, Ø. (2008). Time-course study of injection site inflammatory reactions following intraperitoneal injection of Atlantic cod (*Gadus morhua* L.) with oil-adjuvanted vaccines. *Fish & Shellfish Immunology*, *24*(4), 386–393. https://doi.org/10.1016/j.fsi.2007.08.009

[64] Jiao, X. D., Cheng, S., Hu, Y. H., & Sun, L. (2010). Comparative study of the effects of aluminum adjuvants and Freund's incomplete adjuvant on the immune response to an *Edwardsiella tarda* major antigen. *Vaccine*, *28*(7), 1832–1837. https://doi.org/10.1016/j.vaccine.2009.11.083

[65] Vinay, T. N., Kim, Y. J., Jung, M. H., Kim, W. S., Kim, D. H., & Jung, S. J. (2013). Inactivated vaccine against viral hemorrhagic septicemia (VHS) emulsified with squalene and aluminum hydroxide adjuvant provides long term protection in olive flounder (*Paralichthys olivaceus*). *Vaccine*, *31*(41), 4603–4610. https://doi.org/10.1016/j.vaccine.2013.07.036

[66] León-Rodríguez, L., Luzardo-Álvarez, A., Blanco-Méndez, J., Lamas, J., & Leiro, J. (2012). A vaccine based on biodegradable microspheres induces protective immunity against scuticociliatosis without producing side effects in turbot. *Fish & Shellfish Immunology*, *33*(1), 21–27. https://doi.org/10.1016/j.fsi.2012.03.028

[67] Fredriksen, B. N., & Grip, J. (2012). PLGA/PLA micro-and nanoparticle formulations serve as antigen depots and induce elevated humoral responses after immunization of Atlantic salmon (*Salmo salar* L.). *Vaccine*, *30*(3), 656–667. https://doi.org/10.1016/j.vaccine.2011.10.105

[68] Behera, T., & Swain, P. (2012). Antigen adsorbed surface modified poly-ε-caprolactone microspheres stimulates both adaptive and innate immune response in fish. *Vaccine*, *30*(35), 5278–5284. https://doi.org/10.1016/j.vaccine.2012.05.028

[69] Lövgren Bengtsson, K., Morein, B., & Osterhaus, A. D. (2011). ISCOM technology-based Matrix M™ adjuvant: Success in future vaccines relies on formulation. *Expert Review of Vaccines*, *10*(4), 401–403. https://doi.org/10.1586/erv.11.25

[70] Do Vale, A., Afonso, A., & Silva, M. T. (2002). The professional phagocytes of sea bass (*Dicentrarchus labrax* L.): Cytochemical characterisation of neutrophils and macrophages in the normal and inflamed peritoneal cavity. *Fish & Shellfish Immunology*, *13*(3), 183–198. https://doi.org/10.1006/fsim.2001.0394

[71] Tumbol, R. A., Baiano, J. C., & Barnes, A. C. (2009). Differing cell population structure reflects differing activity of Percoll-separated pronephros and peritoneal leucocytes from barramundi (*Lates calcarifer*). *Aquaculture*, *292*(3-4), 180–188. https://doi.org/10.1016/j.aquaculture.2009.04.030

[72] Afonso, A., Gomes, S., Da Silva, J., Marques, F., & Henrique, M. (2005). Side effects in sea bass (*Dicentrarchus labrax* L.) due to intraperitoneal vaccination against vibriosis and pasteurellosis. *Fish & Shellfish Immunology*, *19*(1), 1–16. https://doi.org/10.1016/j.fsi.2004.09.001

[73] Chaves-Pozo, E., Muñoz, P., López-Muñoz, A., Pelegrín, P., García Ayala, A., Mulero, V., & Meseguer, J. (2005). Early innate immune response and redistribution of inflammatory cells in the bony fish gilthead seabream experimentally infected with *Vibrio anguillarum*. *Cell and Tissue Research*, *320*(1), 61–68. https://doi.org/10.1007/s00441-004-1063-7

[74] Costas, B., Rêgo, P. C. N. P., Simões, I., Marques, J. F., Castro-Cunha, M., & Afonso, A. (2013). Cellular and humoral immune responses of *Senegalese sole*, *Solea senegalensis* (K aup), following challenge with two *Photobacterium damselae* subsp. *piscicida* strains from different geographical origins. *Journal of Fish Diseases*, *36*(6), 543–553. https://doi.org/10.1111/jfd.12033

[75] Jørgensen, J. B., Lunde, H., & Robertsen, B. (1993). Peritoneal and head kidney cell response to intraperitoneally injected yeast glucan in Atlantic salmon, *Salmo salar* L. *Journal of Fish Diseases*, *16*(4), 313–325. https://doi.org/10.1111/j.1365-2761.1993.tb00865.x

[76] Cuesta, A., Rodríguez, A., Salinas, I., Meseguer, J., & Esteban, M. Á. (2007). Early local and systemic innate immune responses in the teleost gilthead seabream after intraperitoneal injection of whole yeast cells. *Fish & Shellfish Immunology*, *22*(3), 242–251. https://doi.org/10.1016/j.fsi.2006.05.005

[77] Korytář, T., Jaros, J., Verleih, M., Rebl, A., Kotterba, G., Kühn, C., & Köllner, B. (2013). Novel insights into the peritoneal inflammation of rainbow trout (*Oncorhynchus mykiss*). *Fish & Shellfish Immunology*, *35*(4), 1192–1199. https://doi.org/10.1016/j.fsi.2013.07.032

[78] Wangkahart, E., Scott, C., Secombes, C. J., & Wang, T. (2016). Re-examination of the rainbow trout (*Oncorhynchus mykiss*) immune response to flagellin: *Yersinia ruckeri* flagellin is a potent activator of acute phase proteins, anti-microbial peptides and pro-inflammatory cytokines *in vitro*. *Developmental & Comparative Immunology*, *57*, 75–87. https://doi.org/10.1016/j.dci.2015.12.017

[79] Attaya, A., Jiang, Y., Secombes, C. J., & Wang, T. (2019). Distinct response of immune gene expression in peripheral blood leucocytes modulated by bacterin vaccine candidates in rainbow trout *Oncorhynchus mykiss*: A potential *in vitro* screening and batch testing system for vaccine development in aquaculture. *Fish & Shellfish Immunology*, *93*, 631–640. https://doi.org/10.1016/j.fsi.2019.08.002

[80] Nunnally, B. K., Turula, V. E., & Sitrin, R. D. (Eds.). (2015). Vaccine Reference Standards. In *Vaccine analysis: strategies, principles, and control* (pp. 563–572). Springer Berlin, Heidelberg, Germany.

11 Mass Vaccination in Aquaculture
Possibilities and Limitations

Hethesh Chellapandian and Sivakamavalli Jeyachandran
National College (Autonomous)

Joseph Selvin
Pondicherry University

Kiyun Park and Ihn-sil Kwak
Chonnam National University

CONTENTS

11.1 INTRODUCTION

Despite various scientific approaches for the development of innovative therapeutics in commercial aquaculture, disease outbreak has a huge impact on the country's economy and causes huge economic losses in the coastal communities who are into aquaculture. Even though chemotherapeutics and antibiotics can be applied for several diseases, they have their own drawbacks/side effects like drug resistance and safety concerns with regard to human consumption if the aquatic organisms, such as shrimp, fish, crabs, etc., are treated with antibiotics. In global aquaculture, vaccination can be used as an effective method for preventing a humungous variety of bacterial, viral, and

DOI: 10.1201/9781003388548-15

other pathogenic diseases that contribute a lot to social, economic, and environmental sustainability. The first instance of vaccination was reported in the late 1940s, followed by many vaccines being developed, which significantly reduces the impact of diseases or prevents certain pathogenic diseases (Snieszko et al., 2019). In certain areas of the world, there has been a transition toward vaccination from antibiotics, which has led to annual vaccination of millions of fishes rather than negative implications of antibiotics. For example, in Norwegian salmon farming, there has been a considerable reduction in the use of antibiotics since the application of vaccines, and vaccines have become the most sustainable and cost-efficient method for preventing/controlling various diseases in aquaculture (Horzinek et al., 1997). Usually, a typical aquaculture vaccine consists of antigens, which stimulate adaptive or innate immune response of an aquatic organism in resistance to specific pathogens (bacteria or virus). During the past two decades, the research activities on fish immunology and fish vaccines have increased in a dramatic way. History, types, advancements, and administration routes of aquaculture-based vaccines along with the prospects and challenges of vaccine development has been described in several researches and review articles by various researchers (Thompson & Roberts, 2016).

Recently, other than injection as a delivery method of vaccine (Plant & Lapatra, 2011), several alternative methods have been developed along with traditional and promising new age adjuvants based on their protective efficacies (Tafalla et al., 2014). In addition, several review articles have focused on present applications of vaccines in large-scale operational fish farms and future perspectives on diverse types of vaccines such as DNA, live attenuated, and inactivated vaccines, but still there is a void for the development of aquaculture vaccine technologies and therefore the vaccine development sector needs comprehensive additional information (Sommerset et al., 2005). The preponderance of licensed vaccines is produced by means of conventional methods that are initiated by culturing target pathogenic organisms. A variety of severe fish diseases has been protected by the abovementioned array of vaccines; aquatic vaccinology can be understood by having a detailed knowledge in two branches of science: immunology and microbiology. The development of vaccines for animals and human use can be obtained by the knowledge advancement in protective antigens and molecular biology (Kim et al., 2016). According to Cimica and Galarza (2017), modern vaccinology targets specific pathogenic components and vaccines that are developed by these approaches, which might include recombinant DNA or subunit vaccines, whereas these vaccines contain novel antigens that are produced through several expression systems. Globally, the developed RNA particle vaccines induce higher level of immunity when compared to other vaccines (Frietze et al., 2016). Even though these advancements are promising, in commercial aqua farming, the implementation/application of mass vaccination is still somewhat limited than in laboratorial conditions due to practical challenges in the aqua environment (Dhar & Allnut, 2011). In this chapter, we describe the application of various molecular approaches and the conventional aquaculture vaccines in large-scale aqua farming systems and challenges, and limitations in mass vaccination and future perspectives of aquaculture vaccine technology and its implication in the aqua environment.

11.2 AQUATIC ORGANISMS' IMMUNE SYSTEM

Unlike vertebrates, shrimps and other crustaceans do not have a distinct lymphatic system; the major transport system is the sole body fluid, which is basically hemolymph. Shrimp hemolymph is made up of cells, soluble inorganic salts, water, and proteins, the most abundant of which is an oxygen transporter copper-based protein. Hemocytes are the unattached cells present inside the hemolymph (Zhi & Fei, 2017). Shrimp, unlike vertebrates, must rely on their innate immune system to defend them from a variety of infections, and hemocytes are essential cells in the innate immune system of shrimp. Hemocytes are formed in the hematopoietic tissue (HPT) surrounding the ocular artery, which has a dorsal position (Tassanakajon et al., 2018). The HPT is made up of thin sheets of relatively dense cells that are grouped into lobules and encircled by connective tissue. Mature hemocytes, prohemocytes, and stem cells are all found in the HPT. Pathogen recognition, encapsulation, phagocytosis, nodulation of pathogens, cytotoxicity, hemolysis, and melanization of pathogens are all actions performed by circulating hemocytes. Hemocytes have two types of secretory granules: big and tiny granules, which carry a variety of defensive molecules.

The morphology and staining features of crustacean hemocytes are used to classify them. Generally, hemocytes are differentiated into three types: granular cells (GC), agranular hyaline cells (HC), and semigranular cells (SGC) (Figure 11.1). Hyaline cells have been studied in the

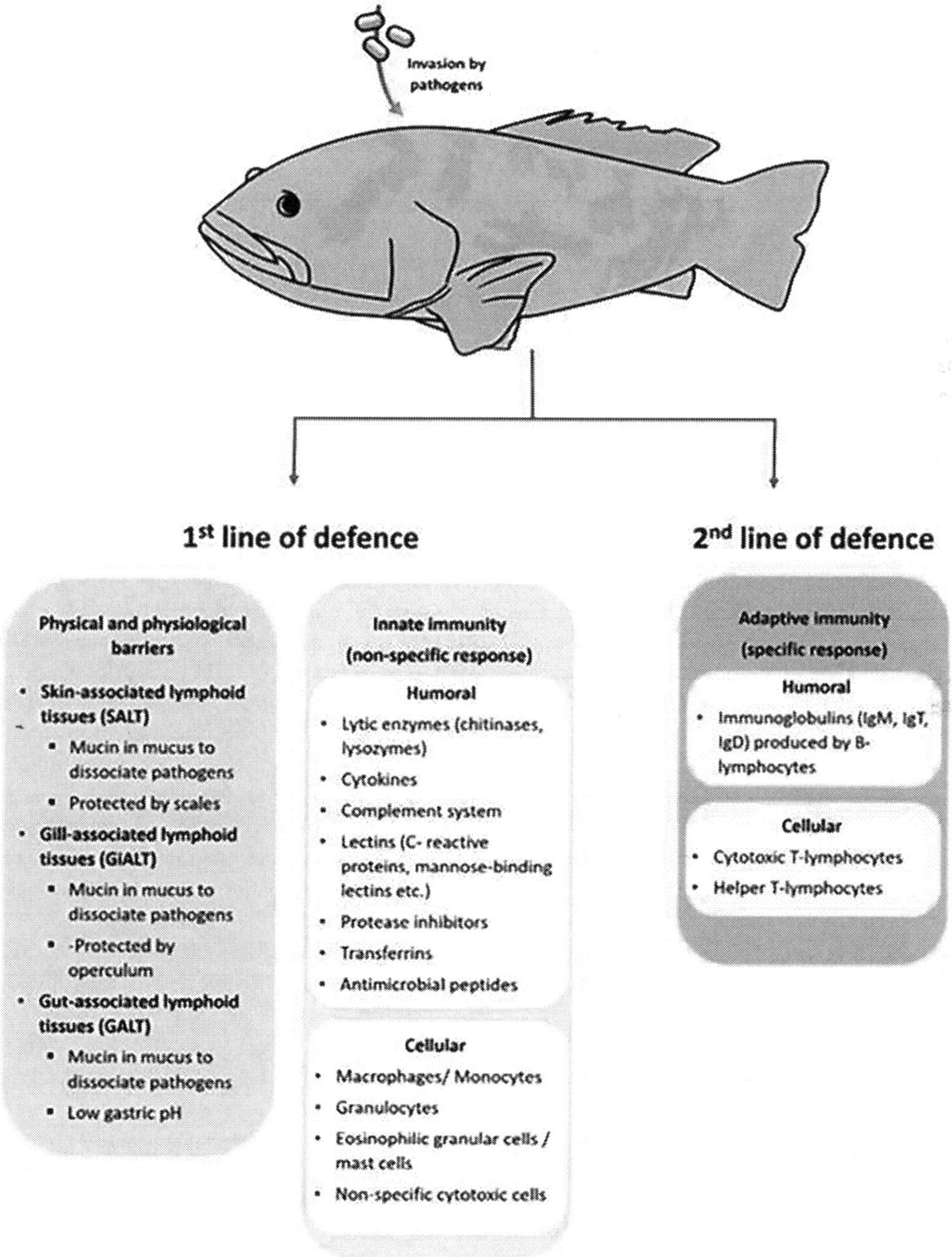

FIGURE 11.1 Defense mechanism in aquatic organisms. [Created by www.biorender.com.]

crayfish, the swimming crab, and the *Penaeus monodon* (Lin & Soderhall, 2011; van de Braak et al., 2002; Hammond & smith, 2002). Smaller globular cells with a bigger central or eccentric nucleus bounded by cytoplasmic granules are known as SGC. The number of tiny eosinophilic granules in SGC varies, although the granules exhibit weak staining characteristics. Granular cells are the biggest hemocytes with a bilobate nucleus, kidney-shaped, and contain differing stain properties for various eosinophilic secretory granules. As a result, certain HCs are eosinophilic, whereas others are partially basophilic and are stainless in some cases. In *P. vannamei*, however, five physiologically active and morphologically different hemocyte subpopulations were extracted using a unique iodixanol density gradient centrifugation approach that differed from the classic Percoll gradient procedure (Dantas-Lima et al., 2013). Hemocytes perform various immune activities and consist of different biochemical properties rather than just morphology. Because of the sensitivity and responsiveness of crustacean hemocytes, functional studies have been difficult, and there have been a number of improvements in methodology and upgrades for extensive investigations of the crustacean distinct hemocytes' features.

Generally, HCs are responsible for phagocytosis in crustaceans, whereas SGCs are responsible for melanization, encapsulation, and coagulation, as well as phagocytosis in certain species (e.g., *Macrobrachium rosenbergii* and *Penaeus japonicas*). The prophenoloxidase (proPO) activation mechanism, cytotoxicity, melanization, and the secretion of antimicrobial peptides are all carried out by GCs containing multiple, large eosinophilic granules. Hemocytes are activated and typically localized in the area of damage or at the surfaces of invading microorganisms during melanization; in order to kill or immobilize invaders, toxic phenol or melanin intermediates are created during the melanization process. According to Liu et al. (2007), melanization and phenoloxidase activity are important in establishing protective immunity in freshwater crayfish against severe pathogens. Since then, RNA-interference suppression of crayfish prophenoloxidase has resulted in greater bacterial growth and mortality, as well as reduced phagocytosis. With the existence of pathogen-associated molecular patterns (PAMPs), such as Gram-negative bacteria's lipopolysaccharides (LPS), Gram-positive bacteria's lipoteichoic acid (LTA), and peptidoglycans (PGN), nucleic acids derived from various microbial pathogens or β-glucans of fungal cells, GC, SGCs, and hemocytes endure hasty degranulation reactions (Amparyup et al., 2012). PAMPs are recognized by pathogen recognition receptors (PRRs) on hemocytes, which then in circulation, release a multitude of effective immune effector molecules near the PAMPs. Antimicrobial peptides (AMPs), lipases, proteases, reactive oxygen species, such as O_2 and H_2O_2, cell adhesion molecules, cytokine-like molecules and cytokines, lysozyme, nitrogen intermediates, complement components (only C1q is known), and clotting proteins (CPs) are all examples of effector molecules (Wang & Wang, 2013). The hemocytes' proPO system is also triggered when PAMPs are recognized. Invertebrates have the most conserved innate immune mechanism that is known as ProPO activation. Pathogens get melanized, and leads to the production of cytokine-like factors, antimicrobial peptides, and other compounds that are considered to assist in iron sequestration and opsonization (Tassanakajon et al., 2017).

The clotting system is a critical component and one of the frontline defenses against invading pathogens in the humoral system. It serves an essential function in limiting hemolymph loss during wound healing and injury when the crustacean exoskeleton barrier is damaged. Shrimp and other crustaceans appear to use the basic coagulation pathway, which involves a calcium-dependent transglutaminase catalyzing the cross-linking of aggregates of CPs (TGase). Many shrimps have been shown to have Tgase, which is a conserved protein that is involved in coagulation as a defensive mechanism in several invertebrates. Aside from acute immunological synapse, hemocytes are key sources of antimicrobial peptides, lectins, proteinase inhibitors, and opsonins such as peroxinectin, a cell adhesion protein. Even though functional studies of hemocytes of crustaceans have been carried out during the 1980s, they have been difficult due to the mechanisms such as sensitivity and reactivity of hemocytes, and there have been many technical breakthroughs over these years that have enabled comprehensive studies and elucidation of immune defense mechanism of hemocytes. The gene silencing/RNA-interference (RNAi) approach (Shockey et al., 2009) was established for studying the involvement of several proteins in hemocyte immune processes such as phagocytosis, virus defense, coagulation, and proPO (Visetnan et al., 2015). Clark and Greenwood attempted to use high-throughput next-generation sequencing technology for immune gene annotation of crustacean in the transcriptome of the American lobster (*Homarus americanus*). Larger invading microorganisms that are too large to be removed by phagocytosis are destroyed by means of nodulation. Multiple layers of hemocytes confine the microbes, and surround them, eventually proPO-mediated melanization end up killing them (Im et al., 2016). By detecting PAMPs and producing effector molecules, multiple conserved signaling pathways are engaged in the frontline of immune defense system to resist pathogen infections. The JAK/STAT, IMD, and toll signaling pathways, which in vertebrates govern the release of pro-inflammatory chemokines, interferons, AMPs, and cytokines, and are thought to exist in shrimp, are likely directly implicated, albeit little is known about their function or activation (Sun et al., 2017).

11.3 HISTORY AND CURRENT PROGRESS IN VACCINE DEVELOPMENT

In 1980's, several countries like Norway, the United Kingdom, and the USA faced a huge and rapid increase of disease outbreak caused by *Aeromonas salmonicida*, *Vibrio species*, and *Yersinia ruckeri* in rainbow trout and Atlantic salmon aquaculture. Consequently, it led to the usage of large quantities of antibiotics; therefore, the concern for antibiotic resistance grew rapidly. As a result the development of fish vaccines was stimulated and the first commercial vaccine was produced against furunculosis, vibriosis, and enteric red mouth (ERM). But, actually the first licensed vaccine was used for salmonid fish against ERM in the USA, in 1976 (Gudding et al., 2014). At present, globally there are 19 major companies that produce fish vaccines including several small-scale companies/industries. Currently, there are almost 30 commercial fish vaccines against both viral and bacterial diseases such as *Yersinia ruckeri* bacterin, *Vibrio anguillarum-ordalii*, *Arthrobacter* vaccine, *A. salmonicida* bacterin, and other immunizations against bacteria in grouper and salmonids, *E. ictaluri* bacterin, and *Flavobacterium columnare* vaccine, respectively, infectious salmon anemia vaccine, infectious pancreatic necrosis virus (IPNV) vaccine, nodavirus vaccine, and other immunizations against viruses in seabass and salmonids, *Streptococcus iniae* vaccine, *Streptococcus agalactiae* vaccine against tilapia, Streptococcosis as well as grass carp hemorrhage disease vaccine, spring viremia of carp vaccine (SVCV), koi herpes vaccine (KHV), and other vaccines against viruses in carps, and similarly a vaccine is commercially available for lobsters (Hang et al., 2021). Live attenuated vaccines are mostly licensed for catfish species in the USA. In Atlantic salmon, against infectious hematopoietic necrosis (IHPN), a DNA vaccine has been licensed in Canada; similarly in Chile, a recombinant vaccine is used against infectious salmon anemia (ISA) and in Norway a subunit vaccine (VP2; peptide) is administrated against IPNV in Atlantic salmon.

Most of the Atlantic salmon vaccines are basically used for microdose applications (i.e., 5 vs 10 μL) that are multivalent in nature. However, compared to salmon farming, the tilapia and carp are widely cultured and established fish, but there are only a handful of vaccines existing for these species. The monovalent furunculosis vaccine market has skyrocketed, eventually decreasing the number of vaccines for trout. There are various significant considerations for using commercial vaccines in fish, such as fish species, farming technology, handling, immune system, stress factors, life history, nutrition, production cycle, environmental factors (pH, salinity, temperature), and even cost benefits. The Responsible Use of Medicines in Agriculture (RUMA) Alliance is the one which provides the guidelines for the utilization of vaccines in fish and other aquatic organisms (https://www.ruma.org.uk/). Majority of vaccines that are commercially available contain adjuvants and are administrated through intraperitoneal injection (Adams and Subashinghe, 2021). In catfish, carp, trout, and tilapia aquaculture, bacterial diseases play a major role in disturbing the production and yield. In addition, those pose noteworthy predicament in marine finfish and salmonids have several efficacious vaccines against viral diseases (Rodger, 2016). Recently, several parasite diseases, such as *Paramoeba perurans* (amoebic gill disease), Ectoparasites, and *Lepeophtheirus salmonis* (sea lice), show potential threat for mass disease outbreaks in Atlantic salmon aqua farming, but still there is a void of vaccines for these parasitic diseases and for fungal and fungus-like diseases. Similarly, facultative parasite (e.g., Aphanomyces and Saprolegnia) that causes disease outbreaks in aquaculture (salmonids and tropical fishes) also does not have any vaccines till date.

11.3.1 Vaccines in Commercial-Scale Fish Farming

The commercially available 28 vaccines protect against 6 viral and 22 bacterial diseases in 17 different species of fish, and these vaccines are available in 40 plus countries (Table 11.1). The large commercial-scale vaccination was first reported in salmon and later, there are vaccines for several other species that are available in most parts of the globe. China is one of the world's largest farmed fish producing countries, and it accounts for around 61% of the total aquaculture production in the world; in common aquaculture practice, vaccination is not applicable. However in China, various intensive researches are done with regard to aquaculture vaccine. This holds a lot of promise in developing and producing an aquaculture vaccine (Wang et al., 2020).

11.4 VACCINE—TYPES OF ADMINISTRATION

The vaccines are administrated through diluted vaccine suspensions using the immersion method, orally through feed, and through the classic method of injecting in the intramuscular (DNA vaccines) or the intraperitoneal route (w/o-based vaccines). When the duration and level of efficacy is taken into account, the injection method is the best mode, but it induces stress while handling fish and shrimp. In the case of small fry which cannot be injected, vaccination can be done through the oral route or the immersion method.

11.4.1 Vaccine Administration—Intraperitoneal/Intramuscular Injection Method

Injection-based vaccination plays a major role in the fortification of aquatic organisms from pathogenic diseases from the production cycle till the harvest period. During the late 1980s, in salmon and trout, water-based vaccines were primarily used against the pathogens such as *Vibrio salmonicida*, *Listonella anguillarum*, and *Yersinia ruckeri*. In Norwegian salmon farming, water-based furunculosis vaccines were not effective against infection caused by *Aeromonas salmonicida*, whereas the application of oil-based and immersion vaccines in the early 1990s significantly reduced the furunculosis disease outbreak (Sommerset et al., 2005). The water-in-oil (w/o) adjuvants had shown prolonged and high efficacy of protection against infections in salmonid culture. For instance, there are several types of oil containing emulsion like water-in-oil-water (w/o/w), oil-in-water (o/w), w/o, and oil-in-water-in-oil (o/w/o), but for long-term protection, the w/o emulsions are used primarily in fish vaccines (Ancouturier et al., 2001). The w/o vaccines have the ability to improve the economy of the aqua farmers by increasing the growth (larger size) and yield till the harvest period. In some cases like during the Atlantic salmon (*Salmo salar*) harvest, the downgrading and reduced growth caused by the injection site reactions of w/o vaccines were obvious (Midtlyng et al., 1996). Several other related studies show that the inflammatory reactions caused by the vaccines may vary in different species of salmonid (Mutoloki et al., 2010). The aftereffects of these vaccines are less vulnerable in some species such as the yellowtail (*Seriola quinqueradiata*), turbot (*Scophthalmus maximus*), sea bass (*Dicentrarchus labrax*), and cod (*Gadus morhua*) (Gravíningen et al., 2008; Afonso et al., 2005; Castro et al., 2008; Maira et al., 2008). Usually the anesthetized fishes are vaccinated by a team of professional vaccinators by following righteous vaccination procedure and a professional vaccinator has the ability to vaccinate up to 3500 fishes per hour. In the past two decades, new automated vaccination machines had been developed in order to reduce the high labor cost for manual vaccination and these automated vaccinators have the capability to vaccinate around 20,000 fishes per hour (Brudeseth et al., 2013). For a matter of fact, the DNA vaccines are also administrated

TABLE 11.1
List of Vaccines Used in Aquatic Organisms

Infectious Viral Pathogens	Viral Family	Vaccine	Experimental Conditions	Vaccines-Dosage (μg)	Relative Percent Survival (RPS)	Reference
White spot syndrome virus (WSSV)	Nimaviridae	Monovalent vaccine, purified VP28 protein	Post-larval staged *Penaeus monodon*, challenged for 3–7 days	25	77	Witteveldt et al. (2004)
WSSV	Nimaviridae	Purified VP24 protein	Advanced post-larval staged *Penaeus monodon*, challenged for 10 days	5 10	64 11	Thomas et al. (2014)
WSSV	Nimaviridae	Transgenic B. subtilis spores, CotB-VP28 CotC-VP26 CotC-VP28	Litopenaeus vannamei oral administration (14 days)	10 each	47 100 50	Nguyen et al. (2014) Ning et al. (2011)
WSSV	Nimaviridae	VP28 protein (Chlamydomonas reinhardtii)	1 g L. vannamei, challenged for 10 days	4 mg	70	Lahn et al. (2021)
WSSV	Nimaviridae	Attenuated envelope protein VP28	Intramuscular injection in L. vannamei for 4 weeks	5	90	Ma et al. (2019)
Infectious hematopoietic necrosis virus	Rhabdoviridae RNA	G Glycoprotein	1 g rainbow trout, challenged for 6 weeks	10	75	Anderson et al. (1996)
Infectious Salmon anemia	Orthomyxoviridae, ssRNA	Hemagglutinin-esterase (HE)	20 g Atlantic salmon pre-smolts, challenge for 9 weeks	15 and two 15 μg boosters	40–60	Mikalsen et al. (2005)
Hemorrhagic septicemia virus (HSV), freshwater	Rhabdoviridae	G	13 g rainbow trout, challenged for 54 weeks	10 50	97 94	Lorenzen et al. (1998)
HSV, Marine	Rhabdoviridae	G	3 g Japanese flounder, challenged for 1 month	10	93–100	Byon et al. (2005)
Hirame rhabdovirus (HIRRV)	Rhabdoviridae	G	2 g Japanese flounder, challenged for 28 days	1 10	70 90	Takano et al. (2004)

(*Continued*)

TABLE 11.1 (*Continued*)
List of Vaccines Used in Aquatic Organisms

Infectious Viral Pathogens	Viral Family	Vaccine	Experimental Conditions	Vaccines-Dosage (μg)	Relative Percent Survival (RPS)	Reference
HIRRV	Rhabdoviridae	Partial G	3 g Japanese flounder, challenged for 21 days	5	95	Seo et al. (2006)
HIRRV	Rhabdoviridae	G	10 g Japanese flounder, challenged for 28 days	10	96	Yasuike et al. (2007)
Spring viremia carp virus (SVCV)	Rhabdoviridae	Mixture of two G plasmids	11 g common carp, challenged for 6 weeks	25 each	48	Kanellos et al. (2006)
Atlantic halibut nodavirus (AHNV)	Nodaviridae, ssRNA	Capsid protein C	3.6 g turbot, challenged for 35 days	5	0	Sommerset et al. (2003)
AHNV	Nodaviridae	C	2.2 g turbot, challenged for 10 weeks	20	7	Sommerset et al. (2005)
Infectious pancreatic necrosis virus	Birnaviridae, dsRNA	Segment A large ORF and VP2 VP2 alone	20 g Atlantic salmon post-smolts, challenged for 69 days	15 each	84	Mikalsen et al. (2005)
Channel catfish virus (CCV)	Herpesviridae, large dsDNA	ORF6 ORF59 ORF6 & 59	6–10 months old channel catfish, challenged for 4–6 weeks	50 50 50	15 38 46	Nusbaum et al. (2002)
Red seabream iridovirus	Iridoviridae, large dsDNA	Major capsid ORF569	5–10 g red seabream, challenged for 30 days	25 each	43–69 48–71	Caipang et al. (2006)

through intramuscular injection in most of the cases; for example, the DNA vaccine used against infectious hemorrhagic necrosis (IHN) is administrated by intramuscular injection.

11.4.2 Administration of Immersion Vaccines in Aquatic Organisms

The immersion vaccines that are commercially available are either live bacterial vaccines or suspension of formalin-inactivated bacteria. In immersion vaccination, the inactivated antigens are exposed for a shorter time in a concentrated antigen suspension or exposed in diluted antigen suspension for a longer duration. The standard concentration of immersion vaccines was 1/10 dilution in the concentrated suspension of vaccine, but the duration of exposure is only between 30 and 60 seconds. For example, in rainbow trout (*Oncorhynchus mykiss*), the application of 1/10 dilution radio labeled *A. salmonicida* vaccine suspension with exposure duration between 5 seconds and 10 minutes does not show any considerable changes in immunological level (Tatner et al., 1986). Whereas when compared the 1/10 dilution to 1/100 dilution for 2 hours has resulted in noteworthy uptake in rainbow trout. Likewise, a 1/500 dilution of inactivated immersion vaccines is administrated in the holding tank for 30 minutes by direct bath for better results. Usually, this method is facilitated by increasing oxygenation and reducing water level in the holding tank. When compared to dip, bath vaccination has less stress, labor intensity, and handling of fish. To improve the immersion vaccine uptake, several techniques have been applied in the past two decades such as the ultrasound-mediated uptake, hyperosmotic dip, and multiple puncture instrument (Huising et al., 2003). Despite a higher uptake of vaccine components during vaccination, none of the vaccines have obtained commercial status to use in the aquaculture industry. Also, it is important to vaccinate fishes with diseases, such as infectious pancreas necrosis (IPN) and rainbow trout fry syndrome, while in the fry stage as soon as possible. The efficacy is reliant on the improvement of immunocompetence of the vaccinated fish. In coho salmon (Oncorhyncus kitsutch), fry below 1 g had poor response, whereas fry of 1–2 g had shown improved protection, with a duration of between 3 and 4 months. Long-term protection is exhibited only in fry size above 2 g. In the USA, live vaccines are used against *Flavaobacterium columnaris* and *Edwardsiella ictaluri* in catfish production, where a frozen vial of vaccine is sufficient to vaccinate catfish of 3–4 kg in approximately 20 L of water. In channel catfish fry (*Ictalurus punctatus*), between 7 and 30 days of post-hatching, the vaccine showed effective and remarkable results.

11.4.3 Oral Administration of Vaccines in Large-Scale Aqua Farming

Oral administration of vaccines through feed along with antigens is the superlative method of oral vaccine delivery. In this case, the commercially available vaccine suspensions are either coated in the feed with antigen or feed is mixed with antigen during production. However, the efficiency of oral vaccines depends on the amount of antigen content present in the feed/uptake in gut/gastric degradation and antigen adsorption (Wang et al., 2020). MSD Animal Health is a US-based company, which has patented encapsulation technology for oral vaccines' production. Similarly, for the prevention of ISA and salmon rickettsial septicemia (SRS), Centrovet Laboratories had patented a MicroMatrix™ Targeted Delivery Systems (MTDS). Likewise in Japan, vaccine developing companies have licensed oral vaccine against disease caused by *Lactococcus garvieae* in *Seriola sp.* weighing around 100–400 g. Against enduring endemic diseases, these oral vaccines can be used as booster or primary vaccine for improved protection with higher feasibility for vaccine delivery.

11.4.4 Vaccine Regimes

In various aquatic species, the reason for the development of vaccine regime is to protect the aquatic species (fish, shrimp, etc.) throughout the production cycle period. In this case, the vaccination can be implicated at three different administrative time-points, which usually initiates with dip vaccination further followed by a booster vaccination (bath/dip or oral) and ends up with injectable vaccine. Subsequent vaccinations will not be encouraged after administration/injection of w/o-based vaccination. But, eventually in Chile, injection vaccination has been performed before oral booster vaccination against piscirickettsiosis. Eventually, to reduce the tedious work intensity of vaccine administration, the booster vaccination is frequently administered by means of bath/oral or dip delivery (mixed in diet) of vaccine components, which also reduces the stress occurrence in fish during vaccine administration. Similarly, like described in the vaccination guide, the immersion vaccine regimes were orally delivered for sea bass against pasteurellosis and vibriosis (You et al., 2018).

11.5 VACCINE DEVELOPMENT AND ITS CHALLENGES

An effective vaccine should consist of potential antigens with promising protective response against the pathogen of interest in the host organisms. An array of approaches is required to identify effective protective antigens. The approaches might depend on the reagent availability, mode of administration, fish species, nature and type of pathogen, and are also based on the development of a challenge model and the respective vaccine candidates' efficacy. Most of the commercially available vaccines are composed of whole-cell killed pathogens and are injected intraperitoneally. Usage of whole pathogens in vaccines can masquerade troubles, if the causative pathogen is expensive or difficult to cultivate and if the pathogen is heterogeneous, complex, and intracellular, and if immunosuppression of epitopes occurs. In some cases, injection is not the optimal delivery route of some vaccines whereas for effective protection mucosal delivery might be administered. However, lack of preliminary knowledge and suitable adjuvants slowed down the progression in mucosal vaccines' development. It is cost-prohibitive in some countries for injecting fish manually by hiring staff and in underdeveloped and developing countries, the convenience and handling plays a vital role in the administration of commercial vaccines. In order to confirm the efficiency of vaccines, the vaccines are exposed in standardized challenge, a model which mimics the natural exposure of vaccines against disease-causing organisms (Hoare et al., 2017). However, the injection challenge method is more challenging to standardize and control, whereas the cohabitation and bath challenge method is easier to procure natural exposure requirement. When there is no or lack of experimental disease challenge method, to test the vaccine efficacy, several proxies can be used such as immunological markers of protection (analyzed by gene expression/immunohistochemistry) and pathogen load (measured by qPCR) and in certain circumstances field trails can be performed.

For instance, a Gram-negative bacterium *Flavobacterium psychrophilum* causing rainbow trout fry syndrome (RTFS) is very problematic to experimentally administrate through cohabitation or bath challenge without inducing handling stress. For testing a mucosal vaccine (dip immersion), intramuscular injection mode of administration is not the right challenge method to administrate in fry. Low levels of hydrogen peroxide (H_2O_2) are required during the pretreatment of fish to succeed the challenge models; however, it is not fully sufficient to attest the efficacy of vaccine with the level of infection induced by this treatment. Therefore, field trails are the better option to confirm the vaccine efficiency (Hoare et al., 2017) (Figure 11.2).

Vaccine development faces difficulties due to the diverse species of fish as we all never knew the whole mechanism behind fish immune system. Each and every fish species requires specific primers and reagents to unveil the mechanisms of host pathogen interactions. For example, in Atlantic salmon, the injection mode of vaccine administration is common but in some species such as pangasius and tilapia, it is not viable against the purpose of administration. Some of the novel vaccines that are developed in the recent researches are protective, but the optimal method for the delivery of these novel vaccines are yet to be determined accompanying with appropriate vaccine strategies (e.g., booster/prime vaccination). The mucosal route of vaccine administration is more feasible because the fish has huge mucosal surfaces such as skin, nasal mucosae, gills, and gut; consecutively, this method is more economically affordable in several sectors over the injection method.

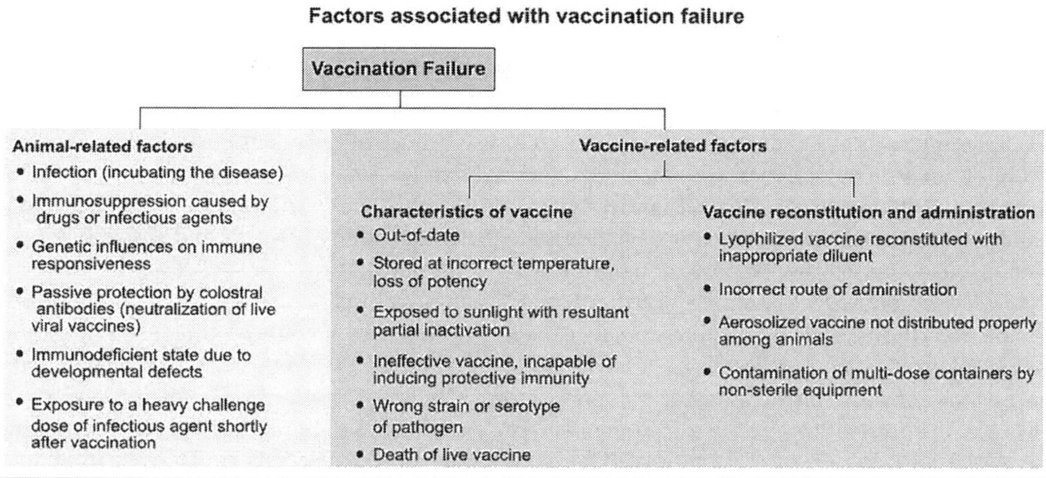

FIGURE 11.2 Factors influencing vaccine efficiency. [Created by www.biorender.com.]

Yet, there are only a limited number of commercial mucosal vaccines that are available (oral and immersion).

As for the vaccine development, the abovementioned factors/challenges had hindered the development of novel vaccines, which also include lack of optimized protection dosage, tolerance, and potential of denaturation of mucosal and oral vaccines in stomach as well as antigens' ability to get access to antigen-presenting cells (APCs) through the mucosal barriers (Munangandu et al., 2012). The APCs that are found in the fish mucosal organs can be activated, as in the mammalian APCs, in order to increase the uptake of antigen, which results in an enhanced adaptive immune system. In some cases, injection vaccines include several adjuvants but not in mucosal vaccination. The adjuvants are basically a bunch of structurally heterogeneous compounds that have ability to modulate the antigen intrinsic immunogenicity (Guy, 2007). These adjuvants are classified based on the immune response that they elicit; signal 1 and 2 facilitators (antigen presentation and secondary signals) activate B and T lymphocytes (Schijns, 2001; Ribeiro and Schijns, 2010). For example, montanides are signal 1 adjuvants obtained from SEPPIC, used in fishes during the injection mode of vaccination. During antigen recognition, co-stimulatory signals are provided by signal 2 adjuvants, such as synthetic oligonucleotides, alums, flagellins, saponins, cytokines, and beta glucans. In a recent study from salmonid pathogen *Yersinia ruckeri*, a recombinant flagellin is isolated, which acted as a potential activator of in vitro antimicrobial peptides, acute phase proteins, inflammatory cytokines, and a mere inflammatory activator than most of other bacterial PAMPs (peptidoglycans, LPS). Recombinant flagellin activates a handful of antimicrobial pathways with higher acute phase proteins expression, which is revealed through in vivo studies; also, it activates complement genes of different tissues and several other antimicrobial peptides. By responding to flagellin stimulation, in trout liver, it exhibits a marked induction of cathelicidin-2, IL-23P19, SAA, IL-11, and IL17C1. So it can be understood that flagellin is one of the potential vaccine adjuvants and immunostimulants in aquaculture. Even though various new discoveries are emerging in vaccinology, there is still a void for efficacious commercial adjuvants (immersion) in aquaculture (Wangkahart et al., 2019).

In teleost fish, it is essential to measure the responses induced through vaccination and to facilitate development of existing and new vaccines. Following vaccination, the adaptive immune response has the responsibility to provide protection, mediated with B and T lymphocytes. Still, there is a lack of knowledge regarding the mucosal vaccines and its protection against disease (Mutoloki et al., 2015); however, a lot of progress can be seen in recent days in this sector. For example, in salmonids, protocol for the isolation of GALT cells has been developed, which expresses a range of dendritic cell markers, B&T-cells, and are responsive to a group of PHA, cytokines, and PAMPS (Attaya et al., 2018). For an efficient vaccination, effective antigen presentation and recognition are required. Usually, the antigens in vaccine are mediated through T-cells from MHC-11 molecules; however, the MHC-1 route is the one mediated in DNA vaccination. It is noteworthy that as the immune response develops, both the responses will be simultaneously triggered. Despite all these findings on cytokine expression, protection markers for T-cell responses still have difficulties (Secombes and Belmonte, 2016). In order to demonstrate the antibodies' protective effect, passive immunization can be used, and it doesn't conclude that association of B-cells and antibody production results in protective response.

Following booster vaccination in fish, there are no boost in antibody level like contrast in the mammals. During memory response, many cells are produced by an increase in the antibody level, but implication of booster vaccination doesn't show any remarkable increase in the antibody level. In various scientific publications, it is reported that, following injectable vaccination, there is an increase in specific IgM, whereas in some fish species, application of immersion vaccines leads to increase in IgT. For example, in rainbow trout, the immersion vaccine leads to increase in IgT cells in the kidney (Hoare et al., 2017). According to Munangandu, in several mucosal organs, there is a compartmentalization of IgM and IgT on physiological distribution and most of the IgT is distributed in the gill lamellae outer surface. This shows that these IgT cells act as a concierge for pathogen entry. On the contrary, IgM is hugely distributed in the arterioles, it acts when the mucosal IgT fails to prevent the entry of pathogen, and it is said to be a secondary defense system (Bayliss et al., 2017).

11.5.1 Types of Vaccination in Aquaculture

11.5.1.1 Live Attenuated Vaccines

In the 1770s, a live vaccination was developed to protect humans from smallpox (Stewart & Devlin, 2006). Although such vaccines have the benefit of stimulating both humoral and cellular activity, they are not permitted to be employed in European aquaculture due to safety concerns. This is due to the possibility of live attenuated isolates reverting to virulent forms. Previously, attenuation was done by repeatedly passing the pathogen in vitro, resulting in random alterations. Molecular technologies may now be used to target specific genes and manufacture genetically modified diseases, but these are designated as GMOs, and their use in aquaculture is governed by tight laws, much more than for traditional live vaccinations. According to reports, defined genetic alterations in viable bacterial immunizations allow for better control and safety than random mutations; therefore, these vaccines may be approved in the future (Frey, 2007).

11.5.1.2 Nanoparticle-Based Vaccine Delivery

Virus-like particles (VLPs), nondegradable nanospheres, immunostimulating complexes (ISCOMs), polymers, and liposomes are reported to have prospects as vaccination antigen delivery vehicles; they stabilize vaccine antigens while also serving as adjuvants (Gregory et al., 2013). They have the potential to modulate the immune response in a number of ways, and they may be crucial in the development of protective responses. These devices can be used to administer vaccines mucosally. Nanoparticles act as both antigen delivery vehicles and antigen release vehicles, minimizing the requirement for booster immunizations. PLGA nanoparticles are extensively employed in humans for the regulated administration of nucleic acids, synthetic proteins, and peptides, and are studied in fish for oral vaccination delivery (Behera et al., 2010). VLPs have been tried in fish more recently. Since they are not infectious and elicit neutralizing antibodies, they are regarded as a new vaccination platform. Chien et al. (2018) showed that NNV orange-spotted grouper (NNVOSG) VLPs may be used as oral vaccines in orange-spotted grouper and other grouper species. In *E. coli*, recombinant capsid proteins were produced and are assembled into VLPs automatically without using any cells. *Pichia pastoris* has been shown to be an effective carrier for delivery of oral antigens and might be used as a substitute for *E. coli*. The yeast acts as an adjuvant and can be administered to older fish in nonencapsulated form or to larval fish in bioencapsulated form (Adams et al., 2019).

11.5.1.3 mRNA Vaccines

The mRNA vaccination is increasingly being employed in clinical practice as an alternative to DNA vaccinations. The method was initially introduced in 1990, but it was not pursued at the time because of concerns about high inherent immunogenicity, stability, and poor in vivo administration. Due to the recent advancements in vaccine delivery and stability mechanisms, nowadays they can be produced cheaply and swiftly (Pardi et al., 2018). Clinical studies using mRNA vaccines for infectious diseases are still not in their infancy stage, with such vaccines being employed primarily for preventative and therapeutic uses in cancer. In HIV clinical studies, they were shown to be safe; however, while eliciting specific antigen CD8+ and CD4+ T-cell immune responses, there were no signs of decrease in viral load. In a review by Pardi et al., vaccine effectiveness against the rabies virus was shown to be highly dependent on the dosage and administration mode, with needleless delivery outperforming injecting directly.

11.5.1.4 Prime Booster vs Prime Vaccination

Adjuvant-administered injection vaccinations are often given only once, producing a long-lasting response. On the other hand, immersion immunization without adjuvant is likely to be ineffective, necessitating a booster vaccine. So far, fish vaccines have failed to provide sterile immunity, and it's possible that activating both systemic and mucosal immunity is needed to do accomplish this phenomenon. An early immersion immunization followed by a prime booster vaccination, or a prime booster vaccination followed by an oral booster vaccination, might accomplish this.

11.5.1.5 Autogenous Vaccines

Commercial vaccine development is costly and time-consuming, and there are several fish species being bred that are vulnerable to a range of illnesses. As a result, developing approved vaccines against all fish infections is unrealistic, and it may even be impossible for some pathogens for a number of reasons. Emergency vaccinations, also known as autogenous vaccines, might be a good alternative. For example, to control the diseases transmission in between sea lice in the United Kingdom and Norway, Cleaner fish (e.g., *Ballan wrasse, Labrus bergylta,* and *lump sucker, Cyclopterus lumpus*) are used occasionally. The pathogens recovered from Ballan wrasse tend to differ from location of site, and autogenous vaccines against bacterial pathogens, such as a typical *Aeromonas salmonicida*, have been shown to be successful in preventing at least some of these disorders when provided by injection. Nonetheless, diseases continue to be a problem in fry, and when this species are immunocompetent, only a little amount information is known, and it is to be ascertained, if at all, whether they can be vaccinated successfully by inducing immersion vaccination. Natural IgM levels in the gut of Ballan wrasse have recently been discovered to be extremely high, leading to speculation that this might be a compensating mechanism in stomachless fish (Bilal et al., 2019).

11.6 FUTURE PERSPECTIVES OF VACCINE DEVELOPMENT

Fish genomes are currently accessible for a range of species. Furthermore, as the cost of technology, such as whole genome sequencing, has decreased, genomes for particular diseases are routinely disclosed, allowing for the development of tailored vaccines for a variety of species. For example, Ngo et al. described over about 300 *Flavobacterium psychrophilum* species, mostly from the United Kingdom, and developed a trivalent whole-cell vaccine. But, these findings are also significant from an epidemiological standpoint (Ngo et al., 2017). Because certain epitopes are immunosuppressive, developing a whole-cell vaccination is not always viable. Instead, impending protective antigens must be discovered and vaccines developed, allowing undesired epitopes to be eliminated. Reverse vaccinology is used for identifying possible vaccine candidates for protein subunit vaccine production, which is basically a genome-based technique. Using computerized programs, pathogen protein sequences are examined, and prospective vaccine candidates are selected. They might be chosen because they're recognized vaccine candidates for other infections, or because they have highly immunogenic proteins, among other things. The effectiveness of recombinant subunit vaccinations (also known as DNA vaccines) is then assessed in vivo. In Europe, DNA vaccines are currently approved for usage and many more are expected to be designed in the future (Dalmo, 2017).

11.7 CONCLUSION

Advances in biotechnology, as well as the discovery of new pathogen vaccines, have helped to reduce the probability of disease outbreaks and resultant mortality in aquaculture. Progress has led to the identification of protective antigens and the development of vaccines that are both safe and affordable. While it has explored a cost-effective technology of monitoring a substantial threatening pathogen, aquaculture vaccination is becoming an important part of the health management. On the other hand, most aquatic animal vaccine development and initiatives are still in their initial phases, and hurdles to multi-component and cost-effective immunization regimens have yet to be overcome. The development and commercialization of vaccines for all economically major fish diseases is governed by technical, scientific, and biological constraints. Although individual fish have a lower production cost than other livestock animals, only low-cost vaccines are cost-effective in protecting fish against diseases. As a result, research should focus on newly discovered immunoglobulins, their roles in immunity, and their relationships with other immunoglobulins in order to develop viable vaccine regimens. Overall, an appropriate aquaculture vaccine must fulfill two crucial criteria for both farmers and vaccine manufacturers. Perhaps, the most crucial category includes three significant roles: these vaccines must provide appropriate immunoprotection against a particular disease in intensive farming circumstances;

provide long-term protection as the animal is most prone to disease; and protect against all serotype variants of the infectious agent; followed by market value and surety for better cointegration of lab performance to farm performance, and the ability to measure this in a proactive manner. Vaccination of small fish also necessitates the development of more appropriate and cost-effective delivery techniques. Vaccination should be considered as part of a balanced fish health management strategy rather than the sole solution to a disease problem. Basic information on fish vaccination could be used for large-scale fish vaccination, and additional development in this sector requires collaboration between basic and application science.

REFERENCES

Adams, A. (2019). Progress, challenges and opportunities in fish vaccine development. *Fish & Shellfish Immunology*, 90, 210–214.

Adams, A., & Subasinghe, R. (2021). Fish vaccines. In *Veterinary Vaccines: Principles and Applications* (pp. 113–118). Wiley, New Jersey, U.S.

Afonso, A., Gomes, S., Da Silva, J., Marques, F., & Henrique, M. (2005). Side effects in sea bass (*Dicentrarchus labrax* L.) due to intraperitoneal vaccination against vibriosis and pasteurellosis. *Fish & Shellfish Immunology*, 19(1), 1–16.

Amparyup, P., Sutthangkul, J., Charoensapsri, W., & Tassanakajon, A. (2012). Pattern recognition protein binds to lipopolysaccharide and β-1, 3-glucan and activates shrimp prophenoloxidase system. *Journal of Biological Chemistry*, 287(13), 10060–10069.

Anderson, E. D., Mourich, D. V., Fahrenkrug, S. C., LaPatra, S., Shepherd, J., & Leong, J. A. (1996). Genetic immunization of rainbow trout (*Oncorhynchus mykiss*) against infectious hematopoietic necrosis virus. *Molecular Marine Biology and Biotechnology*, 5(2), 114–122.

Attaya, A., Wang, T., Zou, J., Herath, T., Adams, A., Secombes, C. J., & Yoon, S. (2018). Gene expression analysis of isolated salmonid GALT leucocytes in response to PAMPs and recombinant cytokines. *Fish & Shellfish Immunology*, 80, 426–436.

Aucouturier, J., Dupuis, L., & Ganne, V. (2001). Adjuvants designed for veterinary and human vaccines. *Vaccine*, 19(17–19), 2666–2672.

Bayliss, S. C., Verner-Jeffreys, D. W., Bartie, K. L., Aanensen, D. M., Sheppard, S. K., Adams, A., & Feil, E. J. (2017). The promise of whole genome pathogen sequencing for the molecular epidemiology of emerging aquaculture pathogens. *Frontiers in Microbiology*, 8, 121.

Behera, T., Nanda, P. K., Mohanty, C., Mohapatra, D., Swain, P., Das, B. K., … & Sahoo, S. K. (2010). Parenteral immunization of fish, Labeo rohita with Poly D, L-lactide-co-glycolic acid (PLGA) encapsulated antigen microparticles promotes innate and adaptive immune responses. *Fish & Shellfish Immunology*, 28(2), 320–325.

Bilal, S., Lie, K. K., Dalum, A. S., Karlsen, O. A., & Hordvik, I. (2019). Analysis of immunoglobulin and T cell receptor gene expression in ballan wrasse (Labrus bergylta) revealed an extraordinarily high IgM expression in the gut. *Fish & Shellfish Immunology*, 87, 650–658.

Byon, J. Y., Ohira, T., Hirono, I., & Aoki, T. (2005). Use of a cDNA microarray to study immunity against viral hemorrhagic septicemia (VHS) in Japanese flounder (Paralichthys olivaceus) following DNA vaccination. *Fish & Shellfish Immunology*, 18(2), 135–147.

Caipang, C. M. A., Takano, T., Hirono, I., & Aoki, T. (2006). Genetic vaccines protect red seabream, Pagrus major, upon challenge with red seabream iridovirus (RSIV). *Fish & Shellfish Immunology*, 21(2), 130–138.

Castro, N., Toranzo, A. E., Núñez, S., & Magariños, B. (2008). Development of an effective Edwardsiella tarda vaccine for cultured turbot (Scophthalmus maximus). *Fish & Shellfish Immunology*, 25(3), 208–212.

Cimica, V., & Galarza, J. M. (2017). Adjuvant formulations for virus-like particle (VLP) based vaccines. *Clinical Immunology*, 183, 99–108.

Chien, M. H., Wu, S. Y., & Lin, C. H. (2018). Oral immunization with cell-free self-assembly virus-like particles against orange-spotted grouper nervous necrosis virus in grouper larvae, Epinephelus coioides. *Veterinary Immunology and Immunopathology*, 197, 69–75.

Dalmo, R. A. (2018). DNA vaccines for fish: Review and perspectives on correlates of protection. *Journal of Fish Diseases*, 41(1), 1–9.

Dantas -Lima, J. J., Tuan, V. V., Corteel, M., Grauwet, K., An, N. T. T., Sorgeloos, P., & Nauwynck, H. J. (2013). Separation of Penaeus vannamei haemocyte subpopulations by iodixanol density gradient centrifugation. *Aquaculture*, 408, 128–135.

Dhar, A. K., & Allnutt, F. (2011). Challenges and opportunities in developing oral vaccines against viral diseases of fish. *Journal of Marine Science: Research & Development*, 1.

Dubovskiy, I. M., Kryukova, N. A., Glupov, V. V., & Ratcliffe, N. A. (2016). Encapsulation and nodulation in insects. *Invertebrate Survival Journal*, 13(1), 229–246.

Frietze, K. M., Peabody, D. S., & Chackerian, B. (2016). Engineering virus-like particles as vaccine platforms. *Current Opinion in Virology*, 18, 44–49.

Frey, J. (2007). Biological safety concepts of genetically modified live bacterial vaccines. *Vaccine*, 25(30), 5598–5605.

Gregory, A. E., Williamson, D., & Titball, R. (2013). Vaccine delivery using nanoparticles. *Frontiers in Cellular and Infection Microbiology*, 3, 13.

Gudding, R., Lillehaug, A., & Evensen, Ø. (Eds.). (2014). *Fish vaccination* (Vol. 624). Fairford, UK: Wiley Blackwell.

Gravningen, K., Sakai, M., Mishiba, T., & Fujimoto, T. (2008). The efficacy and safety of an oil-based vaccine against Photobacterium damsela subsp. Piscicida in yellowtail (Seriola quinqueradiata): A field study. *Fish & Shellfish Immunology*, 24(5), 523–529.

Guy, B. (2007). The perfect mix: Recent progress in adjuvant research. *Nature Reviews Microbiology*, 5(7), 396–397.

Hammond, J. A., & Smith, V. J. (2002). Lipopolysaccharide induces DNA synthesis in a sub-population of hemocytes from the swimming crab, Liocarcinus depurator. *Developmental & Comparative Immunology*, 26(3), 227–236.

Hang, B. T. B., Phuong, N. T., & Kestemont, P. (2014). Can immunostimulants efficiently replace antibiotic in striped catfish (Pangasianodon hypophthalmus) against bacterial infection by Edwardsiella ictaluri? *Fish & Shellfish Immunology*, 40(2), 556–562.

Hoare, R., Ngo, T. P., Bartie, K. L., & Adams, A. (2017). Efficacy of a polyvalent immersion vaccine against Flavobacterium psychrophilum and evaluation of immune response to vaccination in rainbow trout fry (Onchorynchus mykiss L.). *Veterinary Research*, 48(1), 1–13.

Horzinek, M. C., Schijns, V. E. C., Denis, M., Desmettre, P., & Babiuk, L. A. (1997). General description of vaccines. In *Veterinary vaccinology*. Amsterdam, The Netherlands: Elsevier, 131–52.

Huising, M. O., Guichelaar, T., Hoek, C., Verburg-van Kemenade, B. L., Flik, G., Savelkoul, H. F., & Rombout, J. H. (2003). Increased efficacy of immersion vaccination in fish with hyperosmotic pretreatment. *Vaccine*, 21(27–30), 4178–4193.

Kanellos, T., Sylvester, I. D., D'Mello, F., Howard, C. R., Mackie, A., Dixon, P. F., … & Russell, P. H. (2006). DNA vaccination can protect Cyprinus Carpio against spring viraemia of carp virus. *Vaccine*, 24(23), 4927–4933.

Kim, H., Lee, Y. K., Kang, S. C., Han, B. K., & Choi, K. M. (2016). Recent vaccine technology in industrial animals. *Clinical and Experimental Vaccine Research*, 5(1), 12–18.

Lanh, P. T., Nguyen, H. M., Duong, B. T. T., Hoa, N. T., Tam, L. T., Thu, H. T., … & Van Quyen, D. (2021). Generation of microalga Chlamydomonas reinhardtii expressing VP28 protein as oral vaccine candidate for shrimps against White Spot Syndrome Virus (WSSV) infection. *Aquaculture*, 540, 736737.

Lin, X., & Söderhäll, I. (2011). Crustacean hematopoiesis and the astakine cytokines. *Blood: The Journal of the American Society of Hematology*, 117(24), 6417–6424.

Lorenzen, N., Lorenzen, E., Einer-Jensen, K., Heppell, J., Wu, T., & Davis, H. (1998). Protective immunity to VHS in rainbow trout (Oncorhynchus mykiss, Walbaum) following DNA vaccination. *Fish & Shellfish Immunology*, 8(4), 261–270.

Ma, Y., Liu, Y., Wu, Y., Jia, L., Liu, X., Wang, Q., & Zhang, Y. (2019). An attenuated Vibrio harveyi surface display of envelope protein VP28 to be protective against WSSV and vibriosis as an immunoactivator for Litopenaeus vannamei. *Fish & Shellfish Immunology*, 95, 195–202.

Maira, C., Lystad, Y., Schrøder, M. B., Lund, V., & Gudmundsdottir, B. (2008). Side effects of injectable vaccines in cod–Adapting the sideeffect scale to cod. *Fish Farming Xpert*, 3, 57–61.

Nguyen, A. T., Pham, C. K., Pham, H. T., Pham, H. L., Nguyen, A. H., Dang, L. T., … & Phan, T. N. (2014). Bacillus subtilis spores expressing the VP28 antigen: A potential oral treatment to protect Litopenaeus vannamei against white spot syndrome. *FEMS Microbiology Letters*, 358(2), 202–208.

Ning, D., Leng, X., Li, Q., & Xu, W. (2011). Surface-displayed VP28 on Bacillus subtilis spores induce protection against white spot syndrome virus in crayfish by oral administration. *Journal of Applied Microbiology*, 111(6), 1327–1336.

Nusbaum, K. E., Smith, B. F., DeInnocentes, P., & Bird, R. C. (2002). Protective immunity induced by DNA vaccination of channel catfish with early and late transcripts of the channel catfish herpesvirus (IHV-1). *Veterinary Immunology and Immunopathology*, 84(3–4), 151–168.

Midtlyng, P. J., Reitan, L. J., Lillehaug, A., & Ramstad, A. (1996). Protection, immune responses and side effects in Atlantic salmon (Salmo salarL.) vaccinated against furunculosis by different procedures. *Fish & Shellfish Immunology*, 6(8), 599–613.

Mikalsen, A. B., Sindre, H., Torgersen, J., & Rimstad, E. (2005). Protective effects of a DNA vaccine expressing the infectious salmon anemia virus hemagglutinin-esterase in Atlantic salmon. *Vaccine*, 23(41), 4895–4905.

Munang'andu, H. M., Fredriksen, B. N., Mutoloki, S., Brudeseth, B., Kuo, T. Y., Marjara, I. S., … & Evensen, Ø. (2012). Comparison of vaccine efficacy for different antigen delivery systems for infectious pancreatic necrosis virus vaccines in Atlantic salmon (*Salmo salar* L.) in a cohabitation challenge model. *Vaccine*, 30(27), 4007–4016.

Mutoloki, S., Cooper, G. A., Marjara, I. S., Koop, B. F., & Evensen, Ø. (2010). High gene expression of inflammatory markers and IL-17A correlates with severity of injection site reactions of Atlantic salmon vaccinated with oiladjuvanted vaccines. *BMC Genomics*, 11(1), 1–15.

Mutoloki, S., Munang'andu, H. M., & Evensen, Ø. (2015). Oral vaccination of fish–antigen preparations, uptake, and immune induction. *Frontiers in Immunology*, 6, 519.

Ngo, T. P., Bartie, K. L., Thompson, K. D., Verner-Jeffreys, D. W., Hoare, R., & Adams, A. (2017). Genetic and serological diversity of Flavobacterium psychrophilum isolates from salmonids in United Kingdom. *Veterinary Microbiology*, 201, 216–224.

Pardi, N., Hogan, M. J., Porter, F. W., & Weissman, D. (2018). mRNA vaccines—A new era in vaccinology. *Nature Reviews Drug Discovery*, 17(4), 261–279.

Plant, K. P., & LaPatra, S. E. (2011). Advances in fish vaccine delivery. *Developmental & Comparative Immunology*, 35(12), 1256–1262.

Ribeiro, C., & Schijns, V. E. (2010). Immunology of vaccine adjuvants. In *Vaccine adjuvants* (pp. 1–14). Humana Press, Totowa, NJ.

Rodger, H. D. (2016). Fish disease causing economic impact in global aquaculture. In *Fish vaccines* (pp. 1–34). Springer, Basel, Switzerland.

Schijns, V. E. (2001). Induction and direction of immune responses by vaccine adjuvants. *Critical Reviews in Immunology*, 21(1–3), 75–85.

Secombes, C. J., & Belmonte, R. (2016). Overview of the fish adaptive immune system. In *Fish vaccines* (pp. 35–52). Springer, Basel, Switzerland.

Seo, J. Y., Kim, K. H., Kim, S. G., Oh, M. J., Nam, S. W., Kim, Y. T., & Choi, T. J. (2006). Protection of flounder against hirame rhabdovirus (HIRRV) with a DNA vaccine containing the glycoprotein gene. *Vaccine*, 24(7), 1009–1015.

Shockey, J. E., O'Leary, N. A., de la Vega, E., Browdy, C. L., Baatz, J. E., & Gross, P. S. (2009). The role of crustins in Litopenaeus vannamei in response to infection with shrimp pathogens: An in vivo approach. *Developmental & Comparative Immunology*, 33(5), 668–673.

Snieszko, S. F., & Friddle, S. B. (1949). Prophylaxis of furunculosis in brook trout (Salvelinus fontinalis) by oral immunization and sulfamerazine. *The Progressive Fish-Culturist*, 11(3), 161–168.

Sommerset, I., Krossøy, B., Biering, E., & Frost, P. (2005). Vaccines for fish in aquaculture. *Expert Review of Vaccines*, 4(1), 89–101.

Sommerset, I., Lorenzen, E., Lorenzen, N., Bleie, H., & Nerland, A. H. (2003). A DNA vaccine directed against a rainbow trout rhabdovirus induces early protection against a nodavirus challenge in turbot. *Vaccine*, 21(32), 4661–4667.

Sommerset, I., Skern, R., Biering, E., Bleie, H., Fiksdal, I. U., Grove, S., & Nerland, A. H. (2005). Protection against Atlantic halibut nodavirus in turbot is induced by recombinant capsid protein vaccination but not following DNA vaccination. *Fish & Shellfish Immunology*, 18(1), 13–29.

Stewart, A. J., & Devlin, P. M. (2006). The history of the smallpox vaccine. *Journal of Infection*, 52(5), 329–334.

Sun, J. J., Xu, S., He, Z. H., Shi, X. Z., Zhao, X. F., & Wang, J. X. (2017). Activation of toll pathway is different between kuruma shrimp and Drosophila. *Frontiers in Immunology*, 8, 1151.

Tafalla, C., Bøgwald, J., Dalmo, R. A., Munang'andu, H. M., Evensen, Ø. (2014). Adjuvants in fish vaccines. *Fish Vaccination*, 12, 68–84.

Takano, T., Iwahori, A., Hirono, I., & Aoki, T. (2004). Development of a DNA vaccine against hirame rhabdovirus and analysis of the expression of immunerelated genes after vaccination. *Fish & Shellfish Immunology*, 17(4), 367–374.

Tassanakajon, A., Rimphanitchayakit, V., Visetnan, S., Amparyup, P., Somboonwiwat, K., Charoensapsri, W., & Tang, S. (2018). Shrimp humoral responses against pathogens: Antimicrobial peptides and melanization. *Developmental & Comparative Immunology*, 80, 81–93.

Tatner, M. F. (1986). The ontogeny of humoral immunity in rainbow trout, Salmo gairdneri. *Veterinary Immunology and Immunopathology*, 12(1–4), 93–105.

Thomas, A., Sudheer, N. S., Viswanathan, K., Kiron, V., Singh, I. S. B., & Narayanan, R. B. (2014). Immunogenicity and protective efficacy of a major White Spot Syndrome Virus (WSSV) envelope protein VP24 expressed in Escherichia coli against WSSV. *Journal of Invertebrate Pathology*, 123, 17–24.

Thompson, K.D.; Roberts R.J. (2016). *Fish vVaccines*, 1st ed.; Adams, A., Ed.; Springer: Basel, Switzerland: Springer. ISBN : 978--3--0348--0978--8

Van de Braak, C. B. T., Botterblom, M. H. A., Huisman, E. A., Rombout, J. H. W. M., & Van der Knaap, W. P. W. (2002). Preliminary study on haemocyte response to white spot syndrome virus infection in black tiger shrimp Penaeus monodon. *Diseases of Aquatic Organisms*, 51(2), 149–155.

Visetnan, S., Supungul, P., Hirono, I., Tassanakajon, A., & Rimphanitchayakit, V. (2015). Activation of PmRelish from Penaeus monodon by yellow head virus. *Fish & Shellfish Immunology*, 42(2), 335–344.

Wang, X. W., & Wang, J. X. (2013). Pattern recognition receptors acting in innate immune system of shrimp against pathogen infections. *Fish & Shellfish Immunology*, 34(4), 981–989.

Wang, Z., & Zhu, F. (2017). MicroRNA-100 is involved in shrimp immune response to white spot syndrome virus (WSSV) and Vibrio alginolyticus infection. *Scientific Reports*, 7(1), 1–12.

Wang, Q., Ji, W., & Xu, Z. (2020). Current use and development of fish vaccines in China. *Fish & Shellfish Immunology*, 96, 223–234.

Wangkahart, E., Secombes, C. J., & Wang, T. (2019). Studies on the use of flagellin as an immunostimulant and vaccine adjuvant in fish aquaculture. *Frontiers in Immunology*, 9, 3054.

https ://www.ruma.org.uk/.

Witteveldt, J., Vlak, J. M., & van Hulten, M. C. (2004). Protection of Penaeus monodon against white spot syndrome virus using a WSSV subunit vaccine. *Fish & Shellfish Immunology*, 16(5), 571–579.

Yasuike, M., Kondo, H., Hirono, I., & Aoki, T. (2007). Difference in Japanese flounder, Paralichthys olivaceus gene expression profile following hirame rhabdovirus (HIRRV) G and N protein DNA vaccination. *Fish & Shellfish Immunology*, 23(3), 531–541.

You, H., Cai, P., Tebeje, B. M., Li, Y., & McManus, D. P. (2018). Schistosome vaccines for domestic animals. *Tropical Medicine and Infectious Disease*, 3(2), 68.

Section V

Vaccines in Aquaculture

Challenges and Perspectives

12 Fish Vaccination
How Close Are the Fish Farmers?

Saurabh Dubey
Norwegian University of Life Sciences

Kizito K. Mugimba
Makerere University

Jitendra Kumar
Acharya Narendra Deva University of Agriculture and Technology

Hetron M. Munang'andu
Nord University

CONTENTS

DOI: 10.1201/9781003388548-17

12.1 INTRODUCTION

Aquaculture, which is considered as under-water agriculture, is one of the fastest growing food-producing sectors in the world with about 600 farmed aquatic species including more than 200 fish species [1]. It creates employment and generates income for millions of people worldwide. It is expected to continue growing with an estimated production of more than 100 million tons by 2030 [2]. However, disease is one of the main constraints adversely impacting the economic growth of aquaculture in several countries with projected estimated losses of 6 billion USD by 2030 [3]. Diseases do not only cause losses due to mortality, but they also lower fish quality, growth performance, and feed conversion ratio. Several factors contribute to the increase in the disease burden currently faced by aquaculture. The moving of eggs, brood stock, adult fish, and fish products from one place to the other has contributed to transborder transmission of diseases [4,5]. Other factors include overstocking commonly done with a view to increase output, which often leads to poor growth rate, increased stress on fish due to increase in competition for resources, sometimes leading to cannibalism causing open wounds that serve as entry points for pathogens, and is often linked to high disease transmission index [6]. Thus, developing protective vaccines aimed at reducing the disease burden is bound to significantly contribute in making aquaculture sustainable.

12.2 IDEAL FISH VACCINE AND BENEFITS OF VACCINATION

Vaccines are derived from infectious microorganisms containing antigens that stimulate vaccinated individuals to develop protective immunity against the same microorganism in subsequent exposure. As shown in Figure 12.1, an ideal vaccine should have the ability to produce both innate and adaptive immune responses in the immunized individual in order to confer protective immunity [7]. However, for a vaccine to be considered safe, it must be able to stimulate protective immunity without producing disease and side effects in vaccinated fish. In addition, the vaccine should be safe for use in aquatic environments, and should have no adverse effects on consumers [7,8]. An ideal vaccine should have broad neutralizing antibodies against different strains of the same pathogen and should protect close to 100% of the vaccinated individuals [7]. A vaccine is considered ideal if it provides long-term protective immunity covering the duration of the production cycle when fish are most vulnerable to infection. In polyculture, where pathogens have a tropism for a wide range

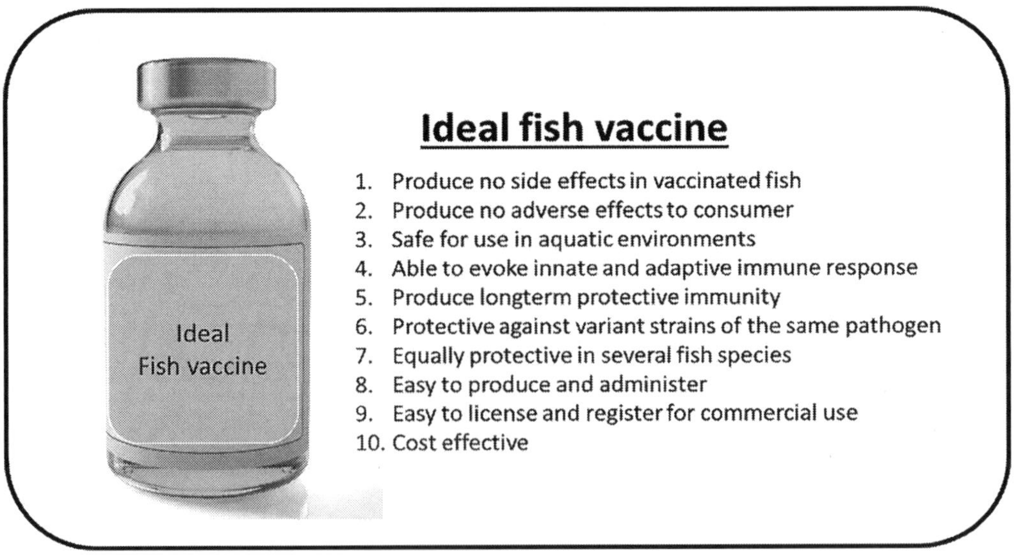

FIGURE 12.1 Ideal fish vaccine.

of fish species, an ideal vaccine should equally protect several fish species [7,8]. Also, an ideal fish vaccine should be easy to produce, administer, and easy to evaluate its efficacy [7,8]. Furthermore, it should be cost-effective by being inexpensive to promote mass vaccination and to provide a reasonable profit margin to stimulate increase in production [8]. Finally, an ideal vaccine should be easy to license and register for commercial use [7]. The major challenge is that most fish vaccines are in the developmental stages and, thus, they do meet the criterion of an ideal vaccine.

The benefits of fish vaccination include promoting a safe environment without antimicrobial deposition and increased feed conversion ratio because fish are healthy during the production cycle. Thus, fish eat more and convert the feed into quality output due to lack of stress caused by infectious diseases unlike diseased fish that grow at a slow rate with poor feed conversion ratio. Vaccines are nontoxic and, thus, are safe for the environment and safe for human consumption too [9]. Vaccination has a positive economic impact not only by reducing mortality but also by improving the quality of fish produced, thereby rendering easy access to market. In addition, vaccination has a positive food safety impact by reducing the transmission of fish foodborne zoonoses [10]. Thus, fish farmers get better returns through vaccination by producing high-quality fish lacking drug residues and adverse effects to consumers.

12.3 VACCINE DEVELOPMENT VERSUS ANTIMICROBIAL TREATMENT IN AQUACULTURE

12.3.1 Global Trend on Antibiotic Use

To control diseases in aquaculture, there is the need for disease surveillance, timely diagnosis, and prevention using scheduled vaccination regimes [5]. However, most fish diseases are still controlled by means of chemotherapeutics in most countries especially for the control of bacterial disease [5], which has increased the risk of antimicrobial resistance in aquaculture. Asian countries that account for 89.4% of global aquaculture production are still dependent on antibiotics for the treatment of diseases [6]. China, Indonesia, India, and Vietnam are the top four aquaculture producers in the world with usage of about 5943, 1160, 882.3 and 512.95 tons of antibiotics annually, respectively (Table 12.1). On the other hand, Japan only uses 179 tons of antibiotics and uses 2.5–3.0 million doses of fish vaccine annually (Table 12.1). South Korea uses 67.71 million doses of fish vaccines annually, which is 20 times higher than the doses used by Japan (Table 12.1). Norway is the only country in the world that uses the least quantity of antibiotics for treatment of fish diseases only amounting to 222 kg. In 2018 and 2019, the Norwegian salmon industry used 369 and 397 million doses of vaccines, respectively (Table 12.1) [11]. On the contrary, Chile which also produces

TABLE 12.1
Aquaculture Producing Countries with Disease Control Measure

Global Aquaculture Position	Country	Production in Million Metric Tons	Antibiotics (tons)	Vaccine Million (Doses)	Licenced Vaccine	Ref.
1	China	63.70	5,943	-	7	[12,13]
2	Indonesia	16.60	1,160	-	1	[9,13]
3	India	5.70	882.3	-	1	[13,14]
4	Vietnam	3.60	512.95	-	2	[9,13]
7	South Korea	1.90	-	67.71	29	[15]
8	Egypt	1.40	>200	-	-	[13]
9	Norway	1.30	0.222	397	21	[11,16,17]
10	Japan	1.10	179	2.5-3.0	29	[18,19]
11	Chile	1.04	393.9	150	56	[20–22]

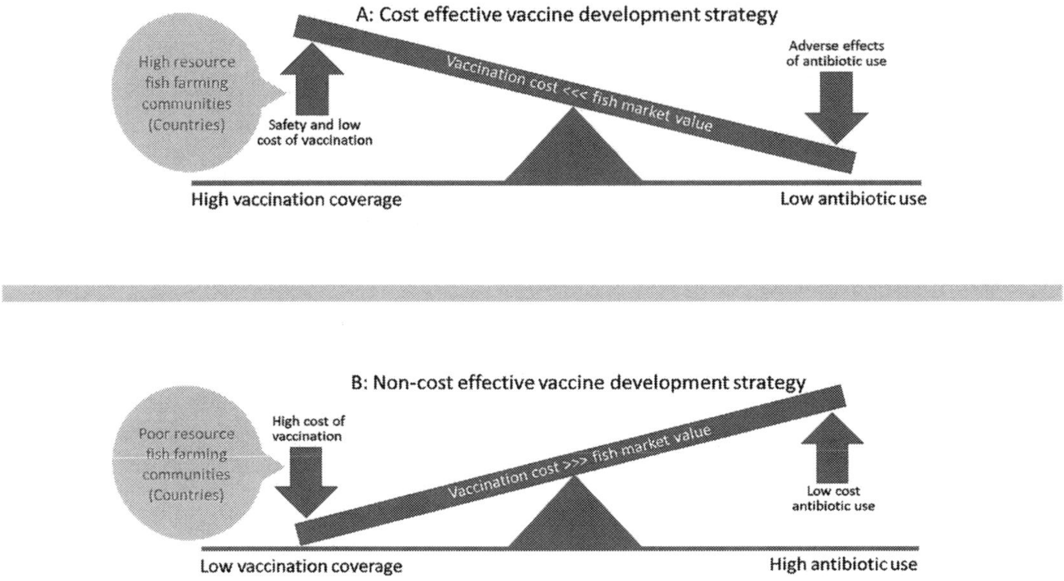

FIGURE 12.2 Comparison of antibiotics use versus vaccination in aquaculture.

Atlantic salmon (*Salmo salar* L) uses 393.9 tons of antibiotics annually and 150 million vaccines for the control of fish diseases (Table 12.1). As shown in Table 12.1, LMICs use <272,000 kg of antibiotics for the control of tilapia and pangasius diseases combined together [3]. Overall, the general trend is that countries such as Norway, Korea, and Japan that use high vaccine doses annually tend to have low antibiotic use in aquaculture while countries with low vaccine coverage use high quantities of antibiotics for the control of fish diseases (Figure 12.2).

12.3.2 Impact of Vaccination on Antibiotics' Use in Aquaculture

Fish vaccination is an important disease prevention method because it also contributes to reduction on the use of antibiotics in the treatment of fish diseases. Lessons from the Norwegian salmon industry provide a good example on how vaccine development contributed to reduction of antibiotics' use in aquaculture. During the 1980s, when Atlantic salmon farming was expanding in Norway, it brought with it increased challenges on the control of fish diseases like furunculosis [23]. It is estimated that the Norwegian salmon industry was using about 48,570 kg of antibiotics per annum in the 1980s to produce 80,000 tons of salmon. However, when vaccination began to increase in the 1990s, the use of antibiotics started to decline by 1996, thereby reducing antibiotics' use to only 1037 kg with an out production of 290,000 tons of Atlantic salmon. Since then, vaccination has increased such that by 2019 the use of antibiotics in Norwegian salmon reached record low levels of only 222 kg of which around 99% of Norwegian farmed Atlantic salmon produced in 2020 were not treated with any form of antibiotics and no residues were found in salmon [23]. This success in the reduction of antibiotics' use in the Norwegian salmon industry is highly attributed to an increase in vaccination against bacteria diseases. Similarly, before 1996, no fish vaccines were approved in Japan, and as such antibiotics were the major way to prevent outbreaks of fish diseases [18]. After introduction of fish vaccines in 1995, economic losses caused by infectious diseases reduced by 70% in subsequent decades such that as fish vaccination increased by >600 fold between 1998 and 2013, antibiotic use reduced by 1000 fold [18]. It is anticipated that other countries like South Korea where vaccination coverage in farmed fish increased from 25.5% in 2014 to 80.6% in 2018 in olive flounder (*Paralichthys olivaceus*) [15] will significantly contribute to reduction of antibiotic use.

FIGURE 12.3 Global distribution of fish producing countries versus vaccination.

12.4 GLOBAL STATUS OF FISH VACCINATION

The global status of fish vaccination can be divided into three groups, namely, the (1) established and developed, (2) established but not fully vaccinating on a large scale, and (3) under development (Figure 12.3), as discussed below.

12.4.1 Vaccination Status of High-Priced Fish in the Established and Developed Category

Fish species classified in the high price category include Atlantic salmon (*Salmo salar* L), rainbow trout (*Oncorhynchus mykiss*), Atlantic cod (*Gadus morhua*), seabass (*Dicentrarchus labrax*), and gilt-head sea bream (*Sparus aurata*) (Table 12.2). Several commercial vaccines have been developed and licensed against diseases infecting these fish species [1,9,24]. Thus, countries engaged in large-scale production and vaccination of these fish species are likely to have reciprocal reduction of antibiotic use in aquaculture (Figure 12.2). The top producers of salmonids are Norway, Scotland, Canada, and Chile. Top countries producing European seabass and gilt-head sea bream include Greece, Spain, Portugal, Croatia, and Turkey. In these countries, several vaccines have been licensed for viral nervous necrosis (VNN) and pasteurellosis. Vaccination is done on a large scale and farmers synchronize their vaccination schedules to ensure broad coverage (Table 12.2) [1].

12.4.2 Vaccination Status of Fish in the Established and Developing Phase

The developing category mainly consists of Asian aquaculture where there has been tremendous improvement in fish vaccination in recent decades compared to early 21st century [1]. Most of the top fish-producing countries have licensed vaccines. However, they still lag behind Western countries in terms of mass vaccination coverage. Currently, the top four aquaculture-producing countries in Asia are China, Indonesia, India, and Vietnam, and they all have different licensed fish vaccines. China, which produces close to 70% of the total farmed fish globally, has different in-house and imported vaccines (Tables 12.1 and 12.2) [12]. Licensed fish vaccines in China are mainly for grass carp (*Ctenopharyngodon Idella*), turbot (*Scophthalmus maximus*), sea bass

TABLE 12.2
Price of Farmed Fish Species and Live Vaccines

Fish Species	Price USD/kg 02.06.22	Harvest Size (kg)	Average Value/ Fish (USD)	Production Cycle (Months)	Licenced Vaccine	Country	Ref
Atlantic salmon	~6–8	4–6	28–42	20–36	Most fish get vaccinated. Several vaccines available	Norway, Chile, UK, Canada, Australia, Faroes	[3,9,25,26]
Rainbow tout	~4–6	3–5	15–25	18–24	Most fish get vaccinated. Several vaccines available Autogenous *Lactococcus gravies* vaccine in Australia	Norway, Chile, UK, Canada, Australia	[3]
Sea bass	~3–4	0.5–0.6	1.75–2.1	15–24	VNN by Pharmaq® VNN Hipra® Pasteurellosis vaccine Hipra® AquaVac IridoV® VibriFishvax by Intervet® in Malaysia Autogenous *S. iniae* vaccine in Australia	Greece, Turkey, Italy, Spain, Croatia, Malaysia Indonesia and Egypt	[3,9,17,20]
Tilapia	~1–2	0.4–1	0.6–1.5	8–10	*Streptococcus agalactiae* vaccine by Pharmaq in Brazil *Flavobacterium columnare* vaccine AquaVac Strep SI® by Intervet® in Malaysia Iridovirus vaccine in Singapore *S. agalactiae* and *S. iniae* vaccine in Indonesia and Veitnam	China, Brazil, Indonesia, Egypt, Thailand, Malaysia	[1,3,9,13,16,17]
Pangasius	~0.7–1.5	~0.8–1.2	0.88–1.32	5–7	Alpha Jet Panga 1 Pharmaq® Alpha Jet 2 *Aeromonas hydrophila* Pharmaq® *Edwardsiella ictaluri* vaccine by Pharmaq® in Vietnam. *E. ictaluri* vaccine in Indonesia	Thailand, Indonesia, Bangladesh, India and Philippines	[3,9,19]
Olive flounder	~2–4	0.8–1.2	2.4–3.6	12–14	*E. tarda* inactivated and Multivalent South Korea *S. iniae* inactivated Multivalent vaccine in South Korea	South Korea, China and Japan	[15]

(*Continued*)

TABLE 12.2 (*Continued*)
Price of Farmed Fish Species and Live Vaccines

Fish Species	Price USD/kg 02.06.22	Harvest Size (kg)	Average Value/ Fish (USD)	Production Cycle (Months)	Licenced Vaccine	Country	Ref
Grouper	~3–4	0.4–0.6	1.4–2.1	7–9	Nodavac-R by CIBA for milkfish, grey mullet and red snapper, India VNN vaccine by Pharmaq® Betanodavirus (RGNNV) vaccine in Australia Iridovirus vaccine in Indonesia	China, India, Taiwan, Indonesia, Malaysia,	[3,14]
Grass carp	~0.5–1.5	~0.8–1.2	0.5–1.5	8–10	Live and cell -cultured inactivated vaccine against Grass carp reovirus in China	China, India	[12]
Silver carp	~0.64–1.75	~0.75–1.5	0.89–1.67	8–10	Inactive vaccine against *A. hydrophila* in China	China, India	[12,21]
Common carp					Spring viremia of carp subunit vaccine by Belgium Spring viremia of carp inactivated vaccine in Ireland Koi herpesvirus vaccine in Indonesia	Belgium, Ireland, China, India	[3,9]
Turbot	~3–6	1.5–2	4.5–6	18–24	Live vaccine against *E. tarda* *Vibrio anguillarum* vaccine in China	China, UK, Portugal, Spain	[12,21]
Koi carp	~2–4	0.8–1.5	2.4–4.5	12–14	Inactivated Koi harpies virus vaccine in Israel Koi herpesvirus vaccine in Indonesia	China, Japan, India, Israel	[3,9]
Indian carp	~0.5–1.5	~0.8–1.2	0.5–1.5	8–12	Several vaccines are in experimental stage. No approval vaccine available	India, Bangladesh	
Channel catfish	~2–3.5	~0.3–0.6	0.82–1.65	18–36	Oral vaccine against enteric septicemia in USA	USA	[22]
Yellow tail	~3–4	3.5–4.5	12.25–15.75	17–20	*Monovalent, multivalent, and trivalent vaccine against* Photobacterium damselae subsp. Piscicida, *L. garvieae, V. anguillarum* in Japan Autogenous *P. damselae* subsp. Damselae vaccine in Australia	Japan, Australia	[3,18]
Cod	~4–6	2–4	10–20	22–24	Vaccine against vibriosis and furunculosis in Norway	Norway, Canada,	[1]

(*Lates calcarifer*), Japanese flounder (*Paralichthys olivaceus*), and softshell turtle (*Trionyx sinensis*) (Table 12.2 and Figure 12.3) [12]. Indonesia is the second largest aquaculture producer but has fewer licensed vaccines [9] than China although it has started developing autogenous vaccines for tilapia (*Oreochromis niloticus*), common carp (*Cyprinus carpio*), and catfish [3]. As for India, which is the third largest aquaculture producer, it has a local vaccine licensed against VNN for milkfish (*Chanos chanos*), gray mullet (*Mugil cephalus*), and red snapper (*Lutjanus campechanus*) [14] while Vietnam has licensed vaccines for catfish against *Aeromonas spp* and *Edwardsiella ictaluri* produced by Pharmaq® (Zoetis) Ltd. [14]. Thailand is one of the major tilapia producers and has several autogenous vaccines [3], while South Korea, which is the seventh largest aquaculture producer in the world, has several licensed fish vaccines against various bacterial, viral, and parasitic diseases. The total number of vaccine doses used in South Korea in 2018 was 67.71 million doses mainly on Oliver flounder, red seabream, and amber jack (*Seriola* spp) of which Oliver flounder is the main farmed fish species in South Korea (Table 12.2) [15]. In the Asia-Pacific region, Australia has several registered commercial vaccines for salmonids, Asian seabass, grouper, and yellow tail fish. In addition, Australia has also allowed some autogenous vaccines under minor permit (Table 12.2) [3]. Overall, although there is considerable progress in licensing of several vaccines, there is considerably a large proportion of vaccines in the developmental stages particularly for diseases endemic in Asian aquaculture that are yet to be commercialized.

12.4.3 Fish Vaccination Status in the Under Development Stage

The group comprised all of Latin America (except Chile), Africa, and the Middle East. Latin America has the highest proportion of tilapia vaccination coverage estimated at 35%, while Africa and Asia have <1% coverage. However, vaccination coverage of farmed tilapia is globally lower than farmed Atlantic salmon that has close to 95% vaccination coverage [27]. Vaccines used in tilapia include ALPHA JECT micro® 1 TiLa used against *S. agalactiae* produced by Pharmaq® [28] and Hipra® in Brazil [29] while in Argentina, Feedvax®, a start-up company, aims at developing oral vaccine for tilapia [30]. The contribution of African countries toward global aquaculture is approximately 2.7% where top producers are Egypt, Nigeria, and Uganda that account for 90% of total aquaculture produced in the region [31]. These countries have no licensed vaccines for the control of fish disease although several fish vaccine trials are going on, especially for tilapia. Aquaculture in the Middle East is still in the early developmental stages, especially in countries like Saudi Arabia, United Arab Emirates, and Oman, which are investing in fish farming [32]. There are no known licensed fish vaccines produced in this region.

12.4.4 Cost of Vaccination per Fish as a Constraint to Vaccine Development

With the exception of Atlantic salmon, other top 10 farmed fish species (Figure 12.4) are still not vaccinated on a large scale. Although most of the top fish-producing countries have licensed vaccines for tilapia, carp, and catfish, mass vaccination is minimal (Table 12.1). Atlantic salmon, rainbow trout, yellow tail, and Atlantic cod are classified as high-value fish species whose production cost is estimated at 28–42 USD/fish, 15–25 USD/fish, 12.25–15.75 USD/fish and 10–20 USD/fish, respectively. Mass vaccination of these fish species is considered cost-effective. For example, the cost of vaccination of Atlantic salmon is 0.17 USD/fish in Norway, which is <150 times lower than the average price per fish [25] (Table 12.2). In 2006, Thorarinsson et al. [26] estimated the cost of vaccinating Atlantic salmon and sea bass to be 0.132 Euro/fish and 0.051 Euro/fish, respectively. On the contrary, although tilapia, carp, and pangasius are produced in large quantities, they are valued at very low prices of 0.6–1.5 USD/fish (Table 12.2). Also, the size of fish at market value is low (>1 kg), while the vaccine price per dose is too high rendering vaccination not cost-effective. To promote mass vaccination coverage of low-priced fish species like tilapia and carp, vaccine

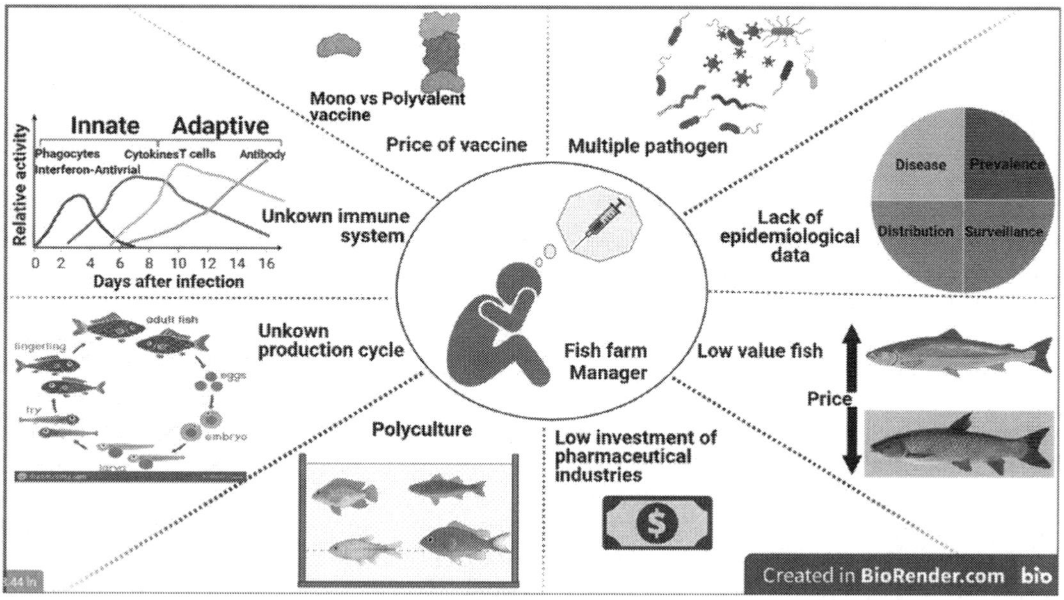

FIGURE 12.4 Fish vaccination managerial challenges.

production efforts should be directed at reducing the vaccine price. This would render vaccination to be attractive and consequently serve as a better alternative to antibiotic use for small-scale fish farmers with low inputs in fish farming. Although tilapia is the second largest farmed fish species after carp, only 5% tilapia are vaccinated [27]. The reason for not vaccinating tilapia at a large scale is that vaccination is considered unprofitable because of the high cost of vaccination compared to the market price of fish. As shown in Figure 12.3, the high cost of fish vaccination has been linked to increase in antibiotics use while increase in vaccination has been linked to reduced antibiotics use in global aquaculture.

12.5 CHALLENGES IN VACCINE DEVELOPMENT

Although vaccination is considered the most effective disease control strategy that does not only reduce the occurrence of disease outbreaks but also has a significant influence in reducing antibiotic use in aquaculture, there are several challenges hindering the success of developing protective vaccines for use in aquaculture as shown in Figure 12.4.

12.5.1 Lack of Epidemiological Data on Disease

Effective disease control in fish requires availability of data regarding distribution, causation, and magnitude of the diseases [10,33]. Such data can only be generated through passive and active surveillance involving the fish farming community. In resource-rich countries, passive surveillance is carried out by daily recording of diseases observed on the farm as part of the fish health management system. Thus, it is easy to identify the source of an outbreak, spread of disease, and it simplifies the task of identifying the causes of vaccine failure in situations where immunized fish show presence of disease. Thus, fish farmers play the leading role in carrying out passive surveillance because their fish health records provide epidemiological data on prevalent of diseases in the ecosystem. Disease trends observed in passive surveillance often call for further investigations using active surveillance to determine etiological agents, emerging novel pathogens or mutants, existing strains/serotypes, and identify factors raising the need for new design. On the contrary, passive surveillance based on record keeping in poor resource fish farming communities is often not implemented as part of fish health management [34]. However, the resource-rich countries have invested heavily in surveillance and research, resource-constrained countries have limited funding for disease surveillance and research [35]. This prohibits the availability of data required to guide rational disease control strategies and vaccine design. Nonetheless, there is a need for epidemiological data to guide in the design of protective vaccines [36–39]. The absence of such data has to a bigger extent limited the use of vaccination, especially in resource limited settings [40,41].

Different species of fish are farmed in the different regions of the world driven mainly by fish adaptability, marketability, and historical/cultural reasons. Relatedly, different pathogens cause disease in the different fish species making vaccine development for specific fish species challenging. It should be noted that some fish species may not be susceptible to certain microorganisms but will serve as reservoirs for the agent [42]. This calls for active surveillance to profile the different pathogens associated with chronic disease carriers or reservoirs [43]. Most of the research has been conducted in high-value fish species like Atlantic salmon, rainbow trout, and European seabass unlike the top three farmed fish species (carp, tilapia, and catfish) that are dominant in resource-poor countries. It is unknown whether these fish carry pathogens that affect the high-value fish species or whether they are a source of emerging diseases resulting from mutations or old pathogens that evolved to infect new host species but pose a challenge to vaccine development. Therefore, there is the need to promote disease surveillance in resource-constrained communities engaged in aquaculture.

12.5.2 Management Challenges in Polyculture and Integrated Farming System

Effective management systems are important in the promotion of fish health in aquaculture. Such management practices would include adequate nutrition, water quality assessment, and maintenance of optimal stocking density. Irregular changes in the management practices cause stress to the fish predisposing them to various diseases [44]. Several studies have underscored the role of nutrition in the maintenance of a robust immune system and response to vaccination [45–48]. A wide range of aquaculture practices are in use in different regions of the world ranging from monoculture in the highly intensive farming systems of high-value fish to the polyculture systems mainly practiced in resource-constrained settings. The proponents of polyculture associate it with maximum exploration of the food web [49,50], and control of some diseases [51]. However, polyculture system can serve as hot spots for emergence and propagation of fish diseases. Different fish species vary in susceptibility to disease pathogens. Consequently, this complicates vaccine development especially where multiple pathogens infect multiple fish species unlike in monoculture, which makes it easy to monitor vaccine performance in a single species. Thus, polyculture practices require multiple assays for measuring immune responses of multiple fish species to vaccination as well as determining the susceptibility of multiple fish species to multiple pathogens. As a result, research in vaccine development for a polyculture system would therefore pursue the path of developing vaccines for multiple fish species but unfortunately such polyculture systems are practiced in resource-constrained countries whose research budget may not support such endeavors [52].

The increase in human population growth with no commensurate increase in global food production, particularly in resource-constrained countries, has led to use of production technologies that may not promote good management practices in aquaculture. One of such production approaches is the integrated aquaculture system where fish is cultured alongside other agricultural activities [53–55]. Such management practices render implementation of fish vaccination programs a challenge especially in situations where aquaculture is integrated with cultivation of crops species like rice. Thus, it can be a challenge to capture all fish cultured in rice pads for vaccination.

12.5.3 Unskilled Personnel and Low Infrastructure Investment

Implementation of effective vaccination programs in aquaculture requires competent staff to choose the correct vaccine for use whether mono- or multivalent, determine the ideal timing for vaccination, and most importantly to administer the vaccines into the fish [56]. There is a big disparity in the distribution of such competent technical personnel between fish farmers engaged in high-priced fish and low-priced fish. This calls for the training of personnel to undertake the vaccination regime. Relatedly, even when the personnel have been trained, there are limitations in the availability of infrastructure to implement the required procedures. Vaccination of fish relies mainly on three major administration routes: injection, oral, and immersion vaccination [57–59]. There is a need to understand the relative advantage and disadvantage of each vaccination method before choosing the method for administering the vaccine in fish. For example, injection is labor-intensive, impacting stress on fish and requires trained personnel to handle fish but has the advantage of ensuring that all fish are injected with the same antigen dose. On the other hand, oral vaccination is less labor-intensive and impacts minimal stress on fish but does not guarantee that all fish will be exposed to the same antigen dose. Although immersion vaccination impacts less stress on fish, it cannot be used for fish cultured in cages in the river or sea.

12.5.4 Several Pathogens with Tropism for Multiple Host Species

Some pathogens infect a wide range of fish species in aquaculture of which some of the infected fish species can carry the pathogen without clinical sign while continuously shedding the pathogen in the environment as a source of infection to susceptible species [60–62]. This is more pronounced in polyculture systems where different species of varying susceptibilities are cultured together [63]. In "co-infection," fish can be infected by multiple pathogens. Co-infection can be synergistic or antagonistic, the outcome of which depends on the virulence of the pathogens involved [64]. Also, co-infection can be homozygous (related pathogens but variant strains) or heterozygous (e.g., pathogens from different genera) [65]. Thus, co-infections pose a challenge to vaccination since they require polyvalent vaccines, which are in most cases not readily available. Bacterial pathogens causing disease in multiple fish species include members of the genera *Aeromonas*, *Edwardsiella*, *Flavobacteria*, and *Vibrio* that are ubiquitously found in aquatic environments used for aquaculture. The common *Aeromonas* spp. causing diseases in fish include *Aeromonas hydrophila*, *A. sobria*, *A. caviae*, and *A. veronii* [66,67]. These species are ubiquitously found in aquatic environments. They cause septicemia, exophthalmos, and ulceration in different fish species. Among the top farmed fish species (Figure 12.5), they have been reported to cause disease in various carp species, tilapia, and catfish. *Vibrio* spp. is known to cause diseases in fish including *Vibrio anguillarum, V. alginolyticus, V. parahaemolyticus, V. ordalii, V. vulnificus*, and *V. anguillarum*. They cause diseases in various fish species that include carp, catfish, salmon, flounder (*Paralichthys olivaceus*), eel (*Anguilla anguilla*), rainbow trout (*Oncorhynchus mykiss*), seabass (*Dicentrarchus labrax*), and seabream (*Sparus aurata*) [68,69]. As for Edwardsiellosis, the major species causing disease in various fish are *E. piscicida* and *E. anguillarum* [70], while in case of *Flavobacterium*, *F. columnare* is the most common species causing columnaris in various species that include catfish, carp, and tilapia [71–73].

Viral pathogens causing disease in multiple fish species include hemorrhagic septicemia virus (VHSV) which causes disease in rainbow trout (*Oncorhynchus mykiss*), steelhead trout, chinook salmon (*O. tshawytscha*), turbot (*Scophthalmus maximus*), yellow perch (*Perca flavescens*), gobies (*Pomatoschistus microps*), emerald shiners (*Notropis atherinoides*), muskies (*Esox masquinongy*), whitefish (*Coregonus clupeaformis*), walleye (*Sander vitreus*), and Japanese flounder [74–76]. Another important emerging viral pathogen infecting multiple fish species is iridovirus of which the most common is infectious kidney and spleen virus (ISKV), which has been reported to infect catfish, carp, tilapia, large yellow croaker, brown-spotted grouper (*Epinephelus malabariacus*), mandarin fish Siniperca (*Siniperca chuatsi*), and several other fish species [77–80].

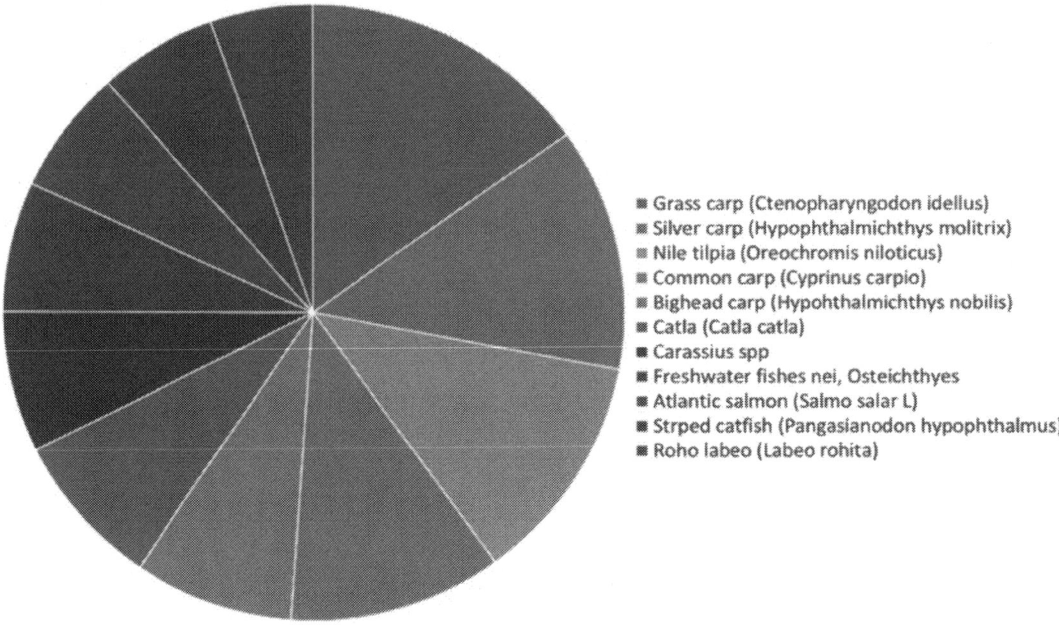

FIGURE 12.5 Top farmed fish species in the world.

The major challenge in developing vaccines infecting multiple fish species is setting priorities on pathogens to select for vaccine development, developing vaccine efficacy test models for different species, developing standard doses that apply across different species, identifying the timing when to vaccinate the multpile species, and several other factors that need synchronizing to suit the multiple fish species susceptible to the same pathogen.

12.5.5 Fish Production Cycle and Vaccination Regimes

Fish life cycles vary among different species even though the developmental stages start with eggs that hatch into larvae for all fish species. When the yolk has been fully absorbed, the larvae develop into fry that develop into juveniles that mature into adults. As fish develop from fry to adults, they are exposed to various pathogens that cause diseases at different stages of the production cycle. Thus, the timing of vaccination has to be done in a way that enables fish to develop protective immunity at the time when they are most vulnerable to infection [42]. However, the major challenge in designing effective vaccination regimes is that stages in the production cycle when fish are most vulnerable to infection have not been determined for most diseases, especially for low-priced fish species like tilapia, carp, and catfish. Thus, a good understanding of stages when fish are most susceptible to infection is a guide for developing effective vaccination regimes. This could involve the design of single or prime-boost vaccination regimes involving multiple immunization programs. For example, in the case of infectious pancreatic necrosis virus (IPNV) where the production cycle of Atlantic salmon (*Salmo salar* L) is well established [81], the most effective timing for vaccination is at the parr/pre-smolt stage before transfer to sea. This enables fish to have protective immunity during smoltification when they are most vulnerable to IPNV infection due to stress induced by transfer from freshwater to seawater [82,83]. Similar timings can be established for other diseases to ensure that fish have protective immunity when they are vulnerable to infection.

12.5.6 Characterizing the Fish Immune System and the Development of Diagnostic Tools

A good understanding of the fish immune systems is crucial for the design of protective vaccines because host response to vaccination is a reliable guide to optimizing vaccine development. There have been more studies on characterization of the immune system of highly priced fish species like Atlantic salmon and rainbow trout than low-priced fish species like carp, tilapia, and catfish, which are among the top farmed fish species in the world (Figure 12.5). This is because carp, catfish, and tilapia lag behind the high-priced species (Atlantic salmon and trout) in attracting research for characterizing immune system induced by vaccination. Consequently, there are few diagnostic tools like polyclonal and monoclonal antibodies (mAbs) developed for the low-priced than the high-priced fish species [82,84–87]. For example, mAbs against MHC-I and II molecules used to evaluate the activation of antigen-presenting cells (APCs) have been developed for salmonids [88] while no similar antibodies have been developed for tilapia, carp, and catfish. Equally, in salmonids, a CD83 mAb used as an immune marker for the activation of macrophages has been developed [89] and yet no similar mAb has been developed for carp, catfish, and tilapia. Similarly, a mAb against CD8α has been developed as an immune marker for activated CD8 T-cell responses in salmonids [90,91] and no similar markers have been developed for tilapia, catfish, and carp. As for humoral responses, IgT mAbs have been developed for evaluating mucosal antibody responses in salmonids [92] while no similar mAbs have been developed for tilapia, catfish, and carp. Equally, there are several anti-salmon polyclonal IgM antibodies that have been in use for a long time and yet most of the anti-tilapia [93–95] and anti-carp [96–98] IgM antibodies are still in the developmental and validation stages. In the absence of such tools, it is difficult to develop immunoassays such as ELISA and immunohistochemistry required for evaluating host responses to vaccination. Also, it is a challenge to determine the mechanisms of vaccine protection and to establish correlates of vaccine protection [99].

12.5.7 Challenges in Development of Vaccines for Intracellular Replicating Bacteria

The majority of vaccines that have contributed to reduction of antibiotics' use in aquaculture are whole cell-inactivated (WCI) bacterial vaccines. The WCI accounts for the largest proportion of vaccines used in aquaculture [100,101]. However, these vaccines are only effective for extracellular replicating bacterial pathogens such as *Aeromonas salmonicida*, *Moritella viscosa*, or *Streptococcus agalactiae*. There is an increasing number of intracellular replicating bacterial pathogens such as members of the genera *Edwardsiella*, *Francisella*, *Piscirickettsia*, *Mycobacteria*, *Renebacteria*, and *Pseudomonas* that are not only difficult to treat using antibiotics but cause granulomatous conditions that lead to high economic losses in infected fish [102]. For example, *Francisella noatunensis* is considered to have contributed to the collapse of the Atlantic cod (*Gadus morhua*) industry in Norway around 2010 [103,104], while *Pseudomonas plecoglossicida* has been reported to adversely impact the large yellow croaker industry in China [105,106]. These bacteria replicate inside fish phagocytic cells like macrophages where they are inaccessible to neutralization by antibodies produced by WCI vaccines. Instead, they require replicative vaccines that evoke the cell-mediated immune response (CMI) able to eliminate infected cells. The major challenge in developing replicative vaccines is the fear of the live attenuated vaccine strains to revert to virulence [18]. Also, genetically attenuated bacteria strains are considered genetically modified organisms (GMOs) that may not be permitted for use as vaccines in countries prohibiting the use of GMOs. In addition, the fear of using avirulent strains is that they can be a source of infection to other fauna in aquatic environments where they can be pathogenic. Thus, various methods are being sought for intracellular delivery of bacterial antigens with the view of developing vaccines able to evoke CMI responses in vaccinated fish [62,107–111]. This far, there are more commercial vaccines against extracellular replicating bacteria than vaccines for intracellular replicating bacteria [1,9,15,112–115].

TABLE 12.3
Vaccine Companies

Vaccine Companies	Country
ScotVax	Australia
Midtjysk Akva ApS	Denmark
Ceva Biovac	France
Vaxxinova International BV	Germany
AniCon Labor GmbH	Germany
Phibro Animal Health Corporation	Israel
Kyoto Biken Laboratories, Inc.	Japan
Pharmaq AS- part of Zoetis	Norway
Choong Ang Vaccine Laboratories (CAVAC)	South Korea
HIPRA	Spain
Eurofish Fish Vaccination	United Kingdom
Aqualife Services Limited	United Kingdom
MARINNOVAC	United Kingdom
Elanco Europe LtD	United Kingdom
Kennebec River Biosciences	USA
MSD Animal Healths	USA
AquaTactic Fish Health and Vaccines	USA

12.5.8 Challenges for the Pharmaceutical Industry in Vaccine Production

The demand for vaccines for tilapia, carp, catfish, large yellow tail, and other fish species is increasing at a large scale corresponding with the increase in intensification of farming systems and high output. As a result, this has attracted several vaccine-producing companies to invest more in these fish species in order to cope with the increasing demand for protective vaccines. However, as shown in Table 12.3, the majority of well-established vaccine-producing companies are based in Western countries, while the demand is highest in Asia, which is the largest aquaculture producer in the world. The major challenge is that vaccine companies based in Western countries would be required to carry out research to develop protective vaccines ideal for Asia. In some cases, issues of compliance with local regulatory policies or agencies regarding vaccine efficacy trials and commercialization are in conflict with ethics of the vaccine companies. Also, the unavailability of epidemiological data makes it difficult to design protective vaccines against pathogens with multiple strains/serotypes prevalent in each ecosystem. These factors, together with several other reasons, have contributed to the slow involvement of the well-established vaccine companies based in Western countries to contribute in developing protective vaccines for the Asian aquaculture industry.

12.6 CONCLUSION

In this synopsis, we have shown that vaccine development is progressing to cope with the global demand for developing preventive measures against several diseases adversely impacting the expansion of the aquaculture industry. However, the progress in vaccine development is highly influenced by the price value of farmed fish with high-priced fish species having more commercial vaccines than low-priced fish species. Consequently, there is significant progress in the reduction of antibiotics in countries with high-resource aquaculture industries than in resource-constrained countries. We have also shown that challenges adversely impacting vaccine development are higher in the resource-constrained countries producing low-priced fish species than in the HRCs producing high-priced fish species. However, there is an increasing trend in the production of fish vaccines as a response to the escalating disease burden in the global aquaculture industry.

DISCLOSURE STATEMENT

The authors declare that they have no conflicts of interest.

REFERENCES

[1] Brudeseth BE, Wiulsrød R, Fredriksen BN, et al. Status and future perspectives of vaccines for industrialised fin-fish farming. *Fish & Shellfish Immunology*, 2013, 35: 1759–1768.

[2] FAO. *The state of world fisheries and aquaculture 2020- sustainability in action*. Food and Agriculture Organization of the United Nations, 2020.

[3] Barnes AC, Silayeva O, Landos M, et al. Autogenous vaccination in aquaculture: A locally enabled solution towards reduction of the global antimicrobial resistance problem. *Reviews in Aquaculture*, 2022, 14: 907–918.

[4] Leaño EM. Transboundary aquatic animal diseases: History and impacts in asean aquaculture. In: *Proceedings of the Aquatic Emergency Preparedness and Response Systems for Effective Management of Transboundary Disease Outbreaks in Southeast Asia: Proceedings of Asean Regional Technical Consultation, August 20–22, 2018, Centara Grand Central Ladprao, Bangkok, Thailand, 2019. Aquaculture Department, Southeast Asian Fisheries Development Center.*

[5] Subasinghe R. Disease control in aquaculture and the responsible use of veterinary drugs and vaccines: The issues, prospects and challenges. *Options Méditerranéennes*, 2009, 86: 5–11.

[6] Dubey S. *Nanoparticle based vaccine delivery of recombinant outer membrane proteins and characterization of aeromonas hydrophila and edwardsiella species in Asian aquaculture*. Norwegian University of Life Sciences, Ås, 2019.

[7] Bedekar MK, Kole S. Fundamentals of fish vaccination. In: *Vaccine design*. Springer, 2022. pp. 147–173.

[8] Ada G. The ideal vaccine. *World Journal of Microbiology and Biotechnology*, 1991, 7: 105–109.

[9] Ma J, Bruce TJ, Jones EM, et al. A review of fish vaccine development strategies: Conventional methods and modern biotechnological approaches. *Microorganisms*, 2019, 7: 569.

[10] Assefa A, Abunna F. Maintenance of fish health in aquaculture: Review of epidemiological approaches for prevention and control of infectious disease of fish. *Veterinary Medicine International*, 2018, 2018: 5432497.

[11] TheFishSite. *28millionmorevaccinedosesdeliveredtoNorwegiansalmonlastyear.* February6,2020. https://thefishsite.com/articles/28-million-more-vaccine-doses-delivered-to-norwegian-salmon-last-year

[12] Wang Q, Ji W, Xu Z. Current use and development of fish vaccines in china. *Fish & Shellfish Immunology*, 2020, 96: 223–234.

[13] Forum FH. *Vaccines provide tilapia producers with a platform for sustainable growth*. The Fish Site, 2021.

[14] Fletcher R. *Ciba launches India's first homegrown fish vaccine*. The Fish Site 2021. The Fish Site Limited, Cork, Ireland. https://thefishsite.com/articles/ciba-launches-indias-first-homegrown-fish-vaccine

[15] Hwang JY, Kwon MG, Seo JS, et al. Current use and management of commercial fish vaccines in Korea. *Fish & Shellfish Immunology*, 2020, 102: 20–27.

[16] Klesius P, Shoemaker C, Evans J. *Tilapia vaccines: Important disease prevention, biosecurity tools*. Global Seafood Alliance, 2007. Global Seafood Alliance, Portsmouth.

[17] Ridzuan MSM, Abdullah A, Ramly R, et al. Current status and advances of fish vaccines in Malaysia. *World*, 2016, 15: 465–482.

[18] Matsuura Y, Terashima S, Takano T, et al. Current status of fish vaccines in Japan. *Fish & Shellfish Immunology*, 2019, 95: 236–247.

[19] McGee MV. *Pangasius culture in Western hemisphere*. Global Seafood Alliance, 2014. Global Seafood Alliance, Portsmouth.

[20] The Fish Site. *New sea bass vaccine approved*. The Fish Site, 2018. Fish Site Limited, Cork.

[21] FAO. *Chinese Fish Price Report - Issue 1/2020*. Product Trade and Marketing Branch, 2020. FAO, Rome.

[22] The Fish Site. *Oral vaccine could save us catfish farmers millions*. The Fish Site, 2019. Fish Site Limited, Cork.

[23] Norwegian Seafood Council. *The success recipe for eradicating antibiotics in norwegian salmon farming*. Norwegian Sea Food Council, 2021.

[24] Kayansamruaj P, Areechon N, Unajak S. Development of fish vaccine in southeast asia: A challenge for the sustainability of se asia aquaculture. *Fish & Shellfish Immunology*, 2020, 103: 73–87.

[25] Merete F. *Key figures from Norwegian aquaculture industry*. Directorate of Fisheries Department of Aquaculture, Oslo, 2020.

[26] Thorarinsson R, Powell DB. Effects of disease risk, vaccine efficacy, and market price on the economics of fish vaccination. *Aquaculture*, 2006, 256: 42–49.

[27] Howell M. *Why are global tilapia vaccination rates so low?* The Fish Site, 2021. The Fish Site Limited, 7/8 Liberty St, Cork.

[28] Damasceno DZ. *Vaccines provide Brazil's fish farmers with a platform for sustainable growth*. Fish Health Forum by PHARMAQ, https://fishhealthforum.com/vaccines-provide-brazils-fish-farmers-with-a-platform-for-sustainable-growth/

[29] Hipra B. *Hipra launches immersion vaccination to boost aquaculture in Brazil. HIPRA, Spain. https://www.hipra.com/en/hipra-launches-immersion-vaccination-boost-aquaculture-brazil*

[30] Fletcher R. *The oral vaccine that aims to tackle streptococcus in tilapia*. The Fish Site, 2021, https://thefishsite.com/articles/the-oral-vaccine-that-aims-to-take-down-streptococcus-in-tilapia

[31] Adeleke B, Robertson-Andersson D, Moodley G, et al. Aquaculture in africa: A comparative review of egypt, nigeria, and uganda vis-a-vis South Africa. *Reviews in Fisheries Science & Aquaculture*, 2020, 29: 167–197.

[32] Berdikeeva S. The rise of fish farming in over-fished gulf nations. 2019, https://insidearabia.com/the-rise-of-fish-farming-in-over-fished-gulf-nations/

[33] Peeler EJ, Taylor NG. The application of epidemiology in aquatic animal health-opportunities and challenges. *Veterinary Research*, 2011, 42: 1–15.

[34] Hedrick RP. Movement of pathogens with the international trade of live fish: Problems and solutions. *Revue Scientifique et Technique*, 1996, 15: 523–531.

[35] McCauley DJ, Jablonicky C, Allison EH, et al. Wealthy countries dominate industrial fishing. *Science Advances*, 2018, 4: eaau2161.

[36] Vinitnantharat S, Gravningen K, Greger E. Fish vaccines. *Advances in Veterinary Medicine*, 1999, 41: 539–550.

[37] Mondal H, Thomas J. A review on the recent advances and application of vaccines against fish pathogens in aquaculture. *Aquaculture International*, 2022, 30, 1971–2000.

[38] Akansha K, Ram R. Vaccines and their role in fish disease management: A review. *Biochemical and Cellular Archives*, 2015, 15: 39–46.

[39] Georgiadis M, Gardner I, Hedrick R. The role of epidemiology in the prevention, diagnosis, and control of infectious diseases of fish. *Preventive Veterinary Medicine*, 2001, 48: 287–302.

[40] Begg N, Miller E. Role of epidemiology in vaccine policy. *Vaccine*, 1990, 8 3: 180–189.

[41] Johannes RE. The case for data-less marine resource management: Examples from tropical nearshore finfisheries. *Trends in Ecology & Evolution*, 1998, 13: 243–246.

[42] Mugimba KK, Byarugaba DK, Mutoloki S, et al. Challenges and solutions to viral diseases of finfish in marine aquaculture. *Pathogens*, 2021, 10: 673.

[43] Murugadas V, Prasad M. *Disease surveillance in aquaculture*. In: ICAR-Central Institute of Fisheries Technology. 2020. ICAR, New Delhi. https://krishi.icar.gov.in/jspui/bitstream/123456789/45985/1/03_Disease%20Surveillance%20in%20aquaculture.pdf

[44] Crane M, Hyatt A. Viruses of fish: An overview of significant pathogens. *Viruses*, 2011, 3: 2025–2046.

[45] Dawood MAO, Koshio S, Esteban MÁ. Beneficial roles of feed additives as immunostimulants in aquaculture: A review. *Reviews in Aquaculture*, 2018, 10: 950–974.

[46] Gatlin III DM. Nutrition and fish health. In: *Fish nutrition*. Elsevier. 2003. 671–702. Elsevier, AmsterdamNederland.

[47] Trichet VV. Nutrition and immunity: An update. *Aquaculture Research*, 2010, 41: 356–372.

[48] Lazard J, Dabbadie L. Freshwater aquaculture and polyculture. *Fisheries and Aquaculture*, 2009, 4: 55.

[49] Wang M, Lu M. Tilapia polyculture: A global review. *Aquaculture Research*, 2016, 47: 2363–2374.

[50] Thomas M, Pasquet A, Aubin J, et al. When more is more: Taking advantage of species diversity to move towards sustainable aquaculture. *Biological Reviews*, 2021, 96: 767–784.

[51] Wang M, Chen Y, Zhao Z, et al. A convenient polyculture system that controls a shrimp viral disease with a high transmission rate. *Communications Biology*, 2021, 4: 1276.

[52] Milstein A. Polyculture in aquaculture. In: *Proceedings of the Animal Breeding Abstracts*, 2005. CABI Publishing: CABI Digital Library.

[53] Ahmed N, Garnett ST. Integrated rice-fish farming in bangladesh: Meeting the challenges of food security. *Food Security*, 2011, 3: 81–92.

[54] Reddy PR, Kishori B. Integrated rice and aquaculture farming. *In: Aquaculture-plants and invertebrates*. IntechOpen. 2018. IntechOpen Limited.

[55] Zhou M, Wan Q, Sarath Babu V, et al. Bacterial features in tilapia (*Oreochromis niloticus*) and environments in a goose-tilapia polyculture model. *Aquaculture*, 2018, 497: 313–319.
[56] Brugere C, Onuigbo DM, Morgan KL. People matter in animal disease surveillance: Challenges and opportunities for the aquaculture sector. *Aquaculture*, 2017, 467: 158–169.
[57] Soltani M, Alishahi M, Mirzargar S, et al. *Vaccination of rainbow trout against streptococcus iniae infection: Comparison of different routes of administration and different vaccines.* Iranian Journal of Fisheries Sciences 2007.
[58] Li J, Ma S, Woo NY. Vaccination of silver sea bream (*Sparus sarba*) against *Vibrio alginolyticus*: Protective evaluation of different vaccinating modalities. *International Journal of Molecular Sciences*, 2015, 17: 40.
[59] Fukuda Y, Kusuda R. Efficacy of vaccination for pseudotuberculosis in cultured yellowtail by various routes of administration. *Bulletin of the Japanese Society of Scientific Fisheries*, 1981, 47: 147–150.
[60] Bandín I, Dopazo CP. Host range, host specificity and hypothesized host shift events among viruses of lower vertebrates. *Veterinary Research*, 2011, 42: 67.
[61] Öztürk RÇ, Altınok İ. Bacterial and viral fish diseases in Turkey. *Turkish Journal of Fisheries and Aquatic Sciences*, 2014, 14: 275–297.
[62] Maiti B, Dubey S, Munang'Andu HM, et al. Application of outer membrane protein-based vaccines against major bacterial fish pathogens in india. *Frontiers in Immunology*, 2020, 11: 1362.
[63] Yu LQJ, Mu Y, Zhao Z, et al. Economic challenges to the generalization of integrated multi-trophic aquaculture: An empirical comparative study on kelp monoculture and kelp-mollusk polyculture in Weihai, China. *Aquaculture*, 2017, 471: 130–139.
[64] Abdel-Latif HMR, Dawood MAO, Menanteau-Ledouble S, et al. The nature and consequences of co-infections in tilapia: A review. *Journal of Fish Diseases*, 2020, 43: 651–664.
[65] Kotob MH, Menanteau-Ledouble S, Kumar G, et al. The impact of co-infections on fish: A review. *Veterinary Research*, 2016, 47: 98.
[66] Wamala SP, Mugimba KK, Dubey S, et al. Multilocus sequence analysis revealed a high genotypic diversity of aeromonas hydrophila infecting fish in Uganda. *Journal of Fish Diseases*, 2018, 41: 1589–1600.
[67] Dubey S, Maiti B, Girisha SK, et al. Aeromonas species obtained from different farmed aquatic species in India and Taiwan show high phenotypic relatedness despite species diversity. *BMC Research Notes*, 2021, 14: 1–8.
[68] Mukherjee S. *Fish diseases in India, their causes and control measures-winter school on recent advances in diagnosis and management of diseases in mariculture, 7th to 27th November 2002, course manual.* Central Marine Fisheries Research Institute, Kochi, India 2002. http://eprints.cmfri.org.in/5453/
[69] Das B. Bacterial diseases of fish and their control. In: Swain P, Sahoo PK, Das BK *Disease Diagnosis and Health Management of Freshwater Fish and Shellfish*, 2004, 45–52. CIFA, Bhubaneswar.
[70] Dubey S, Maiti B, Kim SH, et al. Genotypic and phenotypic characterization of edwardsiella isolates from different fish species and geographical areas in Asia: Implications for vaccine development. *Journal of Fish Diseases*, 2019, 42: 835–850.
[71] Arias C, Welker T, Shoemaker C, et al. Genetic fingerprinting of flavobacterium columnare isolates from cultured fish. *Journal of Applied Microbiology*, 2004, 97: 421–428.
[72] Wakabayashi H. Genotypic diversity of strains of flavobacterium columnare from diseased fishes. *Fish Pathology*, 1999, 34: 65–71.
[73] Decostere A, Haesebrouck F, Devriese L. Characterization of four Flavobacterium columnare (Flexibacter columnaris) strains isolated from tropical fish. *Veterinary Microbiology*, 1998, 62: 35–45.
[74] Meyers TR, Winton JR. Viral hemorrhagic septicemia virus in north america. *Annual Review of Fish Diseases*, 1995, 5: 3–24.
[75] Hedrick R, Batts W, Yun S, et al. Host and geographic range extensions of the north american strain of viral hemorrhagic septicemia virus. *Diseases of Aquatic Organisms*, 2003, 55: 211–220.
[76] Thompson TM, Batts WN, Faisal M, et al. Emergence of viral hemorrhagic septicemia virus in the north american great lakes region is associated with low viral genetic diversity. *Diseases of Aquatic Organisms*, 2011, 96: 29–43.
[77] Figueiredo HCP, Tavares GC, Dorella FA, et al. First report of infectious spleen and kidney necrosis virus in Nile tilapia in Brazil. *Transboundary and Emerging Diseases*, 2021, 69: 3008–3015.
[78] Fu X, Li N, Liu L, et al. Genotype and host range analysis of infectious spleen and kidney necrosis virus (ISKNV). *Virus Genes*, 2011, 42: 97–109.
[79] Jung-Schroers V, Adamek M, Wohlsein P, et al. First outbreak of an infection with infectious spleen and kidney necrosis virus (ISKNV) in ornamental fish in Germany. *Diseases of Aquatic Organisms*, 2016, 119: 239–244.

[80] Ramírez-Paredes JG, Paley RK, Hunt W, et al. First detection of infectious spleen and kidney necrosis virus (ISKNV) associated with massive mortalities in farmed tilapia in Africa. *Transboundary and Emerging Diseases*, 2021, 68: 1550–1563.

[81] Munang'andu HM, Santi N, Fredriksen BN, et al. A systematic approach towards optimizing a cohabitation challenge model for infectious pancreatic necrosis virus in atlantic salmon (*Salmo salar* L.). *PLoS One*, 2016, 11: e0148467.

[82] Munang'andu HM, Fredriksen BN, Mutoloki S, et al. Comparison of vaccine efficacy for different antigen delivery systems for infectious pancreatic necrosis virus vaccines in atlantic salmon (*Salmo salar* L.) in a cohabitation challenge model. *Vaccine*, 2012, 30: 4007–4016.

[83] Munang'andu HM, Mutoloki S, Evensen Ø. Acquired immunity and vaccination against infectious pancreatic necrosis virus of salmon. *Developmental & Comparative Immunology*, 2014, 43: 184–196.

[84] Munang'andu HM, Sandtrø A, Mutoloki S, et al. Immunogenicity and cross protective ability of the central vp2 amino acids of infectious pancreatic necrosis virus in atlantic salmon (*Salmo salar* L.). *PLoS One,* 2013, 8: e54263.

[85] Munang'andu HM, Fredriksen BN, Mutoloki S, et al. Antigen dose and humoral immune response correspond with protection for inactivated infectious pancreatic necrosis virus vaccines in atlantic salmon (*Salmo salar* L). *Veterinary Research*, 2013, 44: 1–16.

[86] Pettersen FE, Fyllingen I, Kavlie A, et al. Monoclonal antibodies reactive with serum igm and leukocytes from atlantic salmon (*Salmo salar* L.). *Fish & Shellfish Immunology*, 1995, 5: 275–287.

[87] Thuvander A, Fossum C, Lorenzen N. Monoclonal antibodies to salmonid immunoglobulin: Characterization and applicability in immunoassays. *Developmental & Comparative Immunology*, 1990, 14: 415–423.

[88] Hetland DL, Jørgensen SM, Skjødt K, et al. In situ localisation of major histocompatibility complex class I and class II and CD8 positive cells in infectious salmon anaemia virus (ISAV)-infected atlantic salmon. *Fish & Shellfish Immunology*, 2010, 28: 30–39.

[89] Braden LM, Rasmussen KJ, Purcell SL, et al. Acquired protective immunity in atlantic salmon salmo salar against the myxozoan kudoa thyrsites involves induction of MHIIβ$^+$ CD83$^+$ antigen-presenting cells. *Infection and Immunity*, 2018, 86: e00556–e00517.

[90] Mikalsen AB, Haugland O, Rode M, et al. Atlantic salmon reovirus infection causes a CD8 t cell myocarditis in atlantic salmon (*Salmo salar* L.). *PLoS One,* 2012, 7: e37269.

[91] Hetland DL, Dale OB, Skjødt K, et al. Depletion of CD8 alpha cells from tissues of atlantic salmon during the early stages of infection with high or low virulent strains of infectious salmon anaemia virus (ISAV). *Developmental & Comparative Immunology*, 2011, 35: 817–826.

[92] Zhang N, Zhang X-J, Chen D-D, et al. Molecular characterization and expression analysis of three subclasses of IgT in rainbow trout (Oncorhynchus mykiss). *Developmental & Comparative Immunology*, 2017, 70: 94–105.

[93] Velázquez J, Rodríguez A, Aragón H, et al. Monoclonal antibody against nile tilapia (Oreochromis niloticus) IgM heavy chain: A valuable tool for detection and quantification of IgM and IgM+ cells. *Fish & Shellfish Immunology*, 2021, 110: 44–54.

[94] Hu H, Zeng W, Wang Y, et al. Development and application of a recombinant protein-based indirect ELISA for detection of anti-tilapia lake virus IgM in sera from tilapia. *Aquaculture*, 2020, 520: 734756.

[95] Al-Harbi AH, Truax R, Thune RL. Production and characterization of monoclonal antibodies against tilapia Oreochromis niloticus immunoglobulin. *Aquaculture*, 2000, 188: 219–227.

[96] Vesely T, Reschova S, Pokorova D, et al. Production of monoclonal antibodies against immunoglobulin heavy chain in common carp (*Cyprinus carpio* L.). *Veterinární Medicína*, 2006, 51: 296.

[97] He Y, Jiang Y, Lu L. Serodiagnosis of grass carp reovirus infection in grass carp Ctenopharyngodon idella by a novel Western blot technique. *Journal of Virological Methods*, 2013, 194: 14–20.

[98] Bag MR, Makesh M, Rajendran K, et al. Characterization of igm of indian major carps and their cross-reactivity with anti-fish IgM antibodies. *Fish & Shellfish Immunology*, 2009, 26: 275–278.

[99] Munang'andu HM, Evensen Ø. Correlates of protective immunity for fish vaccines. *Fish & Shellfish Immunology*, 2019, 85: 132–140.

[100] Munang'andu HM, Mutoloki S, Evensen Ø. Non-replicating vaccines. *Fish Vaccination*, 2014, 22–32. https://onlinelibrary.wiley.com/doi/10.1002/9781118806913.ch3

[101] Gudding R, Lillehaug A, Evensen Ø. *Fish vaccination*. Wiley Online Library, 2014.

[102] Munang'andu HM. Intracellular bacterial infections: A challenge for developing cellular mediated immunity vaccines for farmed fish. *Microorganisms*, 2018, 6: 33.

[103] Puvanendran V, Mortensen A, Johansen LH, et al. Development of cod farming in norway: Past and current biological and market status and future prospects and directions. *Reviews in Aquaculture*, 2022, 14: 308–342.

[104] Nardi G, Prickett R, van der Meeren T, et al. Atlantic cod aquaculture: Boom, bust, and rebirth? *Journal of the World Aquaculture Society*, 2021, 52: 672–690.

[105] Li C, Wang S, Ren Q, et al. An outbreak of visceral white nodules disease caused by Pseudomonas plecoglossicida at a water temperature of 12°C in cultured large yellow croaker (Larimichthys crocea) in China. *Journal of Fish Diseases*, 2020, 43: 1353–1361.

[106] Wang L-Y, Liu Z-X, Zhao L-M, et al. Dual RNA-seq provides novel insight into the roles of *dksA* from *Pseudomonas plecoglossicida* in pathogen-host interactions with large yellow croakers (*Larimichthys crocea*). *Zoological Research*, 2020, 41: 410.

[107] Dubey S, Avadhani K, Mutalik S, et al. Edwardsiella tarda OmpA encapsulated in chitosan nanoparticles shows superior protection over inactivated whole cell vaccine in orally vaccinated fringed-lipped peninsula carp (Labeo fimbriatus). *Vaccines*, 2016, 4: 40.

[108] Maisey K, Montero R, Christodoulides M. Vaccines for piscirickettsiosis (salmonid rickettsial septicaemia, SRS): The chile perspective. *Expert Review of Vaccines*, 2017, 16: 215–228.

[109] Wilhelm V, Miquel A, Burzio LO, et al. A vaccine against the salmonid pathogen Piscirickettsia salmonis based on recombinant proteins. *Vaccine*, 2006, 24: 5083–5091.

[110] Soto E, Wiles J, Elzer P, et al. Attenuated Francisella asiatica iglC mutant induces protective immunity to francisellosis in tilapia. *Vaccine*, 2011, 29: 593–598.

[111] Lagos L, Tandberg JI, Repnik U, et al. Characterization and vaccine potential of membrane vesicles produced by Francisella noatunensis subsp. orientalis in an adult zebrafish model. *Clinical and Vaccine Immunology*, 2017, 24: e00557–e00516.

[112] Sommerset I, Krossøy B, Biering E, et al. Vaccines for fish in aquaculture. *Expert Review of Vaccines*, 2005, 4: 89–101.

[113] Biering E, Villoing S, Sommerset I, et al. Update on viral vaccines for fish. *Developments in Biologicals*, 2005, 121: 97–113.

[114] Toranzo A, Romalde J, Magariños B, et al. Present and future of aquaculture vaccines against fish bacterial diseases. *Options Mediterraneennes*, 2009, 86: 155–176.

[115] Kumar G, Byars TS, Greenway TE, et al. Economic assessment of commercial-scale *Edwardsiella ictaluri* vaccine trials in U.S. catfish industry. *Aquaculture Economics & Management*, 2019, 23: 254–275.

13 Development of Fish Vaccines
Challenges and Future Perspectives

A. P. Desbois and S. J. Monaghan
University of Stirling

CONTENTS

13.1 INTRODUCTION

Veterinary vaccines are environmentally sound tools for disease control and they have contributed to the eradication of several notifiable diseases in terrestrial farming, such as Rinderpest virus of cattle and Aujeszky's disease of swine (Morens et al., 2011; Van Oirschot, 1999). Vaccination is also attracting increasing attention for its potential to address antimicrobial resistance (AMR) in human and animal diseases (Micoli et al., 2021), which is an increasingly urgent global concern (Woolhouse et al., 2015). In fish, perhaps the most successful demonstration of the effectiveness of a vaccination program has been the adoption of vaccines in Atlantic salmon (*Salmo salar* L.) aquaculture in northern Europe to protect against the bacterial diseases of furunculosis (Midtlyng, 2014) and vibriosis (Colquhoun and Lillehaug, 2014), with universal uptake of vaccines leading to dramatic reductions in antibiotics use since their wide-scale introduction in the 1980s (Markestad and Grave, 1997; Sommerset et al., 1995). Thereafter, vaccine programs have been implemented effectively in several other economically important teleost species, including European sea bass (*Dicentrarchus labrax* L.), sea bream (*Sparus aurata* L.), turbot (*Scophthalmus maximus* L.), common carp (*Cyprinus carpio* L.), and striped catfish (*Pangasianodon hypophthalmus* Sauvage, 1878) to protect against vibriosis, pasteurellosis, tenacibaculosis, and motile *Aeromonas* septicemia, among other diseases (Ma et al., 2019). Nevertheless, the challenges for developing and implementing fish vaccines, from discovery through to commercial-scale development and registration, are multifarious and span biological, technical, financial, legislative, and social aspects, and encompass human dimensions of behavior, attitudes, and culture. This suite of challenges can be grouped together broadly into those associated with the market, the pathogen, the host, the formulation, assessing efficacy, and pre- and post-launch (Figure 13.1). The aim of this chapter is to describe the key challenges faced when developing a vaccine for a fish disease and to provide perspective on the ways forward to overcome these barriers.

DOI: 10.1201/9781003388548-18

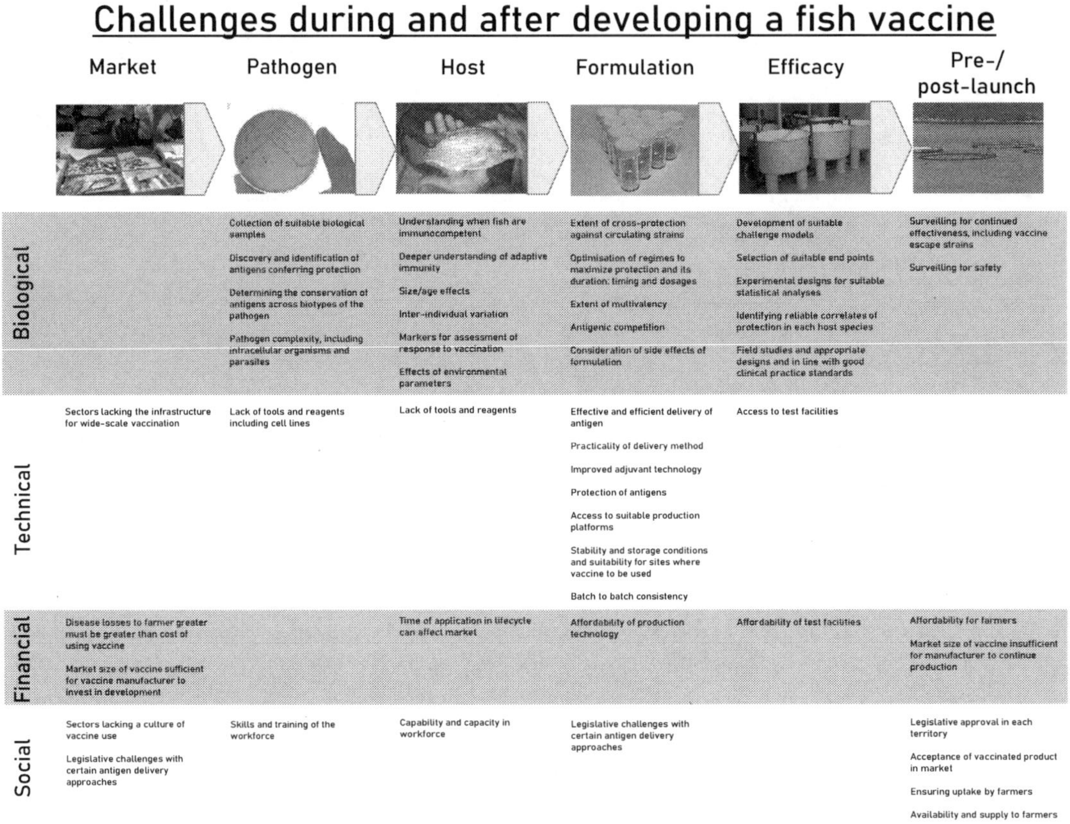

FIGURE 13.1 Summary of the major challenges requiring consideration during and after the development of a fish vaccine. The challenges have been collected together into broad categories depending on the nature of the issue, whether it be biological, technical, financial, or social, and broken down through the development process from initial conception and consideration of the potential market right up to and beyond the launch of a commercial product.

13.2 CONSIDERING THE MARKET FROM THE OUTSET

Several successes have demonstrated the valuable potential of vaccination for many diseases afflicting farmed fish, particularly those caused by microorganisms (Ma et al., 2019), and investment in the development of a vaccine remains driven largely by the financial scale of the problems caused by the disease. In essence, the financial impact of the disease (i.e., mortalities, growth rate retardation, or damage to the fish that reduces saleability and price) and the cost of mitigation measures including treatments, must outweigh the expected financial benefit and incurred expense to using a vaccine, meaning it is more cost-effective to use a vaccine than not (Thorarinsson and Powell, 2006). Factors to be considered here would include the value of the fish, the likelihood of a disease outbreak, the severity of the disease outcome for the fish, and other costs associated with a disease event that may include the need to fallow a site by choice or necessity, such as when the disease is notifiable (Qviller et al., 2020). Yet, even where the application of a vaccine provides a net financial benefit to the fish producer, the scale of the market must be sufficient for the vaccine developer to make the initial investment to explore the possibility to create a vaccine, with justifiable prioritization for larger potential markets, such as sectors culturing species at greater total volume and value.

At early stages of development, public sources of financial support are often used to reduce the risk to the vaccine developer, and this typically facilitates access to expertise in research organizations, including universities, where discovery and proof of principle studies provide a platform for subsequent commercialization.

The demand for vaccines is also influenced by the regulatory framework in the territory where the fish is farmed or intended to be sold, with some countries and sectors far stricter in terms of the types and quantities of chemotherapeutants that can be applied either through legislative controls or the requirements of those purchasing the produce (Okocha et al., 2018). Consequently, territories with more stringent controls on chemical use in aquaculture, or markets where low chemical use is demanded by consumers and implemented through means such as certifying bodies and residue testing programs, offer more favorable conditions for the development of fish vaccines. It is noteworthy that regulatory requirements may be reduced in some circumstances where the application is classified under minor use-minor species limited markets (MUMS) legislation, which exists to support the development of veterinary vaccines for relatively small markets, such as a species produced at low volume or for uncommon diseases of species farmed at greater volume (Cowan et al., 2016). Nonetheless, it is more challenging to justify investment in vaccine development in sectors lacking a culture for vaccination, which therefore will also lack much of the infrastructure necessary for wide-scale adoption of any newly developed vaccines. Such factors must be considered and planned for when embarking upon a vaccine development program and may include efforts to determine the feasibility or acceptability of vaccination to producers and other stakeholders (e.g., Desbois et al., 2021), and the willingness of these parties to invest in local infrastructure such as equipment essential for vaccine storage, transport, and administration.

Looking to the future, vaccination can offer one of the best opportunities to address a particular disease problem, especially where other effective and cost-efficient prophylaxis measures or treatment options are lacking. A notable example is sea lice in Atlantic salmon aquaculture, which remains a perennial and recalcitrant issue for many reasons, including the resistance of these parasites to therapeutic agents, and despite considerable investment in efforts to control the problem through engineering solutions and biological control with cleaner fish (Brooker et al., 2018; Barrett et al., 2020; Guragain et al., 2021). Several recent studies have sought to address this parasitic disease through vaccination, with some success in demonstrating that sea lice antigens may elicit a protective response in Atlantic salmon (Casuso et al., 2022; Contreras et al., 2020), although vaccination against fish parasites remains challenging and has seen very limited commercial success (Shivam et al., 2021). In contrast, considerable strides have been made in the development of vaccines for difficult-to-control notifiable viral diseases, with particular success demonstrated for vaccines to protect against Koi herpesvirus disease (KHVD; caused by the CyHV-3 virus) in carp (Adamek et al., 2022; Boutier et al., 2019), although licensing of the live attenuated vaccines remains a hurdle for registration and commercialization in many places. Still, there will exist many situations where a vaccine is not ultimately the most cost-effective, practical, or efficient means for solving a disease issue, and investment in these circumstances may be directed more usefully to improve infrastructure and other initiatives such as strengthened biosecurity that contribute to reducing the incidence and severity of the disease (Scarfe and Palic, 2020).

Finally, the threat of AMR looms large and may shift the balance further toward encouraging or necessitating the development of more vaccines for fish, and vaccines will become increasingly attractive as diseases become more difficult or expensive to treat. In addition, consumers, retailers, importers, certifiers, and legislators may drive greater investment in finfish vaccine development by insisting on further reductions in the use of chemotherapeutants like antibiotics such as through the introduction of policies and regulations that limit or forbid the use of chemotherapeutants, including by controlling discharge consents, restricting the approval of agents for use in aquaculture, imposing of levies, or incentivizing reduced application via a price premium of the final fish product.

13.3 CHALLENGES WITH THE PATHOGEN

Once the case to develop a vaccine is accepted and investment approved, there is a multitude of biological and technical challenges faced that are associated with the pathogen or parasite to be targeted. First and foremost, it is critical to determine whether the pathogen elicits a response in the host indicative of protective capacity that supports vaccination to be a viable approach. This may be achieved by inoculating the host with the etiological agent and assessing whether a specific immune response is mounted via the production of protective antibodies and cell-mediated immunity. Following this, it is vital to identify immunodominant and conserved antigens associated with the pathogen that elicited the specific response. The dual challenge of seeking to determine both the conservation of the antigen within the population of the etiological agent and its ability to induce long-lasting and specific protection in the host is laborious, costly, and time-consuming, and it often requires the characterization of the population of organisms responsible for the disease by genetic, molecular, and biochemical methods. For example, when developing a vaccine to protect against a bacterium, it is critical to collect 10s to 100s of isolates from diseased hosts that must be identified and authenticated before determining the conservation of antigens across the different biotypes within this collection (e.g., Ni et al., 2010; Ormsby et al., 2016). However, for certain bacteria, not least *Aeromonas* and *Vibrio* species that are responsible for many diseases in farmed fish (Austin and Austin, 2016), genetic heterogeneity is high between the pathogenic strains and this, consequently, can mean antigens are similarly varied and diverse. When taking samples, it is vital to ensure collection of the most important strains in circulation that are truly responsible for the disease problem. Collection of suitable biological material, and the heterogeneity of the disease-causing agent, can pose equally difficult problems for vaccines against viruses, eukaryotic microorganisms, and parasites (Shivam et al., 2021; Mondal and Thomas, 2022).

Each etiological agent comes with its own series of complexities. Ideally, the target organism for the vaccine needs to be propagated in the laboratory and, for parasites in particular, this may require the establishment of an *in vivo* model, which requires full consideration of practical, ethical, and financial issues. Intracellular pathogens, including several bacterial species, can be especially fastidious in their culture requirements (Austin, 2016; Austin and Austin, 2016; Ramírez-Paredes et al., 2017). In addition, viruses require a compatible host cell that permits replication; however, the diversity of fish cell lines available for this task is relatively limited (Ariel et al., 2009). Nevertheless, even if conditions have been determined for culturing the etiological agent of interest, these may be far from optimal, which makes it difficult to generate sufficient material for conventional antigen discovery. Moreover, an additional challenge can be that antigenic profiles may be different between an infection *in vivo* compared to growth in culture medium *in vitro*, such as has been observed for *Aeromonas hydrophila* (Poobalane et al., 2008). In addition, a strain of the protozoan *Paramoeba perurans* (which causes amoebic gill disease) that had become attenuated during subculture *in vitro* had an altered protein profile compared to a recently isolated virulent strain (Dhufaigh et al., 2021). Furthermore, changes in antigens during infection may be observed and, for parasites especially, it is vital to identify the key antigens during different life stages, with an emphasis on those expressed or secreted in the infectious life stage (Hamilton et al., 2018; AmbuAli et al., 2020; Øvergård et al., 2022), or most associated with virulence (Dhufaigh et al., 2021). However, biological understanding for the parasitic life cycle may be rudimentary at best and plugging these fundamental knowledge gaps can be expensive, time-consuming, and often requires a range of specialists working together (Schetters et al., 2016). If sufficient material cannot be generated in controlled settings, such as is often the case with parasites that are rarely culturable *in vitro*, it has to be collected from native hosts in the field but for some diseases achieving this may be unrealistic.

There are several approaches to antigen discovery that aim to save time and money, or can enhance the protective efficacy of current vaccines. For example, reverse vaccinology enables protective antigens to be predicted *in silico* from genetic information of the etiological agent (Bekaert et al., 2021; Ellul et al., 2021; Mondal and Thomas, 2022), thus limiting the need for host-pathogen studies at an early stage of vaccine discovery. Mutagenesis of the pathogen may be a useful way to discover genes important for virulence and this approach has been applied to improve the efficacy and safety of viruses and bacteria in attenuated live vaccines (Pang et al., 2018; Boutier et al., 2019; Zhou et al., 2020). Alternatively, immunoproteomic characterization of immunoreactive antigens can elucidate important antigens for enabling the rational design of cross-protective vaccines (Poobalane et al., 2008; Shahin et al., 2019). These approaches, which can assist the prioritization of antigens for evaluation of protection, have increased in popularity and have the potential to reduce development time and costs but still necessitate the collection of biological samples from diseased hosts. Though predicting and understanding the antigenicity and/or pathogenicity of the target species are crucial steps in rational vaccine design, there remains room for improvement in this process and still the task may be costly to achieve and necessitate expertise in immunoproteomics, genomic technologies, and recombinant biotechnology.

Antigen selection poses a major challenge for vaccine development and the issues raised above ensure that the selection of appropriate antigens for inclusion into a vaccine remains a skillful task, particularly as there are limits on the number of antigens that can be included in a commercial formulation. Many of the key skills in the workforce and access to key technologies that enable antigen selection may be lacking where they are most needed, particularly in low- and middle-income countries where much of global aquaculture is accomplished. This underscores the importance of investing in partnerships and collaborations with relatively wealthier nations to facilitate access and meaningful support, including training and education in fish health (Scarfe et al., 2022).

There remain many important pathogens of fish for which we do not yet have a suitable vaccine and some of the challenges associated with the etiological agents described above go some way to explaining this. Analogous to the situation with antibiotics, where the discovery and development of new antibiotics has become increasingly difficult, many of the diseases for which a vaccine offers a suitable prophylactic solution present complexities that are difficult to tackle, and this hinders the development of a successful commercial vaccine. In addition, the changing climate can be expected to play an important role in the emergence of new pathogens and diseases, where altered conditions extend the normal ranges for etiological agents that bring them into contact with new potential host species (Bricknell et al., 2021; Falconer et al., 2022). Host species may become increasingly susceptible to disease through immune stress induced by unfavorable conditions (Bricknell et al., 2021), and part of the solution to address this unfolding scenario will be the continued development of new fish vaccines to keep producers one step ahead.

13.4 CHALLENGES WITH THE HOST

There remains a major requirement to better understand fundamental aspects of fish immunity, including responses to vaccination, but a serious challenge is presented by fish being the most diverse group of vertebrates (Helfman et al., 2010). Despite possessing the basic features of vertebrate immunity, such as key immune cell types (e.g., B- and T-lymphocytes, granulocytes, neutrophils, and macrophages) and the presence of antigen-detecting receptors (e.g., Toll-like and B- and T-cell receptors) and associated response pathways (e.g., Th1, Th2, and Th17), there is significant variation in immune functionality between and within the three major evolutionary branches of fishes (Rast and Buckley, 2013). An often-cited anomaly is the lack of major histocompatibility complex (MHC) II in Atlantic cod (*Gadus morhua* L.) (Starr and Jentoft, 2012), which is a key component of the adaptive immune system responsible for antigen presentation and important for vaccination, and it means that though cod seem unable to produce a specific antibody response, apparently they can still be protected by vaccination (Mikkalsen et al., 2004; Schrøder et al., 2006). The vast number of fish species cultured across the world complicates efforts to better understand fish vaccination, and this situation contrasts starkly with terrestrial food production that relies on farming just a handful of cornerstone species.

Somatic hypermutation may occur in fish and this enables the production of higher affinity specific antibodies, although there is a limited repertoire of antibodies in fish compared to mammals (Pasquier, 1982). Fish possess three major B-cell lineages of immunoglobulin, where one lineage expresses IgD and a second includes B-cells that co-express IgD and IgM isotypes, with the latter constituting the predominant systemic humoral antibody response. A third B-cell lineage expresses IgT, which is predominantly mucosal, although IgT has yet to be identified in some fish species such as catfishes (Hansen et al., 2005; Zhang et al., 2010). IgD has received relatively little attention, largely because these antibodies are not considered to play an important role in vaccination (Yu et al., 2020). In contrast, fish IgM responses have been well studied in the immune response to vaccination and titers have been shown to correlate to inoculation dose of bacterial whole-cell and surface antigens, such as *Aeromonas salmonicida* in vaccinated Atlantic salmon (Erdal and Reitan, 1992) and *Yersinia ruckeri* in vaccinated rainbow trout (*Oncorhynchus mykiss*, Walbaum 1792) (Riada et al., 2011). Indeed, passive transfer of fish IgM from vaccinated to naïve fish can confer protection in several species to a range of pathogens (reviewed by Munang'andu and Evenson, 2019). Titers of systemic IgM in fish serum are a conventional approach to assess vaccine performance in several fish hosts and enzyme-linked immunosorbent assays (ELISAs) have been developed specifically for this purpose. Research on the importance of IgT in mucosal defense of fish remains in its infancy, but evidence has shown that IgT is associated with protective responses in the skin of rainbow trout infected with the protozoan skin parasite *Ichthyophthirius multifiliis* (Xu et al., 2013), and other studies since have demonstrated IgT responses to mucosal vaccination (Ballesteros et al., 2014; Hoare et al., 2017). Whether IgT responses play a role in protective immunization of fish requires further investigation, but progress in this field remains hampered by a lack of reliable tools to detect IgT levels.

Immune competence (i.e., the ability to respond to a vaccine and produce protective antibodies) is a major challenge for vaccination strategies for cultured fish, especially as it is often imperative to protect against pathogens from hatchery stage rearing. The potential of each fish species to respond to vaccination must be determined through improved understanding of early functional immunogenic ontogenesis. Zebrafish (*Danio rerio* Hamilton, 1822) have been a particularly useful model for this, and it is clear that functional rather than morphological evidence of lymphoid maturation is required to establish immune competence (Zapata et al., 2006). Indeed, the need for this evidence is underscored by observations from several vaccination studies in juvenile fish that have yielded contrasting results. In salmonids, evidence for early adaptive immunity was demonstrated in 1–2.5 g rainbow trout, coho salmon (*Oncorhynchus kisutch*, Walbaum 1792), chinook salmon (*Oncorhynchus tshawytsch*, Walbaum 1792), pink salmon (*Oncorhynchus gorbuscha*, Walbaum 1792), chum salmon (*Oncorhynchus keta*, Walbaum 1792), and sockeye salmon (*Oncorhynchus nerka*, Walbaum 1792) after immersion vaccination with inactivated whole cells of *Vibrio anguillarum* and *Y. ruckeri*; however, differences in protective immunity were evident between the species and protection was shorter in duration compared to individuals vaccinated at >4 g (Johnson et al., 1982). Similarly, protective efficacy of an immersion vaccine against vibriosis was weaker when delivered to 1 g Atlantic cod fry compared to fish at 5 g (Schrøder et al., 2006). Finally, ballan wrasse (*Labrus bergylta* Ascanius, 1767) were not protected against atypical strains of *A. salmonicida* when vaccinated by immersion at <2 g (Papadopoulou et al., 2022), but protection was observed when the vaccine was administered by intraperitoneal injection to 25 g individuals (Ramírez-Paredes et al., 2020).

Being poikilothermic, important disparities exist between cold-water and warm water species regarding the induction of the adaptive response and antibody production, which can have important implications for vaccine prime and boost strategies (Abram et al., 2017). Moreover, appropriate seasonal timing of vaccination must be considered because immunologically permissive temperatures vary across species, and this is especially important for prime vaccinations. While 4°C is permissive in Atlantic salmon, for carp and catfish permissive temperatures are 14°C and 22°C, respectively (Bly and Clem, 1992). Indeed, lower temperatures can suppress antibody production (Raida and Buchmann, 2007; Lorenzen et al., 2009) and this immune suppression is proposed to act primarily via T-helper cells, which renders fish poorly responsive when vaccinated with T-cell-dependent antigens at nonpermissive temperatures (reviewed in Le Morvan et al., 1998; Abram et al., 2017). The incorporation of suitably stimulating adjuvants into the vaccine formulation is especially important when temperature conditions are suboptimal.

Another challenge when vaccinating large populations of fish is the variation in responses between individuals where significant diversity of IgM repertoires exist, that is, the ability to produce high-affinity antibodies to antigens (Mora et al., 2010). Selective breeding for disease resistance has proved effective in Atlantic salmon, including against infectious pancreatic necrosis (Houston et al., 2010), and genetic inheritance of fish with stronger adaptive immune responses offers a promising avenue for future improvement of vaccination strategies. To realize this potential, it is imperative that improved genomic resources for cultured fish species are made available, and this would facilitate better understanding of disease resistance more generally, while also aiding understanding of immune responsiveness to vaccines and adjuvants.

In summary, there is much to be learnt of fish responses to vaccination, not least understanding species and inter-individual differences, including genetic explanations, the influences of environmental conditions, and determining regimens that ensure that the fish are immunocompetent and in an immunopermissive state. Significant investment is required to address these fundamental knowledge gaps, with resources prioritized toward augmenting our knowledge of the main species farmed globally and deemed suitable for the development of future vaccination programs, such as carps, tilapias, catfishes and Atlantic salmon. The availability of genomic information for cultured species requires attention; such resources provide considerable opportunity to better understand immune mechanisms. Given the influence of environmental conditions on the effectiveness of vaccination, especially temperature, the impacts of climate change will need to be considered, monitored, and understood, so that appropriate mitigation steps can be identified as our climate continues to alter. The roles played in effective vaccination by diet and husbandry will provide rich opportunities to improve the efficacy of vaccines, including through the development of immunomodulating diets and probiotics (Mendivil, 2021), while interplay between the innate and adaptive arms of the fish immune system, as well as the roles played by constituents of the microbiota, are primitive areas of research at present but will offer new insights into making vaccination even more effective. Finally, the complexities between each of these factors and how they influence each other will become increasingly apparent as our knowledge base of host immunity is enhanced and such understanding will improve our ability to protect the fish up to harvest against relevant infectious diseases.

13.5 CHALLENGES WITH FORMULATION

There are several challenges in fish vaccine development that may be collected together and considered to be associated with aspects of the formulation, including the different constituents of the vaccine and their concentrations, methods of administration, and the platform technology used for vaccine manufacture.

Like other vaccines, fish vaccines contain the antigen or antigens in various forms, with a decision made early in development on the monovalent or multivalent nature of the formulation, and each approach has its own set of benefits to counterbalance challenges that must be overcome (reviewed by Busch, 1997). Options in this regard include: a subunit vaccine, which contains a protective antigen that may be expressed and produced in a heterologous microbial host and purified; a vaccine relying on delivery of nucleic acid (i.e., DNA or mRNA), which requires the host to express the encoded antigen to elicit protection against the pathogen (Pardi et al., 2018) and this approach has attracted recent attention due to several Covid-19 vaccines utilizing this technology; and a whole cell vaccine, where whole organisms or virus particles are in the vaccine itself, which may be live but attenuated to make them avirulent, or inactivated to render them noninfectious, such as by heat or chemical treatment. Inactivated whole cell vaccines have been, and continue to be, the bedrock approach of fish vaccines that protect predominantly against bacteria and some viruses, since fish prophylaxis was first conceived by Duff (1942). Still, efficacy of whole-cell bacterial vaccines can vary depending on the pathogen, with nonadjuvanted preparations of *Y. ruckeri* protecting against ERM but inactivated whole cells of *A. salmonicida* only providing adequate protection once an appropriate adjuvant is included (reviewed by Midtlyng, 2016 and Bridle and Nowak, 2014). Conventional inactivated whole-cell vaccines may be effective for some intracellular and fastidious bacterial species, such as *Francisella noatunensis* subsp. *orientalis* in Nile tilapia (*Oreochromis niloticus* L.; Shahin et al., 2019), but less efficacious for others (e.g., *Pasteurella atlantica* in Atlantic lumpfish, *Cyclopterus lumpus* L.; Ellul et al., 2019). The delivery of inactivated viruses and bacteria is the approach that underpins the rationale of autogenous vaccines, which are produced using a specific pathogen isolated from an infected fish at a specific site (Barnes et al., 2022). Autogenous vaccines are used widely in fish and they can be highly effective, although they may suffer from offering a relatively low level of cross-protection against other strains of the pathogen that may be in circulation. Indeed, the degree of cross-protection offered against the circulating virulent or infectious strains of the pathogen must be investigated for any commercial fish vaccine, and ultimately this may necessitate the inclusion of further antigens to provide sufficiently broad protection.

Fish vaccines can be administered by injection, usually into the intraperitoneal cavity, or to the mucosa either by immersing the host in the vaccine to target the surface mucosa or constituted in the feed to allow delivery to the gut mucosa. Confirmation and quantification of antigen uptake can be achieved by several methods and may include visualization by microscopy, for example, immunohistochemical detection of specific pathogen or vaccine proteins (Garver et al., 2005), or fluorescently-labeled antigen (Fernandez-Alonso et al., 1999), or bioluminescence offered through *in vivo* imaging (Boutier et al., 2015). The decision on the planned mode of delivery is often taken early, with mucosal delivery, typically by dip vaccination, preferred increasingly because it is noninvasive. However, mucosal delivery presents considerable challenges, particularly to ensure that sufficient antigen is delivered where it is required and that the adaptive response elicited protects sufficiently. In particular, steps may be taken to protect the antigens in vaccines intended for oral mucosal delivery from unwanted degradation in the stomach, and several technologies and approaches may achieve this goal including encapsulation in nanoparticles (Ji et al., 2015). Immune tolerance, where a mucosal vaccine leads to unresponsiveness to otherwise immunogenic antigens, thus eliciting the opposite effect to that desired, is another risk requiring consideration, especially with regard to antigen dose, and this can be especially problematic with vaccines delivered in feed (Mutoloki et al., 2015; Embregts and Forlenza, 2016).

Many vaccines are most effective when a booster dose of the antigen is given, while for some vaccines the administration of a booster is essential to achieve adequate protection as is the case for Atlantic salmon vaccinated against salmon rickettsial syndrome (Tobar et al., 2015). Unlike in mammals, class switching of IgM to higher affinity IgG antibodies does not occur in fish (reviewed in Magadan et al., 2015), meaning booster vaccine strategies require careful consideration for certain vaccines, particularly where systemic IgM antibody levels may wane over time, or where antibodies are produced at low titer or are of only low affinity. Indeed, it is important to establish not only the level of protection conferred by a vaccine but also how this reduces over time. The booster dose may or may not be delivered by the same mechanism as the initial dose; for example, some vaccination regimes for ballan wrasse against bacterial pathogens deliver the antigens by immersion initially, followed by intraperitoneal injection of the vaccine to boost protection shortly before deployment to Atlantic salmon farms (Ramírez-Paredes et al., 2020). Like the timing of the initial vaccine, which needs to account for host factors, including the development of immune competency and ensuring protection is elicited before the host is at risk of contracting the disease, the timing of a booster vaccine must be investigated such that the regimen provides sufficient long-term protection. Oral booster vaccination offers a desirable strategy and in-feed booster vaccines against infectious salmon anemia (ISA) and *Piscirickettsia salmonis* have been shown to maintain antibodies in Atlantic salmon that correlate with protection (Tobar et al., 2015). Approaches to augment a relatively poor immune response to immersion vaccination includes the use of hyperosmotic solutions, immunostimulants, and adjuvants (Bøgwald and Dalmo, 2019).

The stability of antigens and shelf-life require consideration, particularly for vaccines relying on recombinant proteins or other antigen subunits, and this is often overlooked until later stages of development, though there is reasoned argument to consider this issue earlier in development. Approaches to antigen protection, such as nanoparticle encapsulation, may be employed as a means to enhance antigen stability and product shelf-life.

A further issue that must be determined when formulating a vaccine is the concentration(s) of antigens that elicits the desired degree of protection in the fish population, and consideration must be given to the prevalence of associated side effects. After all, production costs may be higher for a vaccine that requires greater quantities of antigens. One novel way to address low titers of an attenuated deletion mutant of CyHV-3 was to use a strain adapted to cell culture conditions (Schröder et al., 2019). Allied to concentration concerns, consideration must be given to the number of antigens that are included in a vaccine, that is, valency. One risk is that protection may be impacted detrimentally by including too many antigens, especially with whole-cell vaccines, as antigenic competition may be encountered, where antigens compete and ultimately the protection elicited in the host differs from one antigen to another (Killie and Jørgensen, 1994; Nikoskelainen et al., 2007; Kuang et al., 2022). Antigenic competition may be influenced by many factors associated with the host and the pathogen and this challenge is attracting increasing research attention. The number of antigens to include in a vaccine will also need to factor in cost considerations and vaccines manufacturers may face practical constraints on the number of antigens that can be produced for a particular vaccine. Similarly, access to, and the availability of, suitable facilities for antigen production must be considered, with access to such sites reduced in many countries in the Global South (Poppe and Koppang, 2016). The introduction of more efficient and inexpensive platforms for vaccine manufacture is a considerable global challenge that is not unique to the production of fish vaccines.

Finally, considering that most fish vaccines are inactivated pathogen formulations, they are relatively weakly immunogenic and require appropriate adjuvants to enhance the immune response to increase the efficacy of protection. Furthermore, enhancement of innate immune responses through adjuvants can drive greater adaptive immunity in fish, for example, the triggering of toll-like receptor TLR-4 with lipopolysaccharide that stimulates naïve T-helper cells to induce T- and B-cell effector immune responses (Schijns and Lavelle, 2011). Despite recent advances in the development of improved adjuvants for fish vaccines that offer greater protection and enhanced safety profiles, further improvements and deeper understanding for their mechanisms of action are necessary to unlock the full potential of vaccination programs (Tafalla et al., 2014). Obtaining a sufficient antigen depot effect (i.e., suitably slow release of the antigen) and inducing cell-mediated and long-lasting antibody responses, while minimizing the induction of adverse side effects such as adhesions, are key goals for future adjuvants. Deserving of special attention in this regard are adjuvants for the delivery of antigens to mucosal surfaces (Tafalla et al., 2014).

Among the challenges grouped together as those associated with vaccine formulation are several that, once addressed successfully, will realize profound benefits for vaccination in fish farming. There is a strong desire to deliver antigens to mucosal surfaces, especially orally (Mutoloki et al., 2015), as this can be, or is perceived to be, more practical and vaccines can be administered at an early stage in the fish life cycle. Generally, mucosal delivery leads to fewer side effects and is not as stressful for the fish as administration by injection, particularly if the vaccine is provided in the feed. To achieve the full potential of mucosal vaccines, advances must be made in technologies to facilitate sufficient antigen uptake, which may include investment in the development of improved adjuvants and approaches that have the potential to enhance and target antigen delivery such as nanoparticles. Moreover, subunit vaccines, including the production of antigens using conventional recombinant expression technology platforms such as yeasts and bacteria, may be an effective way to improve safety profiles of vaccines, while helping to address limits on valency and antigenic competition. Rational design of live attenuated vaccines for fish diseases is still an attractive approach for some pathogens given the better safety data and strong induction of cellular and humoral arms of the immune system (Boutier et al., 2015; Pang et al., 2018; Schröder et al., 2019), but face legislative prohibition in many regions. The opportunities offered by mRNA and DNA vaccines are exciting and this technology opens new possibilities in the future of fish vaccines, not least because DNA vaccines have proven effective for some fish viruses (Dalmo, 2018), even by oral administration (Hølvold et al., 2014). Though more research on the safety of nucleic acid-based vaccines is warranted, this approach to vaccination could hasten the development timeline, allowing for faster translation of a vaccine from one host to another.

13.6 CHALLENGES WITH ASSESSING EFFICACY

Vaccine safety and immunogenicity *in vivo* is relatively simple to accomplish, but a major challenge when embarking on the development of a vaccine is to ensure a suitable challenge model is available to assess efficacy of protection. Challenge models are unavailable for many fish diseases and, even where models have been developed, these are often suboptimal, do not closely mimic the route of infection or progression of the disease in natural conditions, and may suffer a plethora of other shortcomings. Key to any challenge model is reliability and reproducibility, with an ideal model giving the desired experimental outcome (e.g., progression of expected clinical signs or morbidity) in a predictable way. However, with so many factors impinging on the fulfillment of this simple aim, not least the population of fish being used and their life histories, often models are difficult to reproduce reliably even in the same system operating under similar conditions, as highlighted by variations between replicate tanks (Hetland et al., 2011; McBeath et al., 2015). Indeed, many variations in experimental challenges can result in different disease and pathology outcomes, for example, for *A. salmonicida* and CyHV-3, cohabitation, intramuscular, intraperitoneal, and immersion challenge models have all yielded variable infectivity and little reproducibility (Nordmo and Ramstad, 1997; Embregts et al., 2019).

Even where a challenge model exists for a disease, this might not have been established in the species for which the vaccine is being developed, meaning work must be undertaken to develop the model for the species of interest (e.g., Macchia et al., 2022). Indeed, the lack of disease models in target species became a particular challenge with the onset of the mass production of cleaner fish like lumpfish and ballan wrasse for the Atlantic salmon sector, which necessitated the development of several challenge models to support the vaccination programs required for these newly farmed species (Ellul et al., 2019; Papadopoulou et al., 2021). This problem will be encountered when aiming to farm any species commercially for the first time, and it can prove to be a major bottleneck to identifying effective solutions for disease problems. Even where the target species is available and can be acquired, it cannot be taken for granted that fish are available at the desired life stage, with many species only available at certain sizes a limited number of times per year (e.g., Atlantic salmon, which are produced in cohorts up to four times per year). Moreover, the genotypes of the fish used in challenges may be unknown or uncharacterized, which contrasts with many mammalian challenge models, and can impact on the reliability and reproducibility of the model (Herati and Wherry, 2018).

Challenge models aim to closely follow the natural route of infection and progression of disease and the more closely this can be accomplished the more confidence there is that the model will reflect the efficacy of the vaccine in the field. For many diseases, a model where the pathogen is delivered to the host by immersing it in contaminated water is more desirable than one where the pathogen is injected directly into the fish. Indeed, immersion challenge of Atlantic salmon with ISA virus with low and high virulence strains induced differential disease progression and pathogenesis that were indistinguishable by injection challenge (McBeath et al., 2015). Models that include shedders (i.e., fish infected with the pathogen and cohabited with naïve counterparts) often provide an infection route even closer to natural infection conditions.

Typically, ensuring sufficient experimental power in fish challenge studies is difficult to achieve because the tank is the experimental unit. Many trials are run in duplicate or triplicate tanks due to resource and cost limitations meaning that they often fail to reach the power necessary for robust statistical analysis. Also, fish trials often require a minimum number of individuals per tank, and this usually means the total number of fish required is far in excess of requirements in mammalian studies. Tank effects, whereby fish in different tanks respond differently even under the same conditions, can exert major influence on the study outcome (Speare et al., 1995). The removal of individuals from tanks, such as mortalities or just through sampling, may impact on the performance and behavior of the other fish, with biomass density an important factor influencing many aspects of fish biology. Some fish challenge studies have included additional tanks that are used solely for collecting fish for samples throughout a trial to reduce the impact of removing individuals from tanks, but this approach suffers the limitation that samples are collected from a tank that will itself be subject to tank effect influence. For parasite vaccine challenge trials in particular, challenging control and vaccinated fish in the same tanks can provide a good assessment of efficacy, although motile parasites such as sea lice can move between hosts, potentially masking the effects of vaccination at any given point in time (Contreras et al., 2020). Other researchers have conducted trials by housing individuals in separate tanks to improve statistical testing power and reliability of observations (particularly in motile parasite trials), and to reduce overall number of fish required (Hamre et al., 2011). Fish challenge studies may have fewer control groups than experimental groups, which may be justified in ethical terms by reducing the total number of animals required to achieve the aim of an experiment. Clearly, a pragmatic approach must be taken in fish challenge trials to experimental design and statistical analyses, with a balance needing to be achieved that provides confidence in the experimental findings while avoiding the need for the use of excessive numbers of individuals. This skillful task requires careful consideration of many factors and consultation with those knowledgeable in the disease, the challenge methodology, and experimental system to be used. Many of the problems described above are exacerbated in field studies that aim to assess the effectiveness of a vaccine under the conditions where it is expected to be used.

Proof of efficacy is a requirement when obtaining authorization to market a vaccine and this is typically obtained from *in vivo* challenge trials. However, there are continuing efforts to minimize the requirement for *in vivo* disease challenges when establishing vaccine efficacy (Midtlyng et al., 2011) by developing *in vitro* methods that achieve this, and such approaches adhere to the guiding principles of the 3Rs (replacement, reduction, and refinement of the use of animals in research). Indeed, quantification of IgM antibodies in fish serum by ELISA has been used effectively as a correlate of protection in Atlantic salmon against *A. salmonicida* (Romstad et al., 2013) and in carp against CyHV-3 (Schröder et al., 2019). However, the usefulness of antibody abundance as a surrogate for vaccine response must be established for each host and pathogen species interaction, as this is not always a reliable measure (Munang'andu and Evensen, 2019). For example, although Atlantic salmon produce a relatively poor antibody response to lower doses of inactivated ISA vaccine, they still show good protection against the disease (Lauscher et al., 2011), whereas a strong antibody response in Atlantic lumpfish to *Pasteurella* sp. vaccination does not equate to high protection (Ellul et al., 2019). More research is warranted to establish reliable correlates of protection when assessing vaccine protection, with successful validations reducing the need for challenge trials (Munang'andu and Evensen, 2019). Moreover, greater acceptance for *in vitro* assessments of vaccine efficacy is desirable for registration applications, as they may reduce development costs while being more ethically acceptable.

There are several other challenges faced by vaccine developers when assessing the efficacy of a fish vaccine. First, the availability of suitable facilities to test fish vaccines can present a serious limitation, with affordability to perform trials another important factor requiring consideration. Even if facilities can be accessed, the transfer of a challenge model to another facility can be demanding and the expertise and resources needed to achieve this should not be underestimated. Second, field trials, where vaccines are tested in similar conditions to where they will be used commercially, are costly and the data from these trials are often not made available due to understandable commercial sensitivities. However, greater transparency of field trial data, particularly those that failed to demonstrate efficacy, may help to address reasons for failure, including weaknesses in experimental designs, and lead to future improvements.

Looking forward, access to suitable *in vivo* fish challenge models will continue to be crucial when developing vaccines for infectious diseases. As described above, many aspects during the development of a reliable and reproducible challenge model for a particular host and disease can be frustrating and will often require rounds of repeated experiments. For decades, 60% mortality in the control group has been an accepted requirement for vaccine efficacy trials (Amend, 1981), with data still often presented as relative protective survival (RPS), but these concepts require reconsideration. In many countries, mortality is not deemed to be an ethical endpoint for experiments and instead it is encouraged to use the presence and severity of clinical signs or microbiological assessment. Moreover, for some diseases for which vaccines are being developed, it simply has not been possible to achieve such a high level of mortality, such as salmon alphavirus (SAV), where other parameters of efficacy assessment are relied upon including histopathology and viral abundance (Aksnes et al., 2021). The development of replacements to fish challenge models, including *in vitro* technologies and the use of alternative hosts, is a deserving area of research, and such approaches can offer reliable assessments of virulence for fish pathogenic bacteria (McMillan et al., 2015; Djainal et al., 2020). Finally, the determination of reliable correlates of protection for major diseases and hosts, including the advancement of molecular tools derived from a deeper understanding of fish immunology, offers a major opportunity to improve our ability to assess vaccine efficacy.

13.7 CHALLENGES PRE- AND POST-LAUNCH

There are several challenges faced by the vaccine producer just before and after the commercial launch of the product, and consideration may need to be given to the acceptability of the final product to the consumer. Nevertheless, one of the most important hurdles during vaccine development is receiving legislative approval to market and use the vaccine in the territories where it will be applied. This can be a burdensome task and applications need to be prepared carefully with experts familiar in the process, especially because legislation and the requirements of the competent regulatory authorities can vary considerably. In particular, and as highlighted above, some regions do not approve the use of live attenuated vaccines, which may limit the market. For example, though the live attenuated vaccine for KHVD is approved for use in Israel, the USA, and Indonesia, it is not approved in Europe (Dishon et al., 2014), while the DNA vaccine for infectious hematopoetic necrosis virus was approved initially only in Canada (Alonso and Leong, 2013). The major tasks required for attaining the necessary approvals for a vaccine are well described by Holm et al. (2014) and typically an application will consist of a detailed description of vaccine production and quality control measures, in addition to documentation on safety and efficacy.

Of the numerous challenges requiring consideration once a vaccine has been approved and marketed, one of the most important is the need to be vigilant for adverse events that may only become evident during a large-scale roll-out of the new vaccine. Vigilance is especially important if the vaccine technology is associated with particular risks; for example, in theory, a live attenuated vaccine may undergo genetic reversion and cause infections (Boutier et al., 2019). It is key for the producers and end users to work closely together to record all occurrences of adverse events, particularly for a new vaccine or one that has altered in formulation, so that problems are identified early, investigated, and addressed. Examples of adverse events include the formation of adhesions that may occur at the injection site through binding between the gut wall and internal organs; opportunistic infections at the site of injection; mortality in excess of that expected due to handling during vaccine administration; and changes in fish behavior indicative of stress (Evensen et al., 2005; Holm et al., 2014). Vaccine users may encounter the same vaccine inducing more than one type of side effect and they need to be alert to this possibility. Like all other vaccines, some side effects may be acceptable if they are minor in severity or observed only at very low levels, and it is vital to recognize when the side effects are unacceptably high in abundance or severity and thus warrant closer scrutiny. The development and application of scoring schemes, such as the Speilberg scale that is used for adhesions, are helpful for the task (Midtlyng et al., 1996; Tziouvas and Varvarigos, 2021). Adverse events can result from any of the components of the vaccine, including the antigens and the adjuvant.

Monitoring of vaccine effectiveness is important for ensuring continued success and losses in overall protection and will require investigation. Even if a vaccine has been used exactly as intended, many reasons can underlie reduced performance, including genetic changes in the pathogen in the field that affect recognition by the host (e.g., vaccine escape strains) and unrecognized stress in the vaccinated population during administration. Continued surveillance of efficacy and collection of samples from disease outbreaks are important to understand whether changes in the pathogen or parasite are countering the protection offered by the vaccine. Of course, the problem reported may not be associated with the failure of the vaccine at all, but, even if this is the case, doubts over effectiveness can persist unhelpfully in the user's mind and such concerns may be difficult to assuage. In addition, it can be challenging to collect sufficient reliable data to inform effectiveness on farms due to low power or nonblinded experimental designs, and the ways to obtain such data deserve more consideration (Mitchell, 1997).

It is vital that the vaccine is made available to the fish farmers and that steps are taken to make sure supply is affordable. Where there are expected benefits, efforts will be taken to encourage vaccine application and this may require incentivizing and training, particularly where there has not been a culture of vaccine use previously. The infrastructure needs to be in place to deliver the vaccine to the end user and to enable it to be administered correctly. Furthermore, it is essential that sales of the vaccine remain sufficient to justify continued production and, for whatever reason, do not decline to levels where the vaccine becomes financially inviable to manufacture, as has been the case for certain vaccines for rainbow trout in the United Kingdom.

Vaccine producers need to take steps to ensure that the vaccine is used in the correct way by users, as inappropriate use may jeopardize confidence if the vaccine is perceived to be less effective than expected. Provision of training and education is important and materials need to be delivered by an effective means to ensure that the vaccine is used as intended and under conditions likely to provide the desired protection. Practices that will impact vaccine effectiveness include vaccinating fish that are in a stressed condition or have infected individuals in the population; delivering the vaccine to fish of the wrong age or size; diluting the vaccine to decrease the cost per fish; or concomitantly administering antibiotics as a prophylactic for opportunistic injection-site infections. These measures can lead to reduced protection and are ultimately likely to prove false economy. Thus, it is imperative that vaccine producers cultivate relationships with end users as early as possible in the process to build trust, particularly where vaccination is not a common or accepted practice. This process can play a role in providing information that allows for understanding the drivers of vaccine uptake or hesitancy, as this insight can help to increase uptake by the intended end users.

As we look forward, one of the greatest challenges faced is convincing fish farmers who will benefit from a vaccine to invest in using it consistently. Where disease outbreaks do occur in vaccinated stocks, seeds of doubt may be sown in the farmer's mind around vaccine effectiveness, and it is vital to identify the underlying cause to provide reassurances. However, around the world, very often the resources required to achieve a diagnosis are lacking, leaving the farmer uncertain as to the reason for the outbreak and the protection offered by the vaccine. Broadening provision and access to reliable and accurate diagnostics is an area deserving of far greater investment in aquaculture globally. Still, no vaccine can guarantee complete protection against a disease and, beyond occasional cases in the population, where outbreaks are confirmed to be due to the target of the vaccine, these must be investigated to arrive at the reason for failure. There can be many explanations as to why a vaccine did not and the vaccine may not protect universally against all strains of the pathogen, which may necessitate reevaluation of the antigens contained within. Nevertheless, even where the reasons for failure are not intrinsic to the vaccine, disease outbreaks in vaccinated stocks, particularly by the target pathogen, can devastate farmer confidence. Similarly, the side effects induced by the vaccine and its administration may be perceived to be too high, which may also lead to a farmer deciding not to use the vaccine the next time. It is imperative that vaccine suppliers work with customers to provide continued reassurance around the cost-effectiveness of using the product, especially where doubts are raised. At the other end of the spectrum, the complete absence of disease outbreaks in vaccinated stocks over the long term may lull farmers into a false sense of security by incorrectly assuming that the disease is no longer a problem. In these circumstances, farmers may attempt to cut costs by deciding not to apply the vaccine and, though this may provide a short-term benefit, it presents a risk for the reemergence of the pathogen. The relationships between vaccine suppliers and the customers are extremely important to the success of a vaccine and these must be cultivated toward a high level of trust that will deliver mutual benefits. Finally, the Covid-19 pandemic is providing a major insight into attitudes toward vaccines and their use, including the key drivers of uptake and hesitancy, with lessons learned from these studies potentially applicable to improve vaccine use and uptake in aquaculture.

13.8 CONCLUSION AND GENERAL OUTLOOK

Vaccination programs offer an excellent prophylactic option to protect fish against specific disease challenges but they are not a silver bullet and, in many circumstances, more effective and less resource-intensive solutions may exist, such as improved biosecurity, better water quality, breeding for reduced disease susceptibility, use of higher quality feeds, and better all-round husbandry practices. However, the success of fish vaccination programs, as demonstrated most notably by the Atlantic salmon sector of northern Europe, demonstrates considerable potential for effective wide-scale vaccination programs in finfish aquaculture. Nevertheless, many challenges are encountered during the journey from the conception of a fish vaccine through to and beyond its approval and commercial release and the most important of these have been described in this chapter. There are many opportunities to improve the effectiveness and availability of fish vaccines by addressing these challenges, and it is critical that researchers with the key skills needed for each of these tasks continue to be trained and made available to the organizations that need them. Finally, readers are directed to several recent reviews on the development of fish vaccines, most notably Embregts and Forlenza (2016), Adams (2019), and Ma et al. (2019), which provide additional thoughts and perspectives on the main challenges faced in this important area.

REFERENCES

Abram, Q. H., Dixon, B. and Katzenback, B. A. 2017. Impacts of low temperature on the teleost immune system. *Biology* 6:39. https://doi.org/10.3390/biology6040039

Adamek, M., Matras, M., Rebl, A., Stachnik, M., Falco, A., Bauer, J., Miebach, A. C., Teitge, F., Jung-Schroers, V., Abdullah, M., Krebs, T., Schröder, L., Fuchs, W., Reichert, M. and Steinhagen, D. 2022. Don't let it get under your skin! - Vaccination protects the skin barrier of common carp from disruption caused by cyprinid herpesvirus 3. *Frontiers in Immunology* 13:787021. https://doi.org/10.3389/fimmu.2022.787021

Alonso, M. and Leong, J. A. 2013. Licensed DNA vaccines against infectious hematopoietic necrosis virus (IHNV). *Recent Patents on DNA & Gene Sequences* 7:62–65. doi: 10.2174/1872215611307010009

AmbuAli, A., Monaghan, S. J., McLean, K., Inglis, N. F., Bekaert, M., Wehner, S. and Bron, J. E. 2020. Identification of proteins from the secretory/excretory products (SEPs) of the branchiuran ectoparasite *Argulus foliaceus* (Linnaeus, 1758) reveals unique secreted proteins amongst haematophagous ecdysozoa. *Parasites & Vectors* 13:88. https://doi.org/10.1186/s13071-020-3964-z

Ariel, E., Skall, H. F. and Olesen, N. J. 2009. Susceptibility testing of fish cell lines for virus isolation. *Aquaculture* 298:125–130. https://doi.org/10.1016/j.aquaculture.2009.10.003

Aksnes, I., Braaen, S., Markussen, T., Åkesson, C.P., Villoing, S., Rimstad, E. 2021. Genetically modified attenuated salmonid alphavirus: A potential strategy for immunization of Atlantic salmon. *Journal of Fish Diseases* 44:923–937. https://doi.org/10.1111/jfd.13352

Austin, B. 2016. The value of bacterial cultures to modern fish diseases microbiology. *Bulletin of the European Association of Fish Pathologists* 36:2–10.

Austin, B. and Austin, D. A. 2016. *Bacterial fish pathogens*, 6th edition. Springer Nature: Basel.

Ballesteros, N. A., Rodriguez Saint-Jean, S. and Perez-Prieto, S. I. 2014. Food pellets as an effective delivery method for a DNA vaccine against infectious pancreatic necrosis virus in rainbow trout (*Oncorhynchus mykiss*, Walbaum). *Fish and Shellfish Immunology* 37:220–228. https://doi.org/10.1016/j.fsi.2014.02.003

Barnes, A. C., Silayeva, O., Landos, M., Dong, H. T., Lusiastuti, A., Phuoc, L. H. and Delamare-Deboutteville, J. 2022. Autogenous vaccination in aquaculture: A locally enabled solution towards reduction of the global antimicrobial resistance problem. *Reviews in Aquaculture* 14:907–918. https://doi.org/10.1111/raq.12633

Barrett, L. T., Oppedal, F., Robinson, N. and Dempster, T. 2020. Prevention not cure: A review of methods to avoid sea lice infestations in salmon aquaculture. *Reviews in Aquaculture* 12:2527–2543. https://doi.org/10.1111/raq.12456

Bekaert, M., Goffin, N., McMillan, S. and Desbois, A. P. 2021. Essential genes of *Vibrio anguillarum* and other *Vibrio* spp. guide the development of new drugs and vaccines. *Frontiers in Microbiology* 12:755801. https://doi.org/10.3389/fmicb.2021.755801

Bly, J. E. and Clem, L. W. 1992. Temperature and teleost immune functions. *Fish and Shellfish Immunology* 2:159–171.

Bøgwald, J. and Dalmo, R. A. 2019. Review on immersion vaccines for fish: An update 2019. *Microorganisms* 7:627. https://doi.org/10.3390/microorganisms7120627

Boutier, M., Gao, Y., Donohoe, O. and Vanderplasschen, A. 2019. Current knowledge and future prospects of vaccines against cyprinid herpesvirus 3 (CyHV-3). *Fish and Shellfish Immunology* 93:531–541. https://doi.org/10.1016/j.fsi.2019.07.079

Boutier, M., Ronsmans, M., Ouyang, P., Fournier, G., Reschner, A., Rakus, K., Wilkie, G. S., Farnir, F., Bayrou, C., Lieffrig, F., Li, H., Desmecht, D., Davison, A. J. and Vanderplasschen, A. 2015. Rational development of an attenuated recombinant cyprinid herpesvirus 3 vaccine using prokaryotic mutagenesis and *in vivo* bioluminescent imaging. *PLOS Pathogens* 11:e1004690. https://doi.org/10.1371/journal.ppat.1004690

Bricknell, I. R., Birkel, S. D., Brawley, S. H., Kirk, T. V., Hamlin, H. J., Capistrant-Fossa, K. and Moeykens, S. 2021. Resilience of cold water aquaculture: A review of likely scenarios as climate changes in the Gulf of Maine. *Reviews in Aquaculture* 13:460–503. https://doi.org/10.1111/raq.12483

Bridle, A. and Nowak, B. 2014. Vaccination against Yersiniosis, In: *Fish vaccination*, eds. R. Gudding, A. Lillehaug and Ø. Evensen, pp. 226–235. John Wiley & Sons Ltd: Chichester. https://doi.org/10.1002/9781118806913.ch19

Brooker, A. J., Papadopoulou, A., Gutierrez, C., Rey, S., Davie, A. and Migaud, H. 2018. Sustainable production and use of cleaner fish for the biological control of sea lice: recent advances and current challenges. *Veterinary Record* 183:383. https://doi.org/10.1136/vr.104966

Busch, R. A. 1997. Polyvalent vaccines in fish: the interactive effects of multiple antigens. *Developments in Biological Standardization* 90:245–256.

Casuso, A., Valenzuela-Muñoz, V., Benavente, B. P., Valenzuela-Miranda, D. and Gallardo-Escárate, C. 2022. Exploring sea lice vaccines against early stages of infestation in Atlantic salmon (*Salmo salar*). *Vaccines (Basel)* 10:1063. https://doi.org/10.3390/vaccines10071063

Colquhoun, D. J. and Lillehaug, A. 2014. Vaccination against Vibriosis, In: *Fish vaccination*, eds. R. Gudding, A. Lillehaug and Ø. Evensen, pp. 172–184. John Wiley & Sons Ltd: Chichester. https://doi.org/10.1002/9781118806913.ch15

Contreras, M., Karlsen, M., Villar, M., Olsen, R. H., Leknes, L. M., Furevik, A., Yttredal, K. L., Tartor. H., Grove, S., Alberdi, P., Brudeseth, B. and de la Fuente, J. 2020. Vaccination with ectoparasite proteins involved in midgut function and blood digestion reduces salmon louse infestations. *Vaccines (Basel)* 19:32. https://doi.org/10.3390/vaccines8010032

Cowan, G., Smith, P. and Christofilogiannis, P. 2016. Fish vaccines: The regulatory process and requirements from the laboratory bench to a final commercial product, including field trials, In: *Fish vaccines*, ed. A. Adams, pp. 105–118. Springer: Basel, Switzerland.

Dalmo, R. A. 2018. DNA vaccines for fish: Review and perspectives on correlates of protection. *Journal of Fish Diseases* 41:1–9. https://doi.org/10.1111/jfd.12727

Desbois, A. P., Garza, M., Eltholth, M., Hegazy, Y. M., Mateus, A., Adams, A., Little, D. C., Høg, E., Mohan, C. V., Ali, S. E. and Brunton, L. A. 2021. Systems-thinking approach to identify and assess feasibility of potential interventions to reduce antibiotic use in tilapia farming in Egypt. *Aquaculture* 540:736735. https://doi.org/10.1016/j.aquaculture.2021.736735

Dhufaigh, K. N., Dillon, E., Botwright, N., Birlanga, V., Talbot, A., O'Connor, I., MacCarthy, E. and Slattery, O. 2021. Comparative proteomic profiling of newly acquired, virulent and attenuated *Neoparamoeba perurans* proteins associated with amoebic gill disease. *Scientific Reports* 11:6830. https://doi.org/10.1038/s41598-021-85988-8

Dishon, A., Ashoulin, O., Weber III, E.S. and Kotler, M. 2014. Vaccination against koi herpesvirus disease, In: *Fish vaccination*, eds. R. Gudding, A. Lillehaug and Ø. Evensen, pp. 321–333. John Wiley & Sons Ltd: Chichester. https://doi.org/10.1002/9781118806913.ch27

Ellul, R. M., Walde, C., Haugland, G. T., Wergeland, H. and Rønneseth, A. 2019. Pathogenicity of *Pasteurella* sp. in lumpsuckers (*Cyclopterus lumpus* L.). *Journal of Fish Diseases* 42:35–46. https://doi.org/10.1111/jfd.12905

Ellul, R. M., Kalatzis, P. G., Frantzen, C., Haugland, G. T., Gulla, S., Colquhoun, D. J., Middelboe, M., Wergeland, H. I. and Rønneseth, A. 2021. Genomic analysis of *Pasteurella atlantica* provides insight on its virulence factors and phylogeny and highlights the potential of reverse vaccinology in aquaculture. *Microorganisms* 9:1215. https://doi.org/10.3390/microorganisms9061215

Embregts, C. W. E. and Forlenza, M. 2016. Oral vaccination of fish: Lessons from humans and veterinary species. *Developmental & Comparative Immunology* 64:118–137. https://doi.org/10.1016/j.dci.2016.03.024

Embregts, C. W. E., Tadmor-Levi, R., Vesely, T., Pokorova, D., David, L., Wiegertjes, G. F. and Forlenza, M. 2019. Intra-muscular and oral vaccination using a Koi Herpesvirus ORF25 DNA vaccine does not confer protection in common carp (*Cyprinus carpio* L.). *Fish and Shellfish Immunology* 85:90–98. https://doi.org/10.1016/j.fsi.2018.03.037

Evensen, O., Brudeseth, B. and Mutoloki, S. 2005. The vaccine formulation and its role in inflammatory processes in fish--Effects and adverse effects. *Developmental Biology (Basel)*. 121:117–125.

Falconer, L., Telfer, T. C., Garrett, A., Hermansen, Ø., Mikkelsen, E., Hjøllo, S. S., McAdam, B. J. and Ytteborg, E. 2022. Insight into real-world complexities is required to enable effective response from the aquaculture sector to climate change. *PLOS Climate* 1:e0000017.

Fernandez-Alonso, M., Alvarez, F., Estepa, A., Blasco, R. and Coll, J. M. 1999. A model to study fish DNA immersion vaccination by using the green fluorescent protein. *Journal of Fish Diseases* 22:237–241. https://doi.org/10.1046/j.1365-2761.1999.00161.x

Garver, K. A., Conway, C. M., Elliott, D. G. and Kurath, G. 2005. Analysis of DNA-vaccinated fish reveals viral antigen in muscle, kidney and thymus, and transient histopathologic changes. *Marine Biotechnology* 7:540–553. https://doi.org/10.1007/s10126-004-5129-z

Guragain, P., Tkachov, M., Båtnes, A. S., Olsen, Y., Winge, P. and Bones, A. M. 2021. Principles and methods of counteracting harmful salmon–arthropod interactions in salmon farming: Addressing possibilities, limitations, and future options. *Frontiers in Marine Science* 8:701793. https://doi.org/10.3389/fmars.2021.701793

Hamilton, S., McLean, K., Monaghan, S. J., McNair, C., Inglis, N. F., McDonald, H., Adams, S., Richards, R., Roy, W., Smith, P., Bron, J., Nisbet, A. J. and Knox, D. 2018. Characterisation of proteins in excretory/secretory products collected from salmon lice, *Lepeophtheirus salmonis*. *Parasites & Vectors* 11:294. https://doi.org/10.1186/s13071-018-2885-6

Hamre, L. A. and Nilsen, F. 2011. Individual fish tank arrays in studies of *Lepeophtheirus salmonis* and lice loss variability. *Diseases of Aquatic Organisms* 97:47–56. https://doi.org/10.3354/dao02397

Helfman, G., Collette, B., Facey, D. and Bowen, B. 2010. *The diversity of fishes*, 2nd edition. Wiley Blackwell: Hoboken, NJ.

Herati, R. S. and Wherry, E. J. 2018. What is the predictive value of animal models for vaccine efficacy in humans? Consideration of strategies to improve the value of animal models. *Cold Spring Harbor Perspectives in Biology* 10:a031583. https://doi.org/10.1101/cshperspect.a031583

Hetland, D. L., Dale, O. B., Skjødt, K., Press, C. M., Falk, K. 2011. Depletion of CD8 alpha cells from tissues of Atlantic salmon during the early stages of infection with high or low virulent strains of infectious salmon anaemia virus (ISAV). *Developmental and Comparative Immunology* 35:817–826. https://doi.org/10.1016/j.dci.2011.02.001

Hoare, R., Ngo, T. P. H., Bartie, K. L. and Adams, A. 2017. Efficacy of a polyvalent immersion vaccine against *Flavobacterium psychrophilum* and evaluation of immune response to vaccination in rainbow trout fry (*Onchorynchus mykiss* L.). *Veterinary Research* 48:43. https://doi.org/10.1186/s13567-017-0448-z

Holm, A., Rippke, B. E. and Noda, K. 2014. Legal requirements and authorization of fish vaccines, In: *Fish vaccination*, eds. R. Gudding, A. Lillehaug and Ø. Evensen, pp. 128–139. John Wiley & Sons Ltd: Chichester. https://doi.org/10.1002/9781118806913.ch11

Hølvold, L. B., Myhr, A. I. and Dalmo, R. A. 2014. Strategies and hurdles using DNA vaccines to fish. *Veterinary Research* 45:21. https://doi.org/10.1186/1297-9716-45-21

Houston, R. D., Haley, C. S., Hamilton, A., Guy, D. R., Mota-Velasco, J. C., Gheyas A. A., Tinch, A. E., Taggart, J. B., Bron, J. E., Starkey, W. G., McAndrew, B. J., Verner-Jeffreys, D. W., Paley, R. K., Rimmer, G. S. E., Tew, I. J. and Bishop, S. C. 2010. The susceptibility of Atlantic salmon fry to freshwater infectious pancreatic necrosis is largely explained by a major QTL. *Heredity* 105:318–327. https://doi.org/10.1038/hdy.2009.171

Johnson, K. A., Flynn, J. K. and Amend, D. F. 1982. Onset of immunity in salmonid fry vaccinated by direct immersion in *Vibrio anguillarum* and *Yersinia ruckeri* bacterins. *Journal of Fish Diseases* 5:197–205. https://doi.org/10.1111/j.1365-2761.1982.tb00474.x

Killie, J-E. A. and Jørgensen, T. Ø. 1994. Immunoregulation in fish I: Intramolecular-induced suppression of antibody responses to haptenated protein antigens studied in Atlantic salmon (*Salmo salar* L.). *Developmental and Comparative Immunology* 18:123–136. https://doi.org/10.1016/0145-305X(94)90240-2

Kuang, B., Xiao, K., Wang, W., You, F., Xiao, P. and Chen, S. 2022. Different antigen ratio in bivalent vaccine can affect immunological activation and protection against *Aeromonas salmonicida* and *Vibrio anguillarum* in Atlantic salmon. *Fish and Shellfish Immunology* 128:644–650. https://doi.org/ 10.1016/j.fsi.2022.08.033

Lauscher, A., Krossøy, B., Frost, P., Grove, S., König, M., Bohlin, J., Falk, K., Austbø, L. and Rimstad, E. 2011. Immune responses in Atlantic salmon (*Salmo salar*) following protective vaccination against infectious salmon anemia (ISA) and subsequent ISA virus infection. *Vaccine* 29:6392–401. https://doi.org/10.1016/j.vaccine.2011.04.074

Lorenzen, E., Einer-Jensen, K., Rasmussen, J. S., Kjaer, T. E., Collet, B., Secombes, C. J. and Lorenzen, N. 2009. The protective mechanisms induced by a fish rhabdovirus DNA vaccine depend on temperature. *Vaccine* 27:3870–3880. https://doi.org/10.1016/j.vaccine.2009.04.012

Ma, J., Bruce, T. J., Jones, E. M. and Cain, K. D. 2019. A review of fish vaccine development strategies: Conventional methods and modern biotechnological approaches. *Microorganisms* 7:569. https://doi.org/10.3390/microorganisms7110569.

Magadan, S., Sunyer, O. J. and Boudinot, P. 2015. Unique features of fish immune repertoires: Particularities of adaptive immunity within the largest group of vertebrates. In: *Pathogen-Host Interactions: Antigenic Variation v. Somatic Adaptations*, eds. E. Hsu, and L. Du Pasquier, pp. 235–264. Springer: Cham, Switzerland.

Markestad, A. and Grave, K. 1997. Reduction of antibacterial drug use in Norwegian fish farming due to vaccination. *Developments in Biological Standardization* 90:365–369.

McBeath, A., Aamelfot, M., Christiansen, D. H., Matejusova, I., Markussen, T., Kaldhusdal, M., Dale, O. B., Weli, S. C. and Falk, K. 2015. Immersion challenge with low and highly virulent infectious salmon anaemia virus reveals different pathogenesis in Atlantic salmon, *Salmo salar* L. *Journal of Fish Diseases* 38:3–15. https://doi.org/10.1111/jfd.12253

Mendivil, C. O. 2021. Dietary fish, fish nutrients, and immune function: A review. *Frontiers in Nutrition* 7:617652. https://doi.org/10.3389/fnut.2020.617652

Micoli, F., Bagnoli, F., Rappuoli, R. and Serruto, D. 2021. The role of vaccines in combatting antimicrobial resistance. *Nature Reviews Microbiology* 19:287–302. https://doi.org/10.1038/s41579-020-00506-3

Midtlyng, P.J. 2014. Vaccination against Furunculosis, In: *Fish vaccination*, eds. R. Gudding, A. Lillehaug and Ø. Evensen, pp. 185–199. John Wiley & Sons Ltd: Chichester. https://doi.org/10.1002/9781118806913.ch16

Midtlyng, P.J. 2016. Methods for measuring efficacy, safety and potency of fish vaccines, In: *Fish Vaccines*, ed. A. Adams, pp. 119–141. Springer: Basel. https://doi.org/10.1007/978-3-0348-0980-1_6

Midtlyng, P.J., Hendriksen, C., Balks, E., Bruckner, L., Elsken, L., Evensen, O., Fyrand, K., Guy, A., Halder, M., Hawkins, P., Kisen, G., Romstad, A. B., Salonius, K., Smith, P. and Sneddon, L. U. 2011. Three Rs approaches in the production and quality control of fish vaccines. *Biologicals*. 39:117–128. https://doi.org/10.1016/j.biologicals.2011.02.001

Midtlyng, P.J., Reitan, L.J. and Speilberg, L. 1996. Experimental studies on the efficacy and side-effects of intraperitoneal vaccination of Atlantic salmon (*Salmo salar* L.) against furunculosis. *Fish and Shellfish Immunology* 6:335–350. https://doi.org/10.1006/FSIM.1996.0034

Mitchell H. 1997. The pitfalls of field trials in fish vaccinology. *Developments in Biological Standardization* 90:321–332.

Mondal, H. and Thomas, J. 2022. A review on the recent advances and application of vaccines against fish pathogens in aquaculture. *Aquaculture International* 30:1971–2000. https://doi.org/10.1007/s10499-022-00884-w

Mora, T., Walczak, A. M., Bialek, W. and Callan Jr, C. G. 2010. Maximum entropy models for antibody diversity. *Proceedings of the National Acadamy of Sciences* 107:5405–5410. https://doi.org/10.1073/pnas.1001705107

Morens, D. M., Holmes, E. C., Davis, A. S. and Taubenberger, J. K. 2011. Global rinderpest eradication: lessons learned and why humans should celebrate too. *Journal of Infectious Diseases* 204:502–505. https://doi.org/10.1093/infdis/jir327

Munang'andu, H. M. and Evensen, Ø. 2019. Correlates of protective immunity for fish vaccines. *Fish and Shellfish Immunology* 85:132–140. https://doi.org/10.1016/j.fsi.2018.03.060

Mutoloki, S., Munang'andu, H. M. and Evensen, Ø. 2015. Oral vaccination of fish - Antigen preparations, uptake, and immune induction. *Frontiers in Immunology* 19:519. https://doi.org/10.3389/fimmu.2015.00519

Ni, X., Wang, N., Liu, Y. and Lu, C. 2010. Immunoproteomics of extracellular proteins of the *Aeromonas hydrophila* China vaccine strain J-1 reveal a highly immunoreactive outer membrane protein. *FEMS Immunology and Medical Microbiology* 58:363–373.

Nikoskelainen, S., Verho, S., Järvinen, S., Madetoja, J., Wiklund, T. and Lilius, E. M. 2007. Multiple whole bacterial antigens in polyvalent vaccine may result in inhibition of specific responses in rainbow trout (*Oncorhynchus mykiss*). *Fish and Shellfish Immunology* 22:206–217. https://doi.org/10.1016/j.fsi.2006.04.010

Nordmo, R., Ramstad, A. 1997. Comparison of different challenge methods to evaluate the efficacy of furunculosis vaccines in Atlantic salmon, *Salmo salar* L. *Journal of Fish Diseases* 20:119–126. https://doi.org/10.1046/j.1365-2761.1997.d01-114.x

Okocha, R. C., Olatoye, I. O. and Adedeji, O. B. 2018. Food safety impacts of antimicrobial use and their residues in aquaculture. *Public Health Reviews* 39:21. https://doi.org/10.1186/s40985-018-0099-2

Ormsby, M. J., Caws, T., Burchmore, R., Wallis, T., Verner-Jeffreys, D. W. and Davies, R. L. 2016. *Yersinia ruckeri* isolates recovered from diseased Atlantic salmon (*Salmo salar*) in Scotland are more diverse than those from rainbow trout (*Oncorhynchus mykiss*) and represent distinct subpopulations. *Applied and Environmental Microbiology* 82:5785–94. https://doi.org/10.1128/AEM.01173-16.

Øvergård, A. C., Midtbø, H. M. D., Hamre, L. A., Dondrup, M., Bjerga, G. E. K., Larsen, Ø., Chettri, J. K., Buchmann, K., Nilsen, F. and Grotmol, S. 2022. Small, charged proteins in salmon louse (*Lepeophtheirus salmonis*) secretions modulate Atlantic salmon (*Salmo salar*) immune responses and coagulation. *Scientific Reports* 12:7995. https://doi.org/10.1038/s41598-022-11773-w

Pang, H., Qiu, M., Zhao, J., Hoare, R., Monaghan, S. J., Song, D., Chang, Y. and Jian, J. 2018. Construction of a *Vibrio alginolyticus* hopPmaJ (hop) mutant and evaluation of its potential as a live attenuated vaccine in orange-spotted grouper (*Epinephelus coioides*). *Fish and Shellfish Immunology* 76:93–100. https://doi.org/10.1016/j.fsi.2018.02.012

Papadopoulou, A., Garvey, K., Hill, T., Ramirez-Paredes, J. G., Monaghan, S. J., Baily, J. L., Davie, A., Katsiadaki, I., Verner-Jeffreys, D., Wallis, T., Migaud, H. and Adams, A. 2021. Novel atypical *Aeromonas salmonicida* bath challenge model for juvenile ballan wrasse (*Labrus bergylta*, Ascanius). *Journal of Fish Diseases* 44:823–835. https://doi.org/10.1111/jfd.13312

Papadopoulou, A., Monaghan, S. J., Bagwell, N., Alves, M. T., Verner-Jeffreys, D., Wallis, T., Davie, A., Adams, A. and Migaud, H. 2022. Efficacy testing of an immersion vaccine against *Aeromonas salmonicida* and immunocompetence in ballan wrasse (*Labrus bergylta*, Ascanius). *Fish and Shellfish Immunology* 121:505–515. https://doi.org/10.1016/j.fsi.2021.09.044

Pardi, N., Hogan, M. J., Porter, F. W. and Weissman, D. 2018. mRNA vaccines – A new era in vaccinology. *Nature Reviews Drug Discovery* 17:261–279. https://doi.org/10.1038/nrd.2017.243

Pasquier, L. 1982. Antibody diversity in lower vertebrates–Why is it so restricted? *Nature* 296:311–313. https://doi.org/10.1038/296311a0

Poppe, T. T. and Koppang, E. O. 2014. Side-effects of vaccination. In: *Fish vaccination*, eds. R. Gudding, A. Lillehaug and Ø. Evensen, 153–161. Chichester: John Wiley & Sons, Ltd.

Poobalane, S., Thompson, K. D., Diab, A., Ardó, L., Jeney, G. and Adams, A. 2008. Protein expression by *Aeromonas hydrophila* during growth *in vitro* and *in vivo*. *Microbial Pathogenesis* 45:60–69. https://doi.org/10.1016/j.micpath.2008.03.001

Raida, M. K. and Buchmann, K. 2007. Temperature-dependent expression of immune-relevant genes in rainbow trout following *Yersinia ruckeri* vaccination. *Diseases of Aquatic Organisms* 77:41–52. https://doi.org/10.3354/dao01808

Ramírez-Paredes, J.G., Thompson, K.D., Metselaar, M., Shahin, K., Soto, E., Richards, R.H., Penman, D.J., Colquhoun, D.J. and Adams, A. 2017. A polyphasic approach for phenotypic and genetic characterization of the fastidious aquatic pathogen *Francisella noatunensis* subsp. *orientalis*. *Frontiers in Microbiology* 8:2324. https://doi.org/10.3389/fmicb.2017.02324

Ramírez-Paredes, J.G., Verner-Jeffreys, D. W., Papadopoulou, A., Monaghan, S. J., Smith, L., Haydon, D., Wallis, T. S., Davie, A., Adams, A. and Migaud, H. 2020. A commercial autogenous injection vaccine protects ballan wrasse (*Labrus bergylta*, Ascanius) against *Aeromonas salmonicida* vapA type V. *Fish and Shellfish Immunology* 107:43–53. https://doi.org/10.1016/j.fsi.2020.09.040

Rast, J. P. and Buckley, K. M. 2013. Lamprey immunity is far from primitive. *Proceedings of the National Acadamy of Sciences* 110: 5746–5747. https://doi.org/10.1073/pnas.1303541110

Scarfe, D., Padrós, F., Iatridou, D., Bravo, A., Palić, D. and Fabris, A. 2022. Workshop Report: Aquatic Animal Health Education. *Bulletin of the European Association of Fish Pathologists* 41:225–232. https://doi.org/10.48045/001c.36867

Scarfe, A. D. and Palić, D. 2020. Aquaculture biosecurity: Practical approach to prevent, control, and eradicate diseases. In: *Aquaculture Health Management*, eds. F. S. B. Kibenge, and M. D. Powell, pp. 75–116. Academic Press: London.

Schetters, T., Bishop, R., Crampton, M., Kopáček, P., Lew-Tabor, A., Maritz-Olivier, C., Miller, R., Mosqueda, J., Patarroyo, J., Rodriguez-Valle, M., Scoles, G. A. and de la Fuente, J. 2016. Cattle tick vaccine researchers join forces in CATVAC. *Parasites & Vectors* 9:105. https://doi.org/10.1186/s13071-016-1386-8

Schijns, V. E. and Lavelle, E. C. 2011. Trends in vaccine adjuvants. *Expert Review of Vaccines* 10:539–550. https://doi.org/10.1586/erv.11.21

Schröder, L., Klafack, S., Bergmann, S. M., Fichtner, D., Jin, Y., Lee, P. Y., Höper, D., Mettenleiter, T. C. and Fuchs, W. 2019. Generation of a potential koi herpesvirus live vaccine by simultaneous deletion of the viral thymidine kinase and dUTPase genes. *Journal of General Virology* 100:642–655. https://doi.org/10.1099/jgv.0.001148

Schrøder, M. B., Mikkelsen, H., Børdal, S., Gravningen, K. and Lund, V. 2006. Early vaccination and protection of Atlantic cod (*Gadus morhua* L.) juveniles against classical vibriosis. *Aquaculture* 254:46–53. https://doi.org/10.1016/j.aquaculture.2005.10.015

Shahin, K., Shinn, A. P., Metselaar, M., Ramirez-Paredes, J. G., Monaghan, S. J., Thompson, K. D., Hoare, R. and Adams, A. 2019. Efficacy of an inactivated whole-cell injection vaccine for nile tilapia, *Oreochromis niloticus* (L), against multiple isolates of *Francisella noatunensis* subsp. *orientalis* from diverse geographical regions. *Fish and Shellfish Immunology* 89:217–227. https://doi.org/10.1016/j.fsi.2019.03.071

Shivam, S., El-Matbouli, M. and Kumar, G. 2021. Development of fish parasite vaccines in the OMICs era: Progress and opportunities. *Vaccines (Basel)* 20:179. https://doi.org/10.3390/vaccines9020179

Speare, D. J., MacNair, N. and Hammell, K. L. 1995. Demonstration of tank effect on growth indices of juvenile rainbow trout (*Oncorhynchus mykiss*) during an *ad libitum* feeding trial. *American Journal of Veterinary Research* 56:1372–1379.

Star, B. and Jentoft, S. 2012 Why does the immune system of Atlantic cod lack MHC II? *BioEssays* 34:648–651. https://doi.org/10.1002/bies.201200005

Tafalla, C., Bøgwald, J., Dalmo, R.A., Munang'andu, H. M. and Evensen, Ø. 2014. Adjuvants in Fish Vaccines. In: *Fish vaccination*, eds. R. Gudding, A. Lillehaug and Ø. Evensen, pp. 68–84. John Wiley & Sons, Ltd: Chichester.

Thorarinsson, R. and Powell, D.B. 2006. Effects of disease risk, vaccine efficacy, and market price on the economics of fish vaccination. *Aquaculture* 256: 42–49. https://doi.org/10.1016/j.aquaculture.2006.01

Tobar, I., Arancibia, S., Torres, C., Vera, V., Soto, P., Carrasco, C., Alvarado, M., Neira, E., Arcos, S. and Tobar, J. A. 2015. Successive oral immunizations against *Piscirickettsia salmonis* and infectious salmon anemia virus are required to maintain a long-term protection in farmed salmonids. *Frontiers in Immunology* 27:244. https://doi.org/10.3389/fimmu.2015.00244

Tziouvas, H., Varvarigos, P. 2021. Intensity scale of side effects in European Sea Bass (*Dicentrarchus Labrax*) post intraperitoneal injection with commercial oil-adjuvanted vaccines. *Bulletin of the European Association of Fish Pathologists* 41:103–110. https://doi.org/10.48045/001c.28222

Van Oirschot, J. T. 1999. Diva vaccines that reduce virus transmission. *Journal of Biotechnology* 73:195–205. https://doi.org/10.1016/s0168-1656(99)00121-2

Woolhouse, M., Ward, M., van Bunnik, B. and Farrar, J. 2015. Antimicrobial resistance in humans, livestock and the wider environment. *Philosophical Transactions of the Royal Society B: Biological Sciences* 370:20140083. https://doi.org/10.1098/rstb.2014.0083

Yu, Y., Wang, Q., Huang, Z., Ding, L. and Xu, Z. 2020. Immunoglobulins, mucosal immunity and vaccination in teleost fish. *Frontiers in Immunology* 11:567941. https://doi.org/10.3389/fimmu.2020.567941

Zapata, A., Diez, B., Cejalvo, T., Gutiérrez-de Frías, C. and Cortés, A. 2006. Ontogeny of the immune system of fish. *Fish and Shellfish Immunology* 20:126–136. https://doi.org/10.1016/j.fsi.2004.09.005

Zhang, Y., Salinas, I., Li, J., Parra, D., Bjork, S., Xu, Z., LaPatra, S. E., Bartholomew, J. and Sunyer, J. O. 2010. IgT, a primitive immunoglobulin class specialized in mucosal immunity. *Nature Immunology* 11:827–835. https://doi.org/10.1038/ni.1913

Zhou, S., Tu, X., Pang, H., Hoare, R., Monaghan, S. J., Luo, J., Jian, J. A. 2020. T3SS regulator mutant of *Vibrio alginolyticus* affects antibiotic susceptibilities and provides significant protection to *Danio rerio* as a live attenuated vaccine. *Frontiers in Cellular and Infection Microbiology* 10:183. https://doi.org/10.3389/fcimb.2020.00183

14 Fish Vaccine
From Field to Development

R. Cascione, Adrian Astier, and Philippe Mahl
Virbac Asia Pacific Co. Ltd.

Øystein Evensen
Norwegian University of Life Sciences

CONTENTS

14.1 INTRODUCTION

The massive growth of aquaculture over the last two decades has not come without an increased number of disease outbreaks that severely impact on production (1). Annually, disease in finfish is estimated to cause losses between 1 US$ and 9.5 billion US$ (2), depending on the species. However, losses are linked to more than mortality alone, and collateral damages associated with farming and husbandry parameters have to be taken into the equation, often with high impact and often forgotten (3). Growth parameters, such as food conversion ratio (FCR), specific grow rate (SGR), and average daily gain (ADG), are factors that impact on disease-related costs. They are difficult to estimate and will vary from farm-to-farm.

Further to this, other concerns have also been raised, mostly associated with public health and end-user safety, like residues due to the usage of antibiotics/chemicals and associated antimicrobial resistance. Therefore, the implementation of good animal husbandry practices, usage of alternative active compounds, and not the least, the use and development of vaccines (4) have become a core interest point, together with the whole growth of the aquaculture sector.

It has been widely recognized that the application of vaccines has resulted in a drastic drop in the usage of antibiotics in the salmonid aquaculture (5), proving that vaccination is the most economically valuable disease control method in the whole aquaculture industry (6).

Vaccine development is therefore becoming a crucial breaking point for the future of the industry, and it is important that private and public entities keep good relations with the farming industry, because this is the only way in which the fish health industry can respond promptly to the increasing needs of prophylactic methods.

DOI: 10.1201/9781003388548-19

14.2 EPIDEMIOLOGICAL ASSESSMENT AND FIELD NEEDS

14.2.1 Field Needs and R&D Requirements

The perfect fish vaccine is a product that has to be economically affordable and applicable for use at a large production scale. Vaccines need to give quick onset and long-lasting immunity; they have to be applied easily and should meet the needs of the largest possible farming system and operation (7).

Development has to meet strict regulations in order to document the efficacy of a vaccine. Percent survival and relative percentage survival (RPS) are still two of the most important parameters to document efficacy in a challenge test. Other important parameters are antibody levels, onset of immunity, and the immune response itself. Even if those latter parameters are definitely the future of the aquaculture vaccine development strategy, the current stage of knowledge of fish immunology doesn't allow us to rely completely on in vitro test models (8).

It is indeed complicated to combine laboratory methods with industry requirements, which clearly has a wider and more complicated number of parameters to consider. For these reasons, field observations are the most essential for us to understand and define the research objectives, as they help to identify the requirements of a specific farming area and the species needs.

The field observation starts with the veterinarian performing a clinical visit, which needs to assess:

- Mortality patterns: percentage of mortality and seasonality. The outbreak can be associated with the dry (summer) or rainy (winter) season, or during transition from one season to the other. Temperature, oxygen levels, and possible upwelling need to be accounted for and included in the recordings.
- Economic impact: the farmers' perception of the outbreaks must be compared with solid data coming from the loss in biomass at harvest, impact on FCR, SDG, AGD, etc.

All this information will help us to have an idea of economic impact of a specific mortality pattern.

- Local remedies and solutions in place: farmers often apply knowledge-based mitigation to reduce mortality during a disease outbreak, such as: early harvest, increased frys stocking for compensate biomass loss, antibiotics/chemical administration, etc. All this information constitutes valuable background knowledge to better understand what would be expected of a vaccine in terms of providing a positive impact on the production.
- Sampling: The collection of samples is the most powerful tool to understand the nature of the outbreak. The clinical visit and the field observation, combined with results from laboratory examination, are key to define a solution to the disease problem.

All these data will be essential to create an ideal map of the major farming areas combined with an epidemiological assessment, which will help us to identify the most common disease threats and use this to build our future vaccination strategy.

14.2.2 Epidemiology Assessment

Epidemiology is "the branch of medicine which deals with the incidence, distribution, and possible control of diseases and other factors relating to health" (9).

The definition of the Oxford dictionary, clearly defines the wide range of information and application that this discipline embraces. Aquaculture is a booming industry with more than 230 farmed species (10) under extensive, intensive, or semi-intensive production systems.

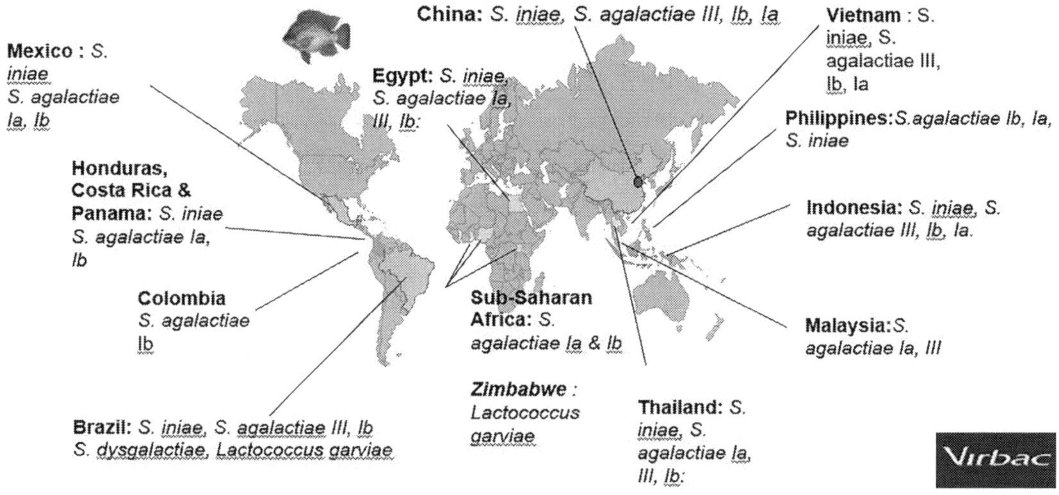

FIGURE 14.1 Streptococcus and other pathogen distribution in the tilapia industry.

The more populated the system is and the more concentrated the production is in specific farming areas, the higher the possibility of introducing and spreading of new pathogens, resulting in disease outbreaks (11).

In addition to this, animal movement from one region to another is a factor that can affect the distribution and spread of disease in different geographic areas. An epidemiological map (Figure 14.1) that aims to identify the major pathogen of a specific species within the area/s where the fish species is farmed is an excellent instrument to identify disease and patterns, which aid in identifying the most prevalent pathogen affecting that specific aquaculture sector.

The epidemiology is not something static; experience tells us that this knowledge needs to be continuously updated, due to the cyclical appearances of diseases. Indeed, the variegate disease scenario, especially of the tropical aquaculture, is one of the major difficulties that the fish health industry has to deal with to provide solutions that are durable and efficient in the long term.

14.2.2.1 Pathogens Choice and Disease

The tilapia (*Oreochromis niloticus*) industry can be used as an example for a vaccine development strategy. With over 6 M tons of production and 20% of global loss from disease outbreaks, the tilapia industry is going through a real upgrading in terms of fish health management. The epidemiological scenario of the species is becoming more and more complex, with the continuous presence of multiple disease patterns and co-infections. Table 14.1 summarizes one of the main diseases observed in tilapia (Streptococcosis), divided into primary or secondary pathogens. The separation into pathogen categories is for the main part based on observations in Asia-Pacific. The dominant pathogen varies by geographic region. For example, this is seen for *Edwarsiella* spp. in East African countries, and *Streptococcus iniae* in South America where both have a clear primary pathogen pattern.

A vaccine development strategy for this species has to consider the impact and importance of each pathogen, once identified in a given production system. The observation of clinical signs, associated environmental parameters, and laboratories' diagnostics are the keys to understand this very complicated puzzle.

The choice of a specific pathogen or class of pathogens for vaccine development is the result of an equation which assembles a multitude of factors, including also an essential economical assessment.

TABLE 14.1
Major Pathogens and Their Association to Clinical Symptoms, Observed in South-East Asia Tilapia Industry

Etiological Agent	Target Size	Season	Geographical Location	Pathogen Type
Streptococcus agalactiae, serotype 1a, 1b and 3	From 1 g till 300 g (higher incidence)	Dry or seasonal change	Worldwide	Primary
Streptococcus iniae	From 1 g till 300 g (higher incidence)	Dry	Worldwide	Primary/ Secondary
Edwarsiella spp	Fingerling stage more susceptible	Dry/Rainy	Africa (++)	Secondary
Francisella orientalis subsp. asiatica	From 1 g till 300 g (higher incidence)	Rainy/Cold season	Worldwide	Primary
Flavobacterium spp	All farming size	Rainy/Cold season	Worldwide	Secondary
Aeromonas spp	All farming size	Dry/Rainy	Worldwide	Secondary
Vibrio spp	All farming size	Dry/Rainy	Worldwide	Secondary
ISKNV	From 1 g till 300 g (higher incidence)	Rainy	Worldwide	Primary
TiLV	From 10 g till harvest size	-	Worldwide	Primary/ Secondary
TiPV	-	-	Asia (++)	Secondary

14.3 VACCINE DEVELOPMENTS

14.3.1 Multifactorial Choice between Industry Demand and Laboratory Answers

Table 14.2 summarizes the major steps that are involved in the development of a commercial vaccine. A combination of scientific, regulatory, and economic factors influences the final decision.

The sample results and the different pathogens detected are the first step of the development process, which of course need to overlap with gross signs' observations and field data. The pathogen or pathogens of choice may spread into new production areas, and therefore the presence in the widest possible geographical area needs to be ascertained (Paragraph 2.0).

A commercial vaccine should be protective and efficient for that specific industry and species, and therefore should be tested so that this can be documented.

Another determining factor in the decision tree is the regulatory framework that will have a very strong influence on the final composition or design of the vaccine. From time to time, regulatory hurdles can hold back the commercialization of a product and delay the development work.

14.3.2 Field Trial and Vaccination Training

Another important step in the development process of a vaccine development is the field trial and field trial application. The R&D challenge tests are essential to obtain initial information, which will be essential for proof of concept and early efficacy studies. For regulatory requirements, the vaccine has to be tested under controlled environmental conditions, that is, in wet laboratories. Then, the field trial is an essential step that will provide information about the efficacy of a vaccine applied under real farming conditions.

By field testing, we will be able to include more data into the efficacy and efficiency analysis, considering all husbandry parameters that will allow for performing a cost-benefit or economic analysis.

TABLE 14.2
Scheme of Decision for Vaccine Development Strategy

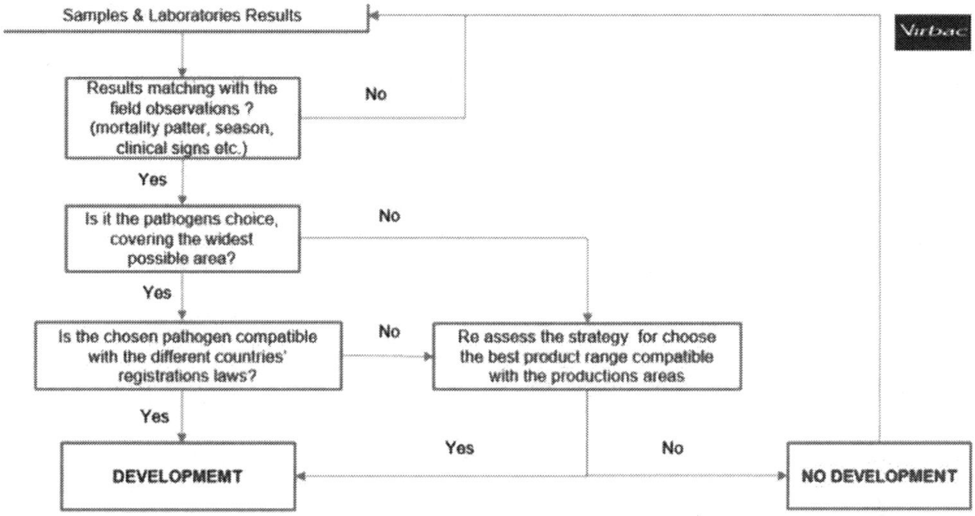

During a field trial testing, it is essential to use a statistically representative fish population and test the efficacy of vaccine under the highest disease pressure possible. Data recording, including mortality, water quality parameters (temperature, ammonia, dry disk), and feed consumption will need to be noted daily for correct data analysis.

The vaccination training is another important part of the field trial. Even if not strictly related to vaccine efficacy, a correct vaccination practice is directly correlated with a positive outcome of the test. A proper vaccination technique will ensure a minor mortality after handling, faster recovery, and adequate or correct onset of immunity.

14.4 ECONOMIC ANALYSIS: MARKET EVALUATION

14.4.1 Field Data Analysis vs Economic Evaluation

The first barrier that a vaccine, which is approaching a market will meet, is the price. The vaccination itself will add an average additional cost on fingerlings of between 10% and 20%. One of the major challenges in the vaccine evaluation and development strategy is to identify how an extra cost can eventually turn into a profitable income. A vaccine needs to create an economical advantage, which is of course a reflection of more sustainable, efficient, and healthy production, where the animal welfare is much more respected.

Many parameters are considered during the development of a vaccine (Table 14.3), but there is definitely a discrepancy between what the field requires and what the laboratory and R&D process takes into account.

This doesn't mean that the R&D does not produce a valuable contribution. It means that the vaccine development is the results of a holistic approach between two data sources, which are both essential for a correct final evaluation and outcome.

For example, an RPS% threshold value for an R&D program is typically between 70% and 80%, and if a vaccine cannot meet this threshold during in vitro efficacy evaluation, the project has to be revised. Differently, an RPS of 40% under field conditions has to be viewed differently than an in vitro result, because we must analyze and take into account what that 40% RPS reflects on biomass,

TABLE 14.3
R&D and Field Parameters Compared

R&D Parameters	Values	Field Parameters	Values
RPS %	➢75%	*RPS %*	➢40%
Survival %	➢85%	*Survival %*	➢75%
DOI	*3 month if possible	*FCR*	1.3-1.4
Onset	15–20 days	*SGR*	*farm-specific
		ADG	*farm-specific
		Harvest ABW	1 kg
		Harvest Biomass	➢ctrl
		Farming time	<ctrl (10 days min)

* All data are based on average and are related to field/R&D condition experience and do not describe the exact situation, but a specific contextualization.

farming period, FCR, and extra costs that are all associated to production, without forgetting to consider the environmental parameters and other causes that might affect mortality in an open system.

Some private financed case studies performed in LATAM and APAC have investigated the economic impact of vaccination under different environments, defining that vaccination can create an clear ROI (return of investment) and a % difference in margins between 8% and 18% in Asia and above 20% in Latin America.

Graph 1 indicates one of the major discriminating factors inducing additional value to the crops under a vaccination program. The vaccinated fish presented both better harvest and higher % of premium size fish, which reflect directly in a better selling price and higher cage biomass.

The more homogenous growth can be related to a general better welfare of vaccinated fish, which are less affected by primary and consequently secondary pathogens that clearly affect growth performances (Figure 14.2).

The analysis and comparison of FCR and other growth-related factors are not easy to evaluate since they are multifactorial, but it is undeniable that vaccinated fish had an average constant better utilization of feed and this has to be compared in each trial to a relevant control.

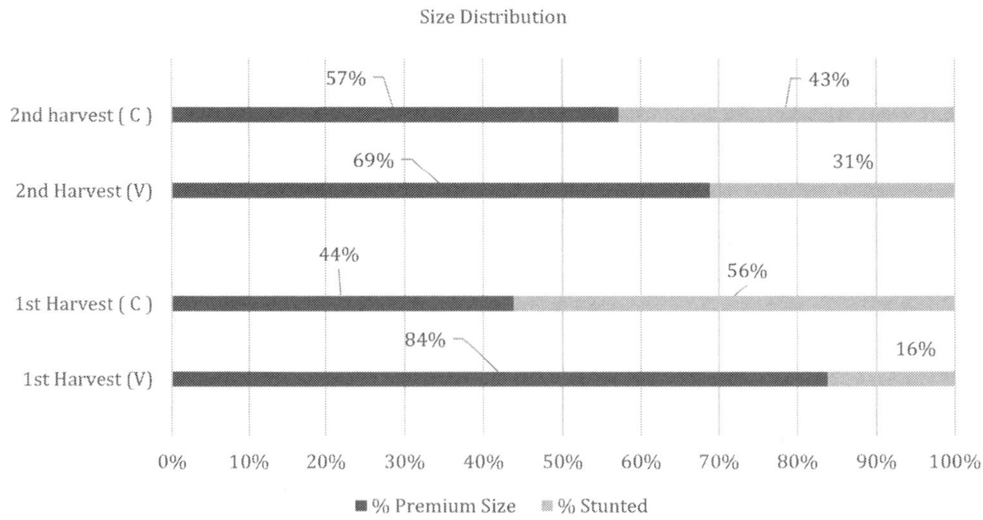

FIGURE 14.2 Comparison of size distribution at harvest between vaccinated fish and their control.

TABLE 14.4
Cost Related and Margins between Vaccinated Cage and Relative Control

	Vacc.		Control	
Cage		**Grow-Out**		**Dif**
Fish, qty.	75,000		75,000	
Survival, %	85		70	15
Harvest weight, kg	0.5		0.5	0
Biomass, kg	28,688		23,625	5,063
Production cost, USD/kg	1.50		1.50	0
Total biomass cost, USD	43,031		35,438	7,594
Selling price, USD/kg	1.84		1.84	0
FCR	1.4		1.40	0.0
Feed cost, USD/kg	0.70		0.70	0.0
Vaccination cost	488		0.00	488
Vaccine cost	3,000		0.00	3,000.0
Mortality cost	3,797		16,875	-13,078.1
Total cost, USD	50,316		52,313	–1,997
Gross income, USD	52,785		43,470	9,315
Total Gross margin, USD	**2,469**		**–8,843**	**11,312**
ROI, %	**4.9**		**–16.9**	**22**

A more detailed economic analysis based on crude expense/income of single cage study has underlined once again the impact on profitability from vaccination. Table 14.4, clearly indicates the costs associated with vaccination and the loss in the control. Even considering all costs associated with vaccine, the final income shows a 21% improved profitability for the vaccinated fish.

14.5 CONCLUSIONS

Diseases in aquaculture have a significant impact on welfare and profitability and prevention methods such as vaccines will have a massive relevance in the future development, especially, of tropical aquaculture. The vaccine development for the aquaculture sector is the result of a constant interaction between the farmers' needs and the support of the fish health industry.

The support of the pharmaceutical and health sector aims to remove the barrier linked to the economical suitability of vaccination, which is still one of the major obstacles to the development of new products.

The field requests are not easy to be met because of complex and sometimes demanding requests and expectations in terms of performance. Vaccination is not a silver bullet and only by applying solid health husbandry procedure, the whole industry will benefit from the use of vaccines. The development of a vaccine is therefore a complex diagram, where epidemiology, veterinary, laboratory expertise, farming knowledge, market assessment, and economical aspects are interacting, which in the end will help to define vaccination as the most economically valuable disease control method in the whole aquaculture industry (6).

REFERENCES

1. Bagum N, Monir MS, Khan MH. Present status of fish diseases and economic losses due to incidence of disease in rural freshwater aquaculture of Bangladesh. *J Innov Dev Strateg.* 2013;7:48–53
2. Monir S, Bagum N, Rahman S, Ashaf-Ud-Doulah M, Bhadra A, Borty SC. Parasitic diseases and estimation of loss due to infestation of parasites in Indian major carp culture ponds in Bangladesh. *Int J Fish Aquat Stud.* 2015;2:118–122.
3. Tavares-Dias M, Martins ML. An overall estimation of losses caused by diseases in the Brazilian fish farms. *J Parasit Dis.* 2017;41(4):913–918. doi:10.1007/s12639-017-0938-y
4. Shinn AJ, Pratoomyot J, Bron J, Paladini G, Brooker E, Brooker A. Economic impacts of aquatic parasites on global finfish production. Global Aquaculture Advocate, September/October, 2015, pp 58–61.
5. Tavares-Dias M, Martins ML. An overall estimation of losses caused by diseases in the Brazilian fish farms. *J Parasit Dis.* 2017;41(4):913–918. doi:10.1007/s12639-017-0938-y
6. Okocha, RC, Olatoye IO and Adedeji OB. Food safety impacts of antimicrobial use and their residues in aquaculture. *Public Health Rev* 2018;39:21.
7. Rodger, HD. Fish disease causing economic impact in global aquaculture. In Fish vaccines, 1st ed; Adams, A., Ed.; Springer: Basel, Switzerland, 2016; pp. 1–34.
8. Horzinek MC, Schijns VECJ, Denis M, Desmettre P, Babiuk LA. General description of vaccines. In Veterinary vaccinology; Pastoret PP, Blancou J, Vannier P, Verschueren C, Eds. Elsevier Press: Amsterdam, the Netherlands, 1997; pp. 132–152
9. Ma J, Bruce TJ, Jones EM, Cain KD. A review of fish vaccine development strategies: Conventional methods and modern biotechnological approaches. *Microorganisms.* 2019;7(11):569.
10. Munang'andu HM, Evensen Ø. Correlates of protective immunity for fish vaccines. *Fish Shellfish Immunol.* 2019;85:132–140. doi:10.1016/j.fsi.2018.03.060
11. Peeler EJ, Taylor NG. The application of epidemiology in aquatic animal health -opportunities and challenges. *Vet Res.* 2011;42(1):94. doi:10.1186/1297-9716-42-94

Index

Made in the USA
Las Vegas, NV
29 October 2024

e0ac5903-1b42-4104-ab24-8152ec9a6a78R01